Nonmetallic Materials and Composites at Low Temperatures

W0042834

CRYOGENIC MATERIALS SERIES

Nonmetallic Materials and Composites at Low Temperatures
Edited by A. F. Clark, R. P. Reed, and G. Hartwig

A Continuation Order Plan is available for this series A continuation order will bring delivery of each new volume immediately upon publication Volumes are billed only upon actual shipment For further information please contact the publisher

Nonmetallic Materials and Composites at Low Temperatures

Edited by
A.F. Clark and R.P. Reed
National Bureau of Standards
Boulder, Colorado

and
G. Hartwig
Nuclear Research Center Karlsruhe
Institute for Technical Physics
Karlsruhe, Federal Republic of Germany

Plenum Press • New York and London

Library of Congress Cataloging in Publication Data
Main entry under title:

Nonmetallic materials and composites at low temperatures.

Proceedings of a conference sponsored by the International Cryogenic
Materials Conference Board.
Includes indexes.
1. Nonmetallic materials – Thermal properties – Congresses. 2. Polymers
and polymerization – Thermal properties – Congresses. 3. Composite
materials – Thermal properties – Congresses. 4. Materials at low tempera-
tures – Congresses. I. Clark, A. F. II. Reed, Richard Palmer, 1934-   III.
Hartwig, Günther. IV. International Cryogenic Materials Conference Board.
TA418.95.N65   620.1'1216   78-26576
ISBN 978-1-4615-7524-5   ISBN 978-1-4615-7522-1 (eBook)
DOI 10.1007/978-1-4615-7522-1

Proceedings of the ICMC Symposium on Nonmetallic Materials and Composites at
Low Temperatures held in Munich, West Germany, July 10–11, 1978

© 1979 Plenum Press, New York
Softcover reprint of the hardcover 1st edition 1979
A Division of Plenum Publishing Corporation
227 West 17th Street, New York, N.Y. 10011

Preface

Cryogenics is an emerging technology filled with promises. Many cryogenic systems demand the use of nonmetallics and composites for adequate or increased performance. Thermal and electrical insulations, potting for superconducting magnets' mechanical stability, and composite structures appear to be some of the most significant applications. Research on nonmetallics at cryogenic temperatures has not progressed to the degree of research on metals. Nor can room temperature research be extrapolated to low temperatures; most polymers undergo a phase transformation to the glassy state below room temperature.

Research by producers, for the most part, has not been practical, because, except for LNG applications, the market for large material sales is not imminent. There are, however, many government-stimulated developmental programs. Research on nonmetallics thus is dictated by development project needs, which require studies oriented toward prototype hardware and specific objectives. As a result, research continuity suffers. Under these conditions, periodic topical conferences on this subject are needed. Industry and university studies must be encouraged. Designers and project research material specialists need to exchange experiences and data. Low-temperature-oriented research groups, such as the National Bureau of Standards and the Institute for Technical Physics - Karlsruhe, must contribute by assisting with fundamentals, interpreting project data, and contributing to project programs through their materials research.

To help meet these needs, the International Cryogenic Materials Conference Board sponsored this special conference, bringing together materials researchers and project designers to discuss properties and applications of nonmetallics and composites at low temperatures. The intensive and successful exchange is evidenced by the papers in this volume. The conference provided a unique forum to explore the full range of research on materials whose design need is urgently present, but whose basic property informa-

tion is quite scarce. It was concluded that nonmetallics and
composites are truly materials whose properties can be tailored to
fit the application, but that much more knowledge about constituent
properties and their combinations is needed.

The editors wish to express their sincere appreciation to
M. Stieg who worked hard to produce a uniform and exceptional
proceedings of this conference.

A. F. Clark
R. P. Reed
G. Hartwig

Contents

POLYMERS - PHYSICAL PROPERTIES AND RADIATION EFFECTS

THERMAL INSULATIONS

COMPOSITES - PROPERTIES OF FIBERS AND COMPOSITES

COMPOSITE AND POLYMER APPLICATIONS

INDEXES

CONFERENCE SUMMARY

NONMETALLIC MATERIALS AND COMPOSITES AT LOW TEMPERATURES

U. T. Kreibich

Ciba-Geigy AG
Basel, Switzerland

The object of the conference, which was attended by representatives of research institutes or industries of most European countries, the United States, and Japan, was a state-of-the-art review of this relatively new field. The conference was held July 10 and 11, 1978 in Munich, Germany and was sponsored by the International Cryogenic Materials Conference. The papers read covered the following:

1. The selection for low-temperature applications of materials, primarily polymers and polymer based composites, with reference to their properties and processing requirements.

2. The mechanical and dielectrical properties, thermal conductivity, and low-temperature thermal expansion of polymers, composites, and foams.

3. The effects of environmental factors, such as gases and radiation, and of incorporate additives and fillers on the properties of polymers and composites.

4. Molecular movement in the base polymers at low temperatures and its assessment and correlation with property variations.

5. Special properties of fibers and composites.

6. Practical applications for nonmetallic materials and composites in thermal or dielectric insulation and in the mechanical anchoring of superconductive cables and superconductive energy storage magnets, the construction of liquefied gas tanks and pipelines, and the manufacture of supports for space-based reflectors and telescopes.

The papers read showed that, for application at low temperatures down to 4 K, the most critical properties of nonmetallic
materials are their flexibility and resultant mechanical behavior,
their coefficients of thermal expansion as compared with that of
the inorganic materials with which they are combined, and their
thermal conductivity. At present, the leading polymeric materials
for low-temperature applications are epoxy resins, polypropylene,
and polyimide strips and films, and polystyrene and polyurethane
based foams. The leading fibers for reinforcement are glass,
graphite, boron, and organic polyaramid.

The types of base materials needed for low-temperature applications can be divided into four main categories:

1. Thermoplastic polymers exhibiting good low-temperature flexi-
    bility. The required property is found primarily in polymers,
    such as polyethylene, polycarbonate, or polyimide. Other
    polymers having moderate low-temperature flexibility (e.g.,
    polypropylene) exhibit relatively good elongation also at 4 K
    after undergoing orientation by stretching.

2. Laminating and bonding resins exhibiting good strength at low
    temperatures. At present, practical use is being made of
    conventional epoxy, polyurethane, and unsaturated polyester
    systems. These more or less meet the requirements, but papers
    and discussions showed that superior materials could probably
    be found or developed.

3. Crosslinking casting and laminating resins with good ductility
    at cryogenic temperatures. To date, the requirements of this
    category are largely unfilled. Rubber-elastic polyurethanes
    only partly meet the requirements, since they are very soft
    at room temperature. It might, however, prove possible to
    improve the low-temperature flexibility of crosslinking sys-
    tems by the introduction of flexible segments.

4. Fibers and fillers compatible with the matrix materials pro-
    viding the low temperature strength or the change in thermo-
    physical properties needed for structural composite applica-
    tions. These constituents must also retain their properties
    during manufacture of the composite and fabrication for the
    application.

High filler loadings and embedded graphite fibers in the
polymers can be used to impart to their composites low coefficients
of thermal expansion or coefficients corresponding more closely to
those of the materials with which they are combined. These techniques reduce the stresses that arise when a structure composed
of materials having different expansion coefficients is cooled to

cryogenic temperatures. They make possible constant dimensions
and stable structures that are vital in space-based optical systems,
for example.

It has been found that the properties of polymers, e.g., the
dielectric loss factor, can be markedly affected by the release of
low-molecular compounds, such as additive or degradation products
following oxidation or exposure to radiation at low temperatures.

Environmental factors must also be taken into consideration.
At low temperatures, some gases near their condensation points may
have a plasticizing effect on polymers and modify their mechanical
properties. And on warming up the material to room temperature,
gases that have diffused into a polymer at low temperature may
expand and change the material properties.

The application of nonmetallics and composites is rapidly on
the increase simply because designers are demanding properties that
can be met only by these new materials. The potential of designing
a material to meet the property needs is also being realized for
modulus, damping, residual stress, thermal expansion, thermal conductivity, and the directionality of these properties. This capability is also leading to innovative design and fabrication of
components, as evidenced by the papers read showing applications
from power lines to pipelines and flexible bellows to rigid magnet
supports.

Considerable uncertainty, however, still prevails regarding
the permissible loading of nonmetallic materials or composite
structures and the necessary design tolerances. The lack of information on these factors poses problems, particularly when designing
large structures. There is a call for materials made to standard
specifications and exhibiting identical quality, although produced
by different manufacturers. There is also a need for standard
testing procedures for these new composite materials.

Cost plays an important role, but, in compiling estimates for
advanced technology applications, the special advantages to be
gained by using a given material, e.g., simplified design, are
always taken into account.

The conference on "Nonmetallic Materials and Composites at Low
Temperatures" gave participants a good overall picture of the present
state of research on low-temperature applications for these needed
materials. The most interesting papers were those describing the
reader's own experiments and experiences. The papers and discussions revealed that materials science for low-temperature applications is a multieffort field rather than a narrow specialty.
Although polymers and composites are already finding industrial

application in spacecraft and the manufacture of liquefied gas
tanks, other applications, such as insulation and structural support for superconducting devices are still in the development stage.

The general impression gained was that an even better exchange
of information between engineers and polymer and composite manufacturers was necessary and would benefit both. The engineer
would have better access to advice on the selection of a material
for a given application, and the manufacturer would get a clearer
idea of the direction his efforts should take to evolve new materials. Cooperation and the choice of the right materials would de-
cide whether engineering projects or products intended to meet
extreme requirements ever go beyond the theoretical or small-model
stage. And cooperation would provide a solid base for efficient
development and research work.

POLYMERS IN LOW TEMPERATURE TECHNOLOGY

U. T. Kreibich, F. Lohse, and R. Schmid

Ciba-Geigy AG
Basel, Switzerland

INTRODUCTION

Polymers for low-temperature applications have received increasing interest with the advances in space research and the technological exploitation of physical phenomena, such as super- conductivity. Crosslinked polymers already find application in laminates for the construction of liquid gas tanks and pipelines. They are used for insulation of superconducting cables, impregna- tion of superconducting energy storing magnets, and supporting devices of optical systems in space applications. The castings have to be processed free of voids and may not release volatile compounds at low temperatures.

Polymers for low-temperature applications have to satisfy three basic requirements. They should (1) be processable by the casting or impregnating techniques, (2) exhibit adequate mechani- cal strength at room temperature, and (3) be flexible and tough at low temperatures.

In the first section, a review is given on the synthesis of crosslinked polymers that might be suitable for low-temperature applications. In the second part, some mechanical properties of polymers at low temperatures are discussed, followed by our own results on the relaxation behaviour and material properties of different flexibilized epoxy resin systems at temperatures down to 77 K.

CROSSLINKING POLYMERS FOR LOW-TEMPERATURE APPLICATIONS

Only a selected group of polyaddition and polymerization systems meet the above mentioned processing requirements. Foremost
among these are relatively low-viscosity resin systems, which, due
to their structural and chemical properties, form macromolecules
with the desired properties. These low-viscosity systems include:

  1. Polyaddition systems based on
        epoxy compounds
        isocyanates

  2. Polymerization systems based on
        acrylic resins
        unsaturated polyesters
        siloxanes

In the following parts, the basic principles of the chemical
formation of these polymers are shown.

Epoxy Resin Systems

Since our experiments were carried out primarily with the
epoxy resin systems,[1-3] we shall deal first with the principles
underlying the synthesis of these polyaddition systems in Fig. 1.
Two components, a resin and a hardener, are needed to produce a
network. One must be at least bifunctional and the other more than
bifunctional. Of the epoxide compounds, or epoxy resins as they
are often called, technical bisphenol-A-diglycidyl ether, novolak
glycidyl ether, hexahydrophthalic acid diglycidyl ester, N,N'-
diglycidyl hydantoine derivatives, and triglycidyl isocyanurate
are the most important. In addition to these, cycloaliphatic
epoxies are used which contain the epoxy group in a cyclic structure, e.g.,

Araldit CY 175[(R)]*

The most widely used hardeners are dicarboxylic acid anhydrides
(e.g., phthalic acid anhydride, tetrahydro and hexahydrophthalic
acid anhydride), long-chain hardener segments with carboxylic end

*(R) - registered trademark.

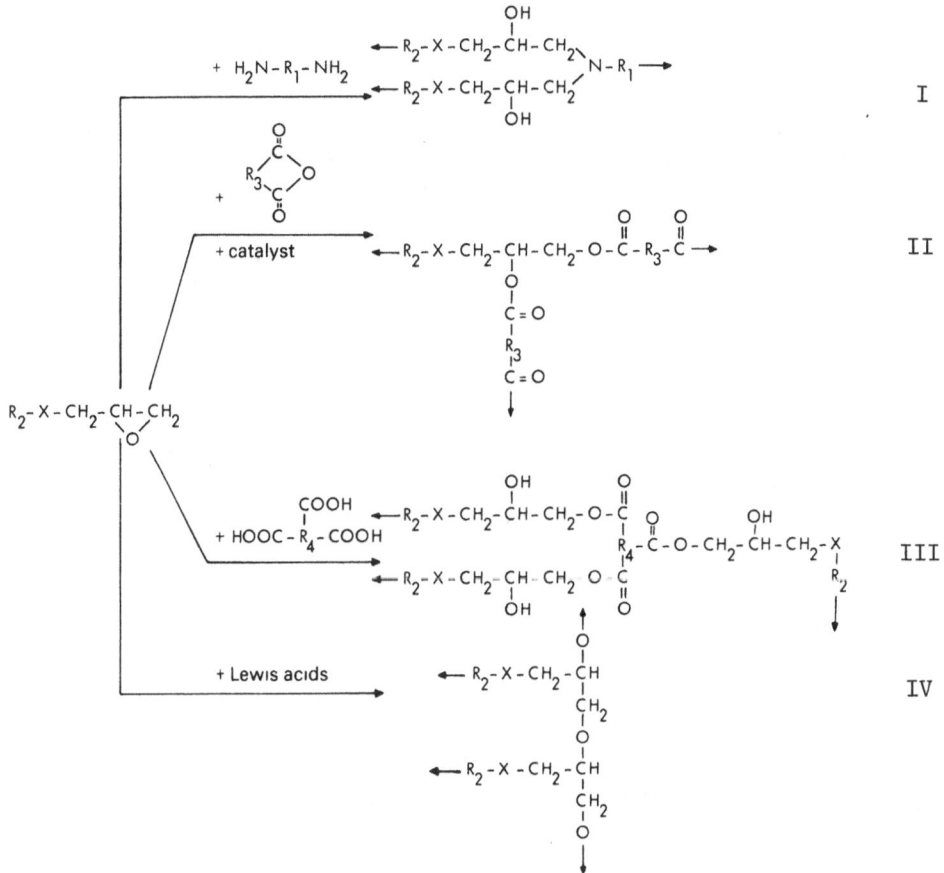

Fig. 1. Fragments of the network formed by the reaction
         of glycidyl compounds with amines, anhydrides,
         segments with carboxylic end groups, and Lewis
         acids.

groups, and amines (e.g., triethylene tetramine, isophorone diamine, diaminodiphenyl methane). The addition process results in
reaction products that have the atomic structure indicated in
Fig. 1 for the reactive groups.

The active hydrogen atoms in an amino group react with glycidyl groups to form bis-(β-hydroxy propyl)-amine structures, as
illustrated by structural fragment I for a primary amine group.

This fragment becomes a characteristic part of the network, the nitrogen atom being the point at which branching or crosslinking occurs.

In anhydride-induced crosslinking, on the other hand, a glycidyl group is attacked, especially in base-catalysed systems, by two moles of dicarboxylic acid anhydride, which form an ethylene glycol diester structure of Type II. The central C atom of the glycidyl residue then becomes the branching point and the

$$- CH - CH_2 - O -$$
$$|$$
$$O$$
$$|$$

residue becomes a structural part of the network.

If polyaddition is induced without using basic catalysts, lower proportions of anhydride must be used, since, depending on the conditions under which the reaction takes place and the acidity of the intermediate ester carboxylic acids, some of the glycidyl groups will form ether groups according to Fig. 1 structural fragment III.

The use of Lewis acids as catalysts results in the direct polymerization of the epoxy groups and formation of structural fragment IV.

So, a fragment of the network formed by an industrial bisphenol-A-diglycidyl ether cured with phthalic acid anhydride is shown schematically in Fig. 2. Amine curing of the same epoxy resin results in the idealized fragment shown in Fig. 3.

For low-temperature applications, special interest attaches to the curing of epoxy resins with long-chain hardener segments, as shown schematically in Fig. 4.

The long-chain hardener segments are built up by a simple condensation reaction between a moles of a dicarboxylic acid and a-1 moles of a glycol, as shown in Fig. 5. Further reaction of the carboxylic groups with diepoxide groups then produces long-chain, flexible epoxide compounds, which require the use of corresponding hardeners. The best results have been obtained with anhydride hardeners.

Isocyanate Addition Systems

Isocyanate polyaddition systems are also of considerable interest in view of the way the properties of the polymer can be varied. A description of the characteristic two-step network

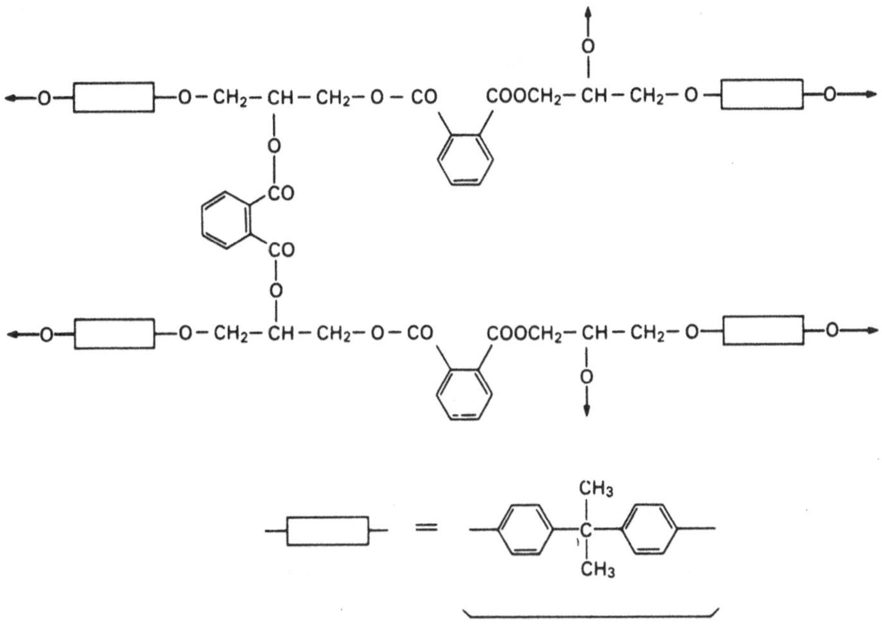

Fig. 2.   Schematic fragment of the network formed by a
          bisphenol-A-diglycidyl ether cured with phthalic
          acid anhydride.

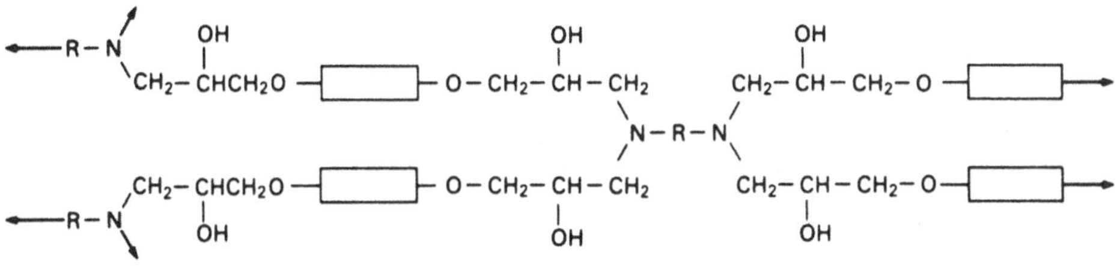

Fig. 3.   Schematic fragment of the network formed by
          bisphenol-A-diglycidyl ether cured with a dipri-
          mary amine.

formation of isocyanates would, however, call for a detailed explanation, which would exceed the scope of this paper. Only the
basic principle, therefore, is illustrated by means of the simple
structure of a polyurethane network (Fig. 6).

     Di- and tri-isocyanates are generally cured with di- or tri-
hydroxy compounds, which can be synthesized as oligoether or
oligoester from low-molecular units. This allows control of the
synthesis to obtain the desired polymer properties. The formation

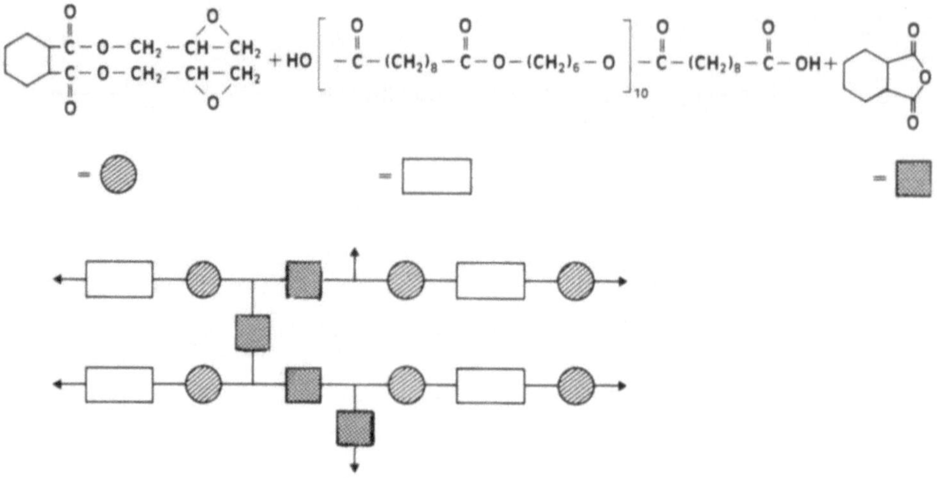

Fig. 4.   Reaction of a bivalent epoxy resin with a bivalent
          hardener and a bivalent oligoester with carboxylic
          end groups (idealized network structure).

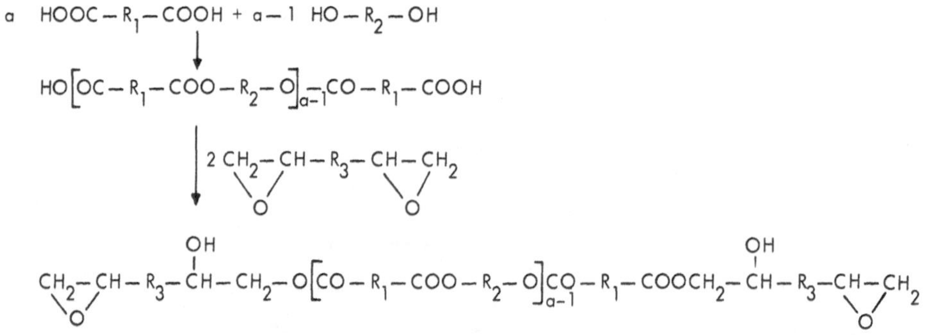

Fig. 5.   Polycondensation reaction of a dicarboxylic acid
          with a glycol and formation of a prepolymer with
          an epoxy resin.

of hydrogen bonds, however, could reduce the flexibility at low
temperatures. This should be proved by further investigations.

     The polymerization systems, the second group of polymers
mentioned before, are not as readily tailored to requirements as
the polyaddition systems described above, where presynthesized
elements can be used to determine network structures. Nevertheless,
polymerization systems could also be suitable materials for low-
temperature applications, since it would appear possible to impart

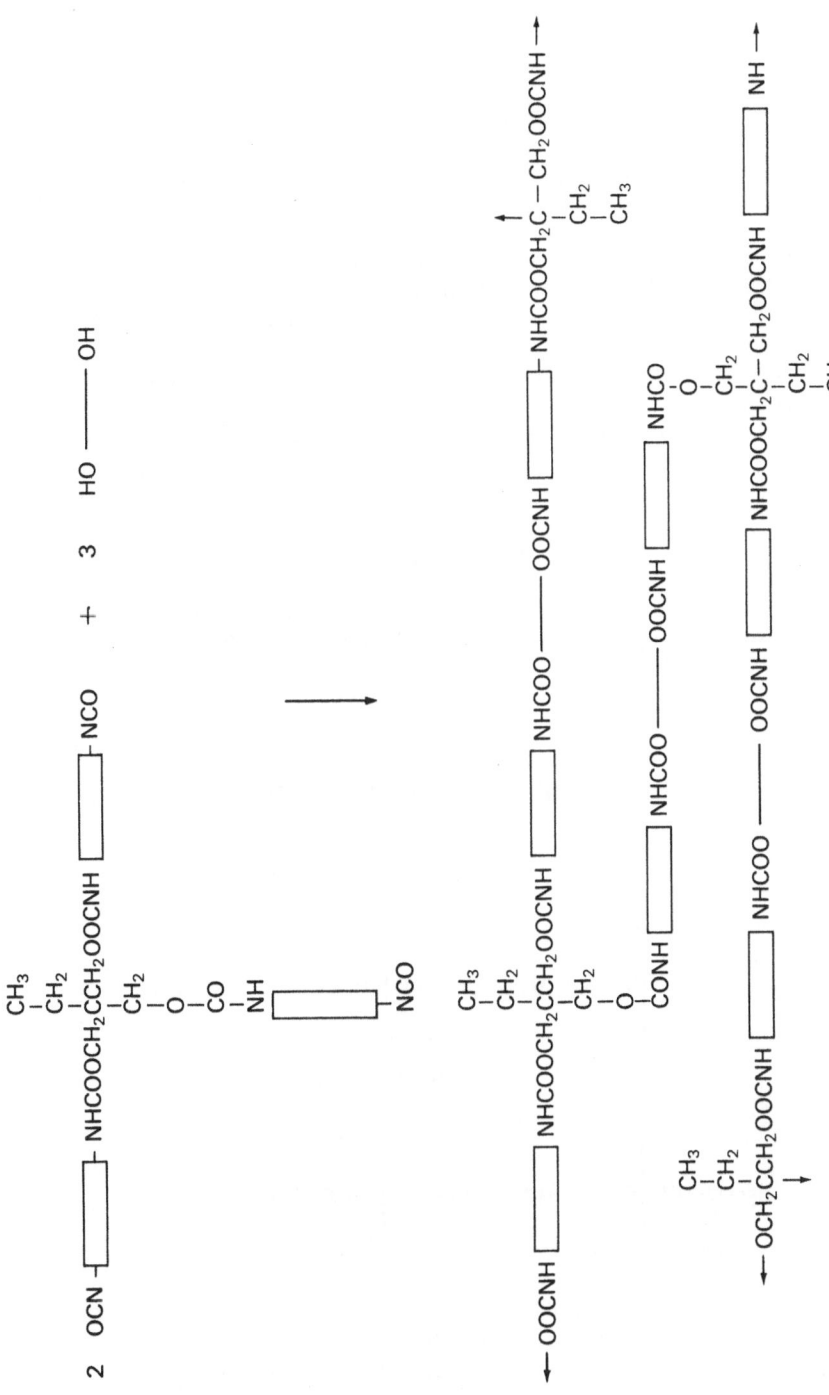

Fig. 6. Schematic structure of a polyurethane network.

the desired low-temperature flexibility by incorporating long
flexible chains.

Polymerization of Acrylic Resins

Oligoesters with acrylic ester end groups can be polymerized
either directly or in blends with other monomers. The synthesis
of the oligoesters from dicarboxylic acid and glycols in Fig. 7
has already been described with similar molar relationships for
epoxide and isocyanate curing agents. The hydroxylic end groups
are saturated with acrylic acid esters and then polymerized.

The reaction of acrylic acid with epoxide pre-adducts (e.g.,
the end product of Fig. 5) also results in useful raw materials
that can be crosslinked by polymerization (Fig. 8). The structure
of such a network is shown schematically in Fig. 9.

Unsaturated Polyesters

In crosslinked polymers based on unsaturated polyesters, two
components determine the properties of the end product: (1) the
structure of the unsaturated polyester, and (2) the structure of
the copolymerization component.

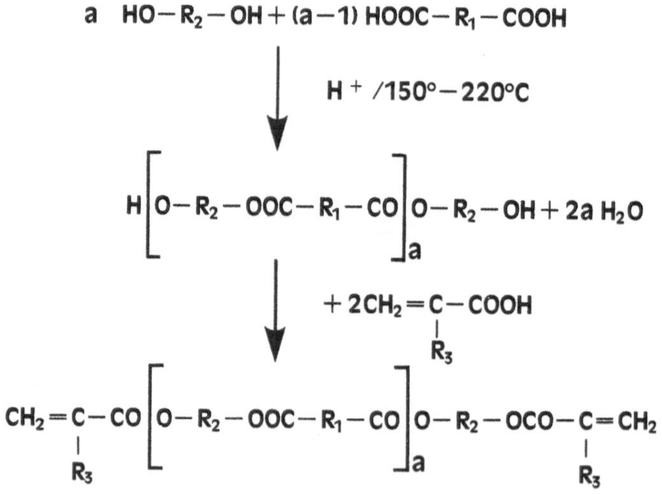

$$\text{a} \quad HO-R_2-OH + (a-1)\,HOOC-R_1-COOH$$

$$\downarrow \quad H^+ / 150^\circ - 220^\circ C$$

$$H\left[O-R_2-OOC-R_1-CO\right]_a O-R_2-OH + 2a\,H_2O$$

$$\downarrow \quad + 2CH_2 = C-COOH \atop \qquad\qquad R_3$$

$$CH_2 = C-CO\left[O-R_2-OOC-R_1-CO\right]_a O-R_2-OCO-C=CH_2 \atop R_3 \qquad\qquad\qquad\qquad\qquad\qquad R_3$$

Fig. 7.   Synthesis of oligoesters with acrylic ester end
          groups by esterification with acrylic acid of
          oligoesters with hydroxylic end groups.

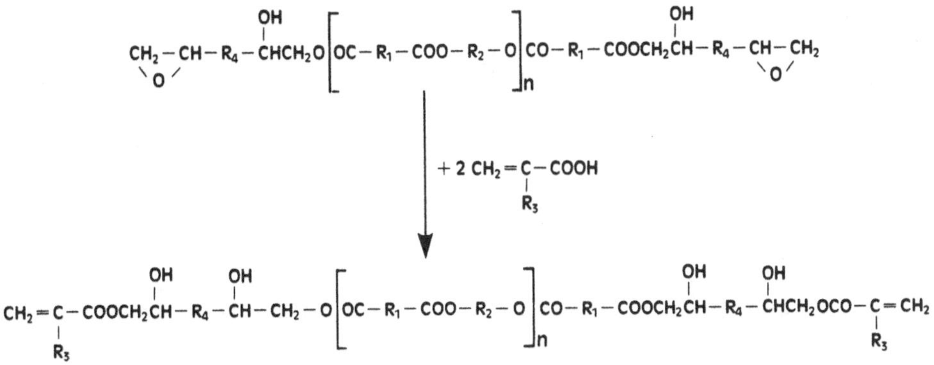

$$CH_2-CH-R_4-\underset{\underset{OH}{|}}{CHCH_2O}\left[OC-R_1-COO-R_2-O\right]_n CO-R_1-COOCH_2\underset{\underset{OH}{|}}{CH}-R_4-CH-CH_2$$

$$+\,2\;CH_2=\underset{\underset{R_3}{|}}{C}-COOH$$

$$CH_2=\underset{\underset{R_3}{|}}{C}-COOCH_2\underset{\underset{OH}{|}}{CH}-R_4-\underset{\underset{OH}{|}}{CH}-CH_2-O\left[OC-R_1-COO-R_2-O\right]_n CO-R_1-COOCH_2\underset{\underset{OH}{|}}{CH}-R_4-\underset{\underset{OH}{|}}{CHCH_2OCO}-\underset{\underset{R_3}{|}}{C}=CH_2$$

Fig. 8. Synthesis of oligoesters with acrylic ester
         end groups from an epoxy prepolymer and
         acrylic acid.

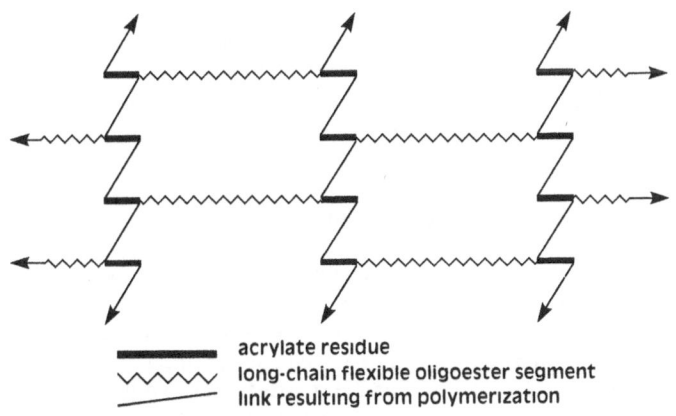

━━━━━ acrylate residue
〜〜〜 long-chain flexible oligoester segment
───── link resulting from polymerization

Fig. 9. Schematic: idealized network of a polymerized
         oligoester acrylate.

     The synthesis of an unsaturated polyester with a molecular
weight that should not be too high because of the many double
bonds present (a molecular weight of 1000 to 3000 suffices) is
illustrated in Fig. 10.

     Maleic acid and phthalic acid are used as oligoester components
together with bivalent alcohols (e.g., ethylene glycol, 1,4-
butanediol) as co-condensation partners. Crosslinking is achieved
by copolymerization of styrene, vinyl acetate, diallyl phthalate,
acrylonitrile or acrylic ester. Cyclic structures impart a higher
glass transition temperature. The aliphatic glycols in the

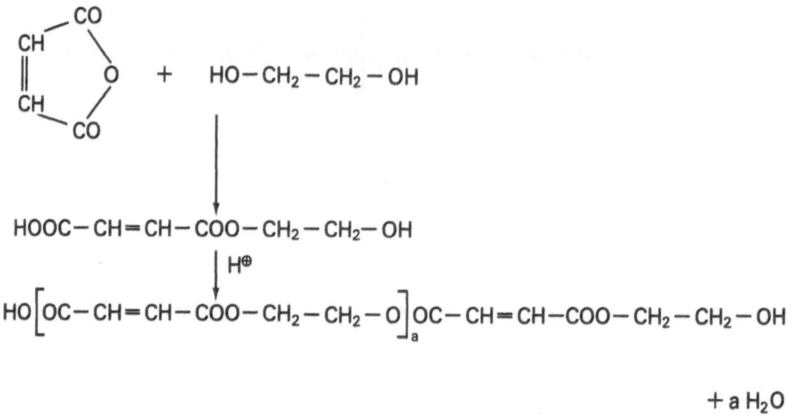

Fig. 10.   Synthesis of unsaturated polyester.

oligoester give flexibility and toughness to the product.   The polar groups, in this case ester groups, impart good adhesion to glass fibres.   Following addition of a monomer, e.g., styrene with radical initiators R–R, copolymerization is performed in the usual way and the initial mixture a is converted into the network b (Fig. 11).

Polysiloxanes

The physical properties of polysiloxanes are relatively un- affected by temperature owing to their exceptionally weak inter- molecular forces.   High-molecular chain-like compounds of this series are always liquids, and their glass transition temperatures may be as low as 150 K.   They fail to meet the requirement of sufficient mechanical strength at room temperature.   This weak- ness can be overcome by crosslinking the chains by "hot vulcani- zation."   Peroxides of the usual type are added to the polydimethyl siloxanes followed by curing at 70 to 120°C (343 to 393 K); see Fig. 12.

The radical crosslinking mechanism links the chains via the $Si-CH_3$ groups, according to the reaction scheme shown in Fig. 12. The distances between the individual links of the resulting net- work are relatively large.

The cold curing method commonly used to crosslink polymethyl silicones is a polycondensation reaction involving the splitting off of low-molecular units.   Such systems cannot fulfil the basic requirements for low-temperature applications.

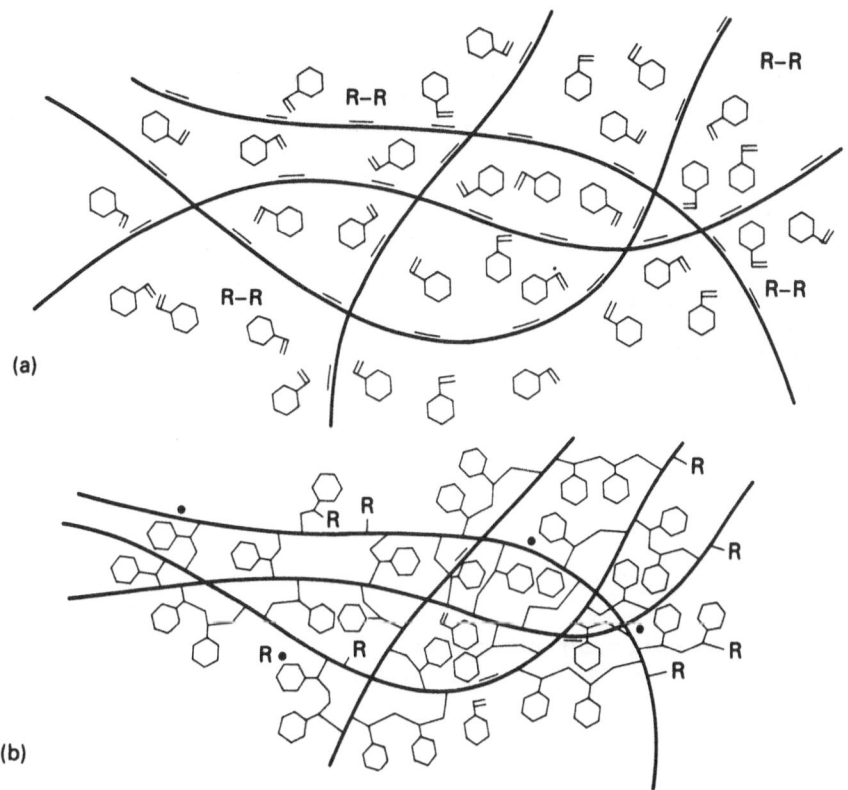

Fig. 11.   Schematic network formation by reaction of an
           unsaturated polyester and styrene.[4]

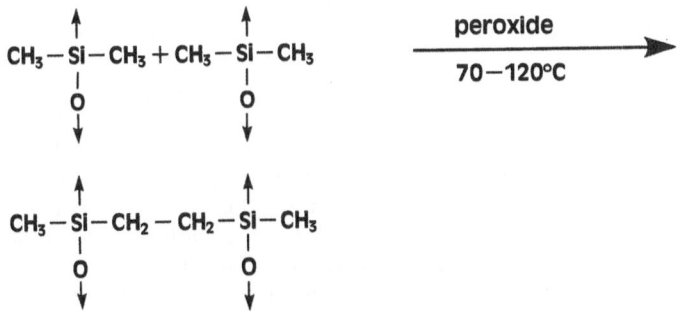

Fig. 12.   Radical crosslinking of polydimethyl silox-
           anes with peroxides.

MATERIAL PROPERTIES AT LOW TEMPERATURES

In the following, some typical properties of polymers at low temperatures[5,6] are described and related to the chemical structure. In theory, falling temperatures always promote a progressive increase in

    modulus
    flexural strength
    tensile strength
    creep resistance
    fatigue strength
    adhesive strength
    dielectric strength and resistance

and a progressive decrease in

    elongation
    deflection
    fracture toughness
    impact strength
    compressive strength
    coefficient of linear expansion
    permittivity and dielectric loss factor

In practice, a decrease in strength is frequently observed with decreasing temperatures, since low temperatures cause a loss of flexibility resulting in premature brittle failure before reaching the yield point.

Mechanical Properties

The effect of temperature on material properties is illustrated by the mechanical properties of some plastics. With decreasing temperatures, the yield stress increases and the elongation decreases (Figs. 13 and 14). A comparison of the tensile tests of polyethylene and polyamide 6.6 at different temperatures shows that, at a temperature of 163 K, double the room temperature load is required to induce failure. In both cases, elongation decreases with temperature. Polyamide 6.6, however, fails at 163 K before its yield point is reached. A comparison of failure as a function of temperature in Figs. 15 and 16 shows that, at low temperatures, the yield stress of polyethylene is markedly lower than its tensile strength. This fact is due to the weak interaction of the CH_2 groups with one another, there being only relatively low shear resistance to overcome. The molecule

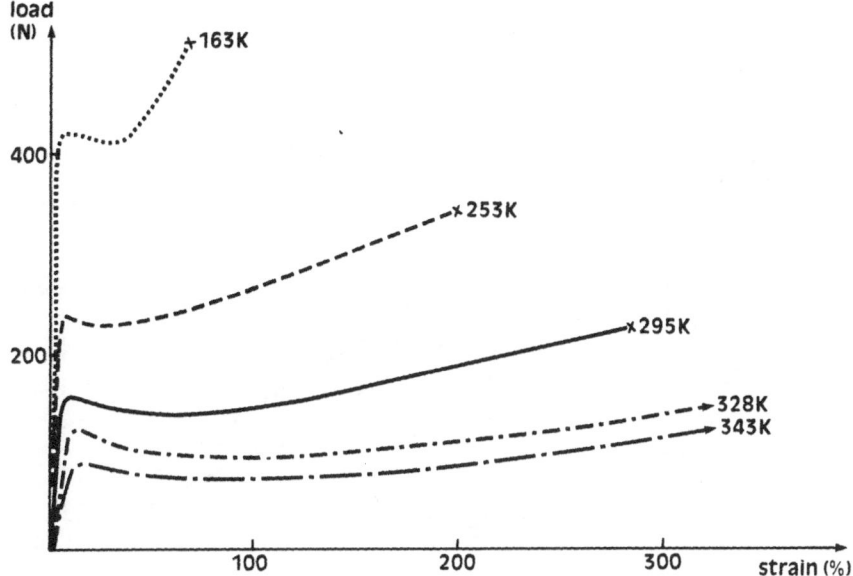

Fig. 13.   Tensile test of high-density polyethylene
           (T_{melt} = 404 K).

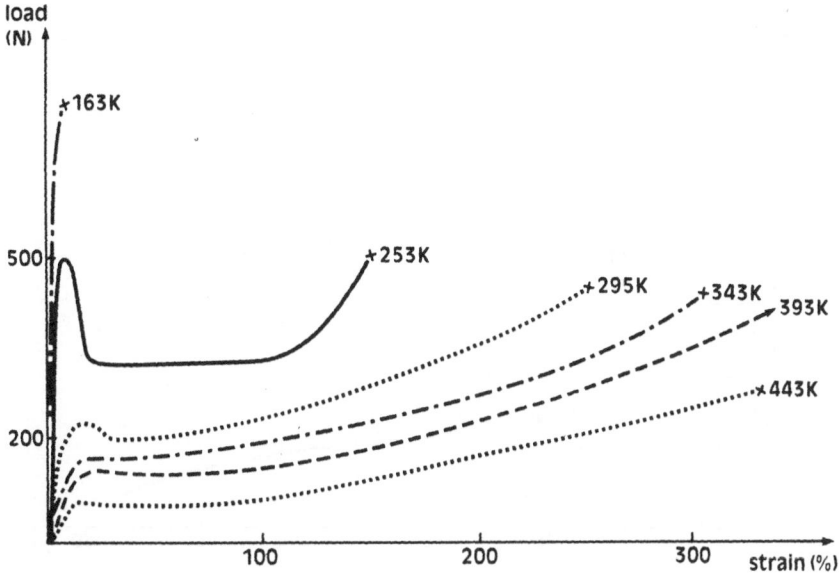

Fig. 14.   Tensile test of polyamide 6.6.

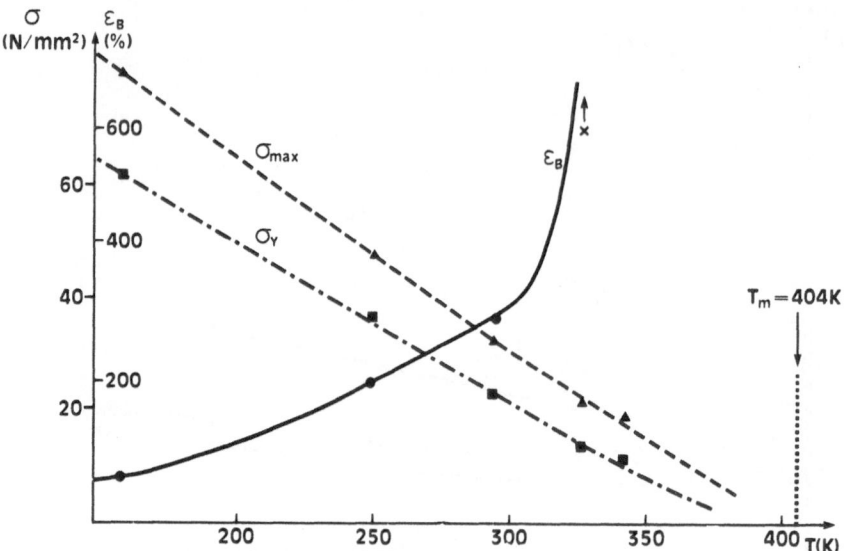

Fig. 15.   Tensile strength, σ_{max}, yield stress, σ_y, and
elongation at break, ε_B, of high-density polyethylene as a function of temperature.

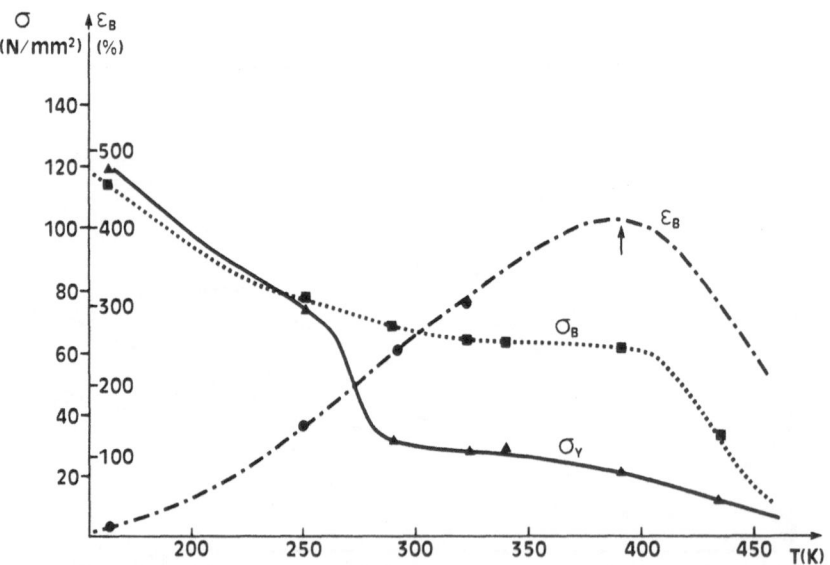

Fig. 16.   Tensile strength, σ_B, yield stress, σ_y, and
elongation at break, ε_B, of polyamide 6.6 as
a function of temperature.

chains are mobile even at low temperatures and can still absorb considerable tension after shearing off and before failure finally occurs. In polyamide 6.6, however, in the glassy state, the high cohesive strength prevents adjacent polymer chains from shearing off. But as soon as the glass transition is reached, there is a dramatic drop in yield stress (Fig. 16).

Failure at very low temperatures is also determined to a considerable degree by the surrounding medium. Just as solvents at high temperatures frequently promote crazing in polymers in their glassy state, gases, such as N_2, Ar, O_2, or CO_2 in a thermodynamically highly active state, that is, near their condensation point, promote crazing and a loss of yield strength in plastics at low temperatures.[7-9] Under such conditions, greater elongation will be observed near 77 K than under vacuum or in the presence of helium, particularly at high testing speeds. The absorbed gas reduces the surface energy of the polymer and promotes the formation of new surface area in the crazes. The absorbed gas also acts as a plasticizer.

Figure 17 provides a comparison of elongation at break in various polymers and shows a relatively good flexibility for polyethylene even at low temperatures. Polyvinyl chloride and the crosslinked epoxy resin system are brittle. High-density

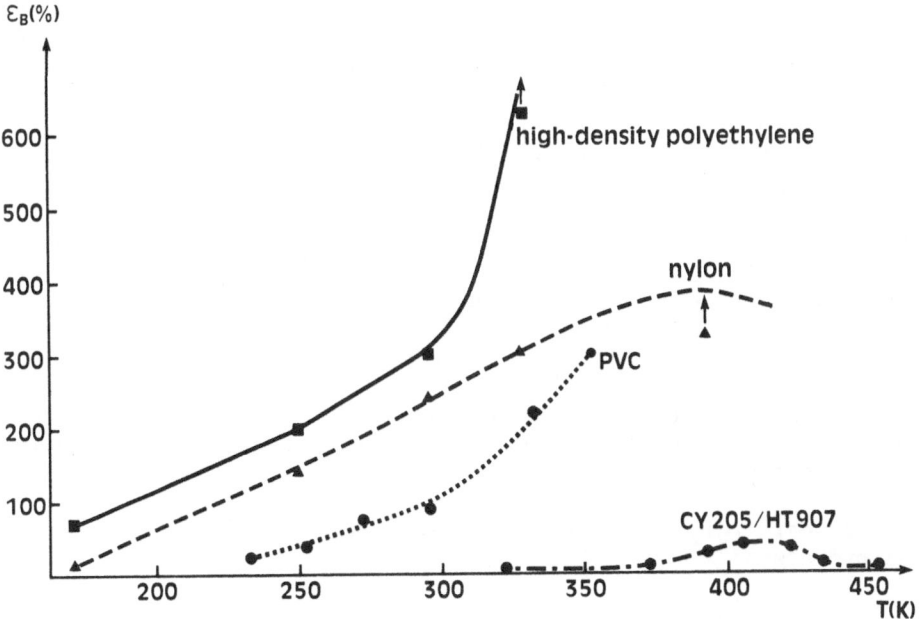

Fig. 17. Elongation at break as a function of temperature
          for various polymers.

polyethylene also displays markedly better impact strength at low
temperatures than other polymers[10] (Fig. 18). Even multiphase systems such as ABS and impact-resistant polystyrene embrittle relatively rapidly at low temperatures. They lose their characteristic
impact strength as soon as the phase with the lower glass transition
temperature, T_2, is freezing-in.

Relaxation Properties

    The application of polymers at low temperatures is determined
by their flexibility and toughness. Both properties are closely
linked to molecular movements in plastics in their glassy state,[11-13]
which are shown as specific secondary relaxations in the mechanical
vibration spectrum.

    The effect of molecular movement on mechanical properties is
dramatically apparent at the glass transition. Toughness and
frequently elasticity attain maximum values at the beginning of
freezing-in of Brownian movement, as has also been shown for epoxy
resin systems.[14] Relatively high toughness persists down to 50 to
150 K below the glass transition temperature.

    The relationship between relaxation properties and deformation
and failure at low temperatures has been the subject of various
papers.[5,6,11,12] Nearly all polymers that are tough and ductile in
their glassy states, having high impact strength and low notch sensitivity, exhibit clearly defined low-temperature relaxations. The
effect of the β relaxation on impact strength, owing to the vibration of aliphatic segments with about 3 to 10 units, is well known.
In the temperature range at which the β relaxation occurs, impact
strength is changing directly.[15,17] A correlation has been found
between the β relaxation and the frequency dependence of fatigue
crack propagation in polymers.[18] Changes in the coefficient of
linear expansion parallel the beginning of molecular movement in
the glassy state well below the glass transition.[19,20] A connection
has been found to exist between the γ relaxation in polyethylene
terephthalate and polycarbonate at 50 K and the beginning of the
nonelastic behaviour in the stress-strain diagram.[9]

    Movement has even been shown to occur in the 1 to 4.2 K range
in polyethylene, polyamide, polyethylene terephthalate, and fluori-
nated hydrocarbons.[21,22,23]

    Secondary relaxation in homopolymers may be induced by several
local processes:[6,24]

    1. Local movements of the main chain, these being weaker
        cooperative movements than at the glass transition,
        e.g., the crankshaft movement of the CH_2 sequence.[25]

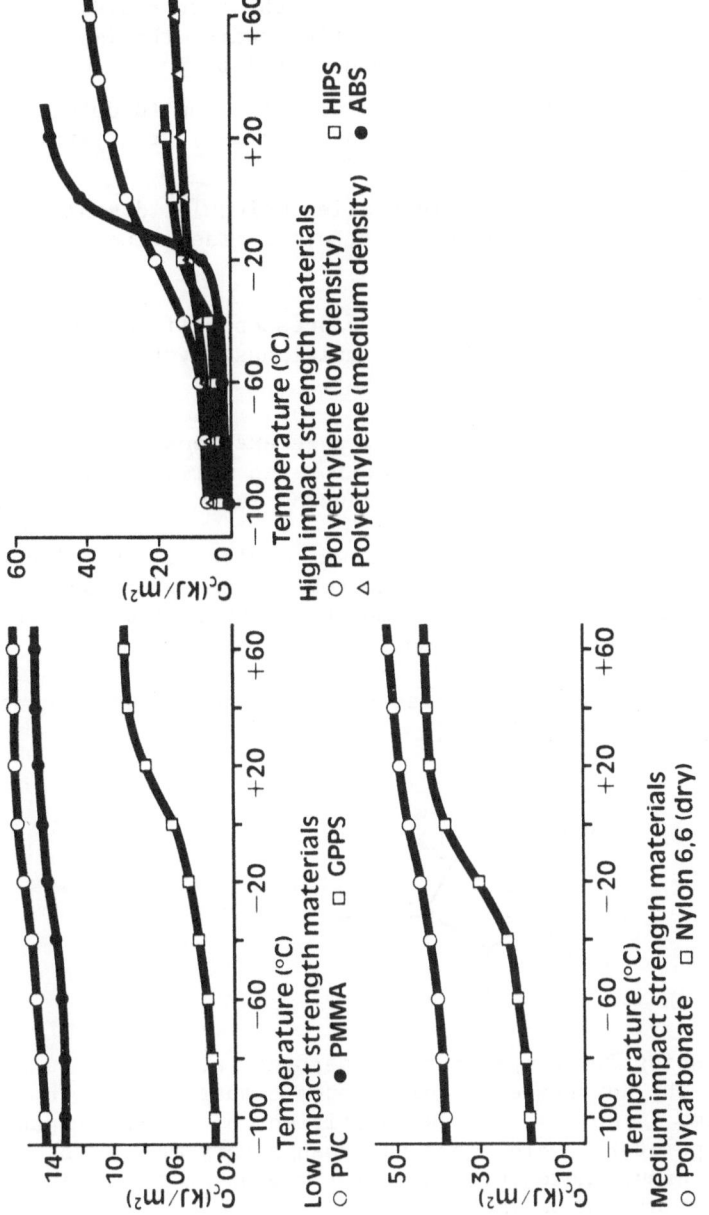

Fig. 18. Critical strain release rate, G_c, from impact tests as a function of temperature (from Ref. 10).

2. Relaxation of side groups involving the main chain
    to some extent.

3. Movement within the side chain without effects on
    the main chain. (Movements in the side chain
    usually contribute very little towards the improve-
    ment of impact strength and ductility.)[15]

4. Movements in the main or side chain due to defects
    in packing or conformity in the glassy or crystal-
    line state.

5. Movements by or within smaller molecules dissolved
    or embedded in the polymer, e.g., plasticizers or
    additives.

Relaxation induced by any of these processes differs theoretically
from low-temperature relaxation in multiphase systems due to the
glass transition of one of the phases.

In highly crosslinked systems, β relaxations may be limited
by the antiplastification effect of foreign molecules that occupy
part of the available free volume causing an increase in modulus.[26]

In the following section, our own results will be discussed
concerning the correlation between chemical structure and the relaxation behaviour, as well as some other physical properties of
epoxy resin polymers at low temperatures.

Of the epoxy resin polymers, systems based on hydantoine segments with fragments such as

have interesting low-temperature properties e.g., Aracase 350[(R)]/
HY 924[(R)]. The advantage of this system compared with other epoxy
resin systems is due to its low coefficient of thermal expansion
at room temperature, which does not change very much with decreasing temperatures. Moreover, the polar groups of the resin cause
good adhesion of the polymer to glass fibers.

Since a good low-temperature flexibility has been found for
polyethylene, the introduction of long aliphatic chains into crosslinking polymers is expected to improve their flexibility, too.[2,27]

The crosslinking density of the flexibilized epoxy resin systems
is variable over a wide range, and the chemical incorporation of
the flexibilizer precludes the migration frequently observed with
plasticizers. Figure 19 shows part of an idealized network structure for epoxy-polyester copolymers.

The flexibilizer markedly modifies the relaxation behaviour
of the epoxy resin systems shown in Figs. 20 and 21. Its incorporation results in crosslinked two-phase systems. But the two phases
are compatible and therefore not clearly separate. The T_g of the
resin-hardener matrix (α_1 relaxation) is smoothly passing into the
T_g of the incorporated oligoester (α_2 relaxation). The maximum of
the α_1 relaxation shifts to lower temperatures as the flexibilizer
content of the system is increased. The α_2 relaxation always occurs
at nearly the same characteristic temperature, as is evident from
the modulus decay at about 240 K in Figs. 20 and 21. The reduction
in modulus observed in the rubber-elastic state shows the decrease
in crosslinking density caused by an increase in flexibilizer content.

As the oligoester content is increased, the pronounced β
relaxation of the unflexibilized system at about 230 K decreases
in intensity. Another damping peak then appears in the 140 K range,
with an intensity that increases as more oligoester is added; this
γ relaxation is due to the vibration of the oligoester segment.
Similar relaxations of aliphatic chain segments have been observed
in styrene-crosslinked polyesters,[28] nylon polymers,[29] and poly-
urethanes.[30,31] The type of movement occurring has not been clearly
identified. For instances, it could be a crankshaft[25] or double-
kink[32] movement.

The effect of the T_g of the resin-hardener matrix on the relaxation properties of the flexibilized product is also of interest.
In a system comprising a resin-hardener matrix with a high T_g
(Fig. 21), the incorporation of an oligoester results in an exceptionally wide temperature interval for the mechanical damping, Λ,
compared with a system with a matrix having a lower T_g (Fig. 20).
There is relatively little shift (about 10 K) in the initial temperature of the α_2-relaxation specific for the flexibilizer or in
the maximum for the γ relaxation despite the very appreciable change
in the T_g of the rigid matrix. Thus, temperature interval and damping intensity can be set within the main relaxation range by suitable
choice of a resin-hardener matrix.

Varying the proportion of flexibilizer added only modifies the
intensity of the γ-relaxation peak, but changes in the chemical
structure of the oligoester segment result in marked changes in
the effect of temperature on shear modulus (Fig. 22) and mechanical
damping (Figs. 23 and 24) at low temperatures. With an increasing
number of CH_2 groups in the acid component at equal content of

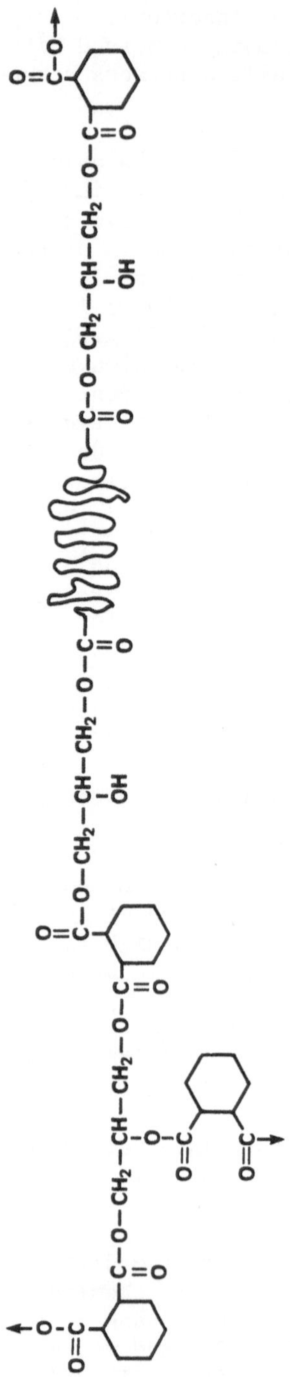

Hexahydrophthalic acid diglycidylester (HHPD) /
hexahydrophthalic anhydride (HHPA)/Oligoester

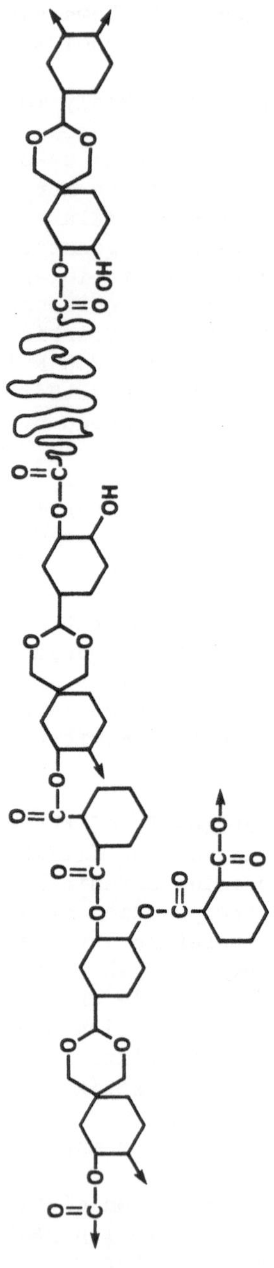

Araldit CY 175$^{(R)}$/HHPA/Oligoester

Fig. 19. Idealized network structure of epoxy-polyester copolymers.

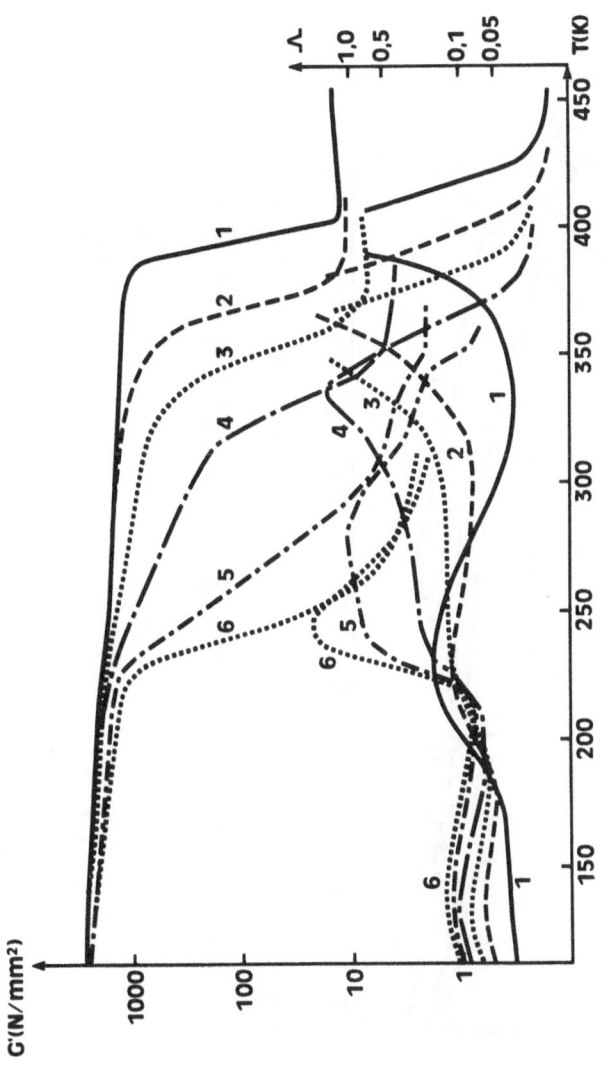

Fig. 20. Shear modulus, G', and mechanical damping, Λ, as a function of the amount of flexibilizer.

Curve	eq. HHPD/eq. HHPA/eq. sebacic acid neopentyl glycol (11:10)		
1	1.0	1.0	–
2	1.05	1.0	0.05
3	1.1	1.0	0.1
4	1.2	1.0	0.2
5	1.5	1.0	0.5
6	2.0	1.0	1.0

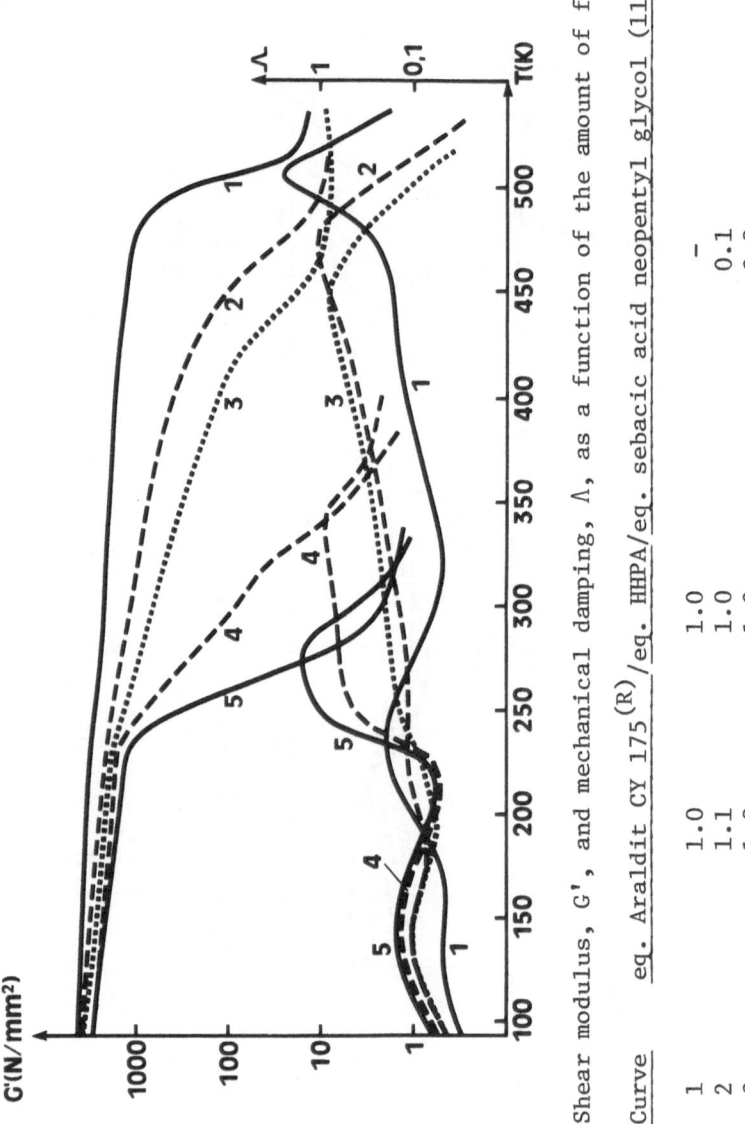

Fig. 21. Shear modulus, G', and mechanical damping, Λ, as a function of the amount of flexibilizer.

Curve	eq. Araldit CY 175$^{(R)}$	/eq. HHPA	/eq. sebacic acid neopentyl glycol (11:10)
1	1.0	1.0	−
2	1.1	1.0	0.1
3	1.2	1.0	0.2
4	1.5	1.0	0.5
5	2.0	1.0	1.0

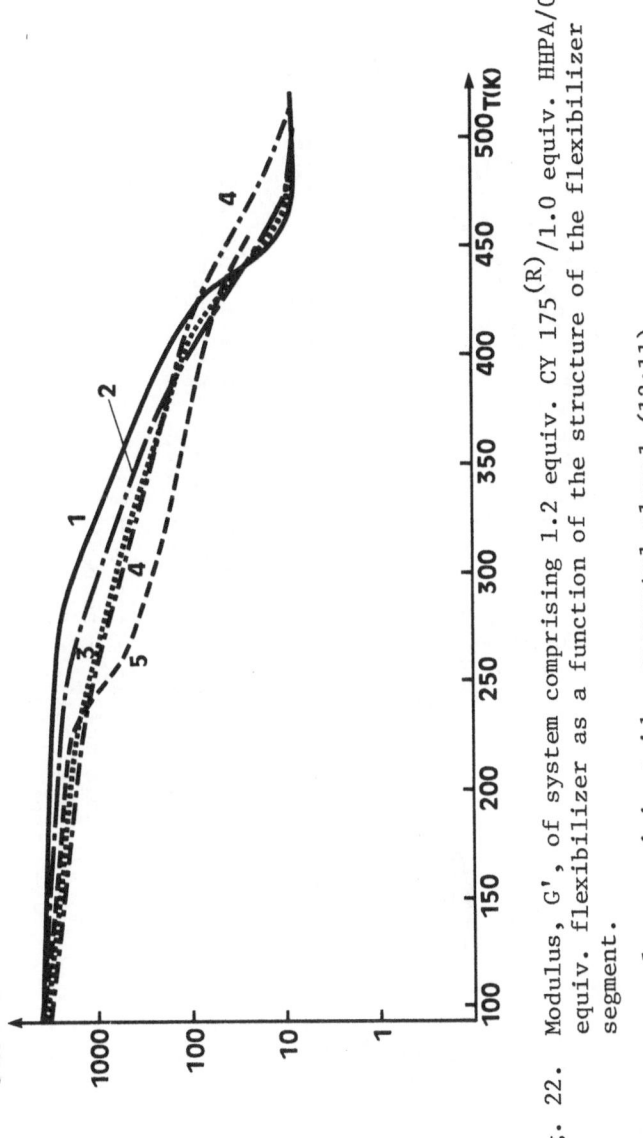

Fig. 22. Modulus, G', of system comprising 1.2 equiv. CY 175$^{(R)}$/1.0 equiv. HHPA/0.2 equiv. flexibilizer as a function of the structure of the flexibilizer segment.

1 – succinic acid – neopentyl glycol (12:11)
2 – adipic acid – neopentyl glycol (11:10)
3 – sebacic acid – neopentyl glycol (11:10)
4 – Empol$^{(R)}$ 1024 – diethylene glycol (3:2)
5 – Hycar$^{(R)}$ CTBN

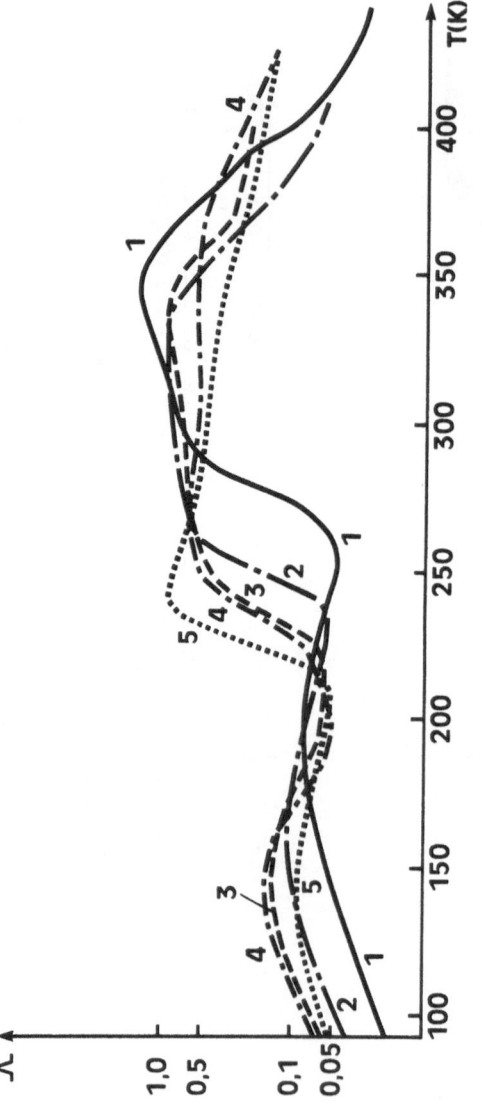

Fig. 23. Mechanical damping, Λ, of system comprising 1.5 equiv. CY 175$^{(R)}$ /
1.0 equiv. HHPA/0.5 equiv. flexibilizer as a function of the
structure of the flexibilizer segment.

1 – succinic acid – neopentyl glycol (12:11)
2 – adipic acid – neopentyl glycol (11:10)
3 – sebacic acid – neopentyl glycol (11:10)
4 – Empol(R) 1024 – diethylene glycol (3:2)
5 – Hycar(R) CTBN

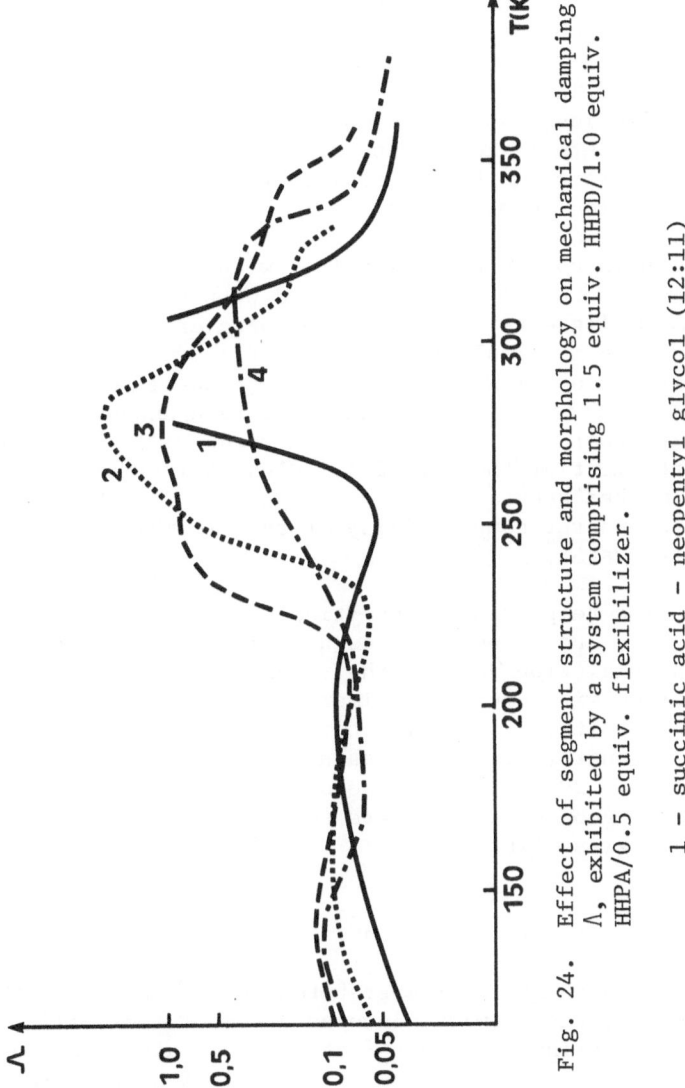

Fig. 24. Effect of segment structure and morphology on mechanical damping, Λ, exhibited by a system comprising 1.5 equiv. HHPD/1.0 equiv. HHPA/0.5 equiv. flexibilizer.

1 – succinic acid – neopentyl glycol (12:11)
2 – adipic acid – neopentyl glycol (11:10)
3 – sebacic acid – neopentyl glycol (11:10)
4 – sebacic acid – hexandiol (11:10)

flexibilizer, the beginning of the α_2 relaxation in the succinic acid-adipic acid-sebacic acid series was found to shift about 40 K to lower temperatures. The temperature of the γ maximum also decreases from about 200 K in the succinic acid system to about 140 K in the sebacic acid system.

The morphology of the multiphase system changes on switching from sebacic acid-neopentyl glycol oligoester to sebacic acid-hexandiol oligoester. The previously completely amorphous epoxy resin polymer goes into a partly crystalline crosslinked state,[33] since crystallization of the long-chain ester segment is no longer inhibited by the methyl side groups of the neopentyl glycol (Fig. 24). Yet, in spite of marked crystallinity, a weaker but clearly defined γ relaxation is preserved.

Relaxation in polymers flexibilized with Empol differs only slightly from that in the system modified with sebacic acid-neopentyl glycol oligoester, even though there are considerable differences in the chemical structures of the two flexibilizers. The incorporation of flexible, long side groups like those in Empol-diethylene glycol oligoester can, therefore, enhance toughness, whereas the incorporation of rigid side groups will cause considerable embrittlement of the material. Flexibilization with Hycar results in a marked separation into two phases, this being a consequence of the poorer compatibility of the flexibilizer with the resin-hardener system.

The good low-temperature properties of the flexibilized epoxy resin systems are due to the increased mobility of the segments from the main relaxation down to the γ relaxation. Epoxy resin polymers modified with Empol[R] (dimerized fatty acid) or Hycar[R] (butadiene-acrylonitril copolymer, terminated by carboxyl groups) show notable strength, impact resistance, and temperature-shock resistance down to 20 K.[34] Yet none of these systems is ductile below 77 K. The good properties are retained when the systems are reinforced with glass fibre.[35,36]

Bonding

Bonding strength at low temperatures is another property that can be enhanced by the chemical incorporation of flexibilizer (Fig. 25).[37] Comparison of the effect of temperature on the shear modulus and the torsional adhesive strength of a bonded aluminium-to-aluminium joint reveals that the incorporation of a flexibilizer reduces the shear modulus but, by imparting greater flexibility, increases the torsional adhesive strength of the bond at low temperatures.

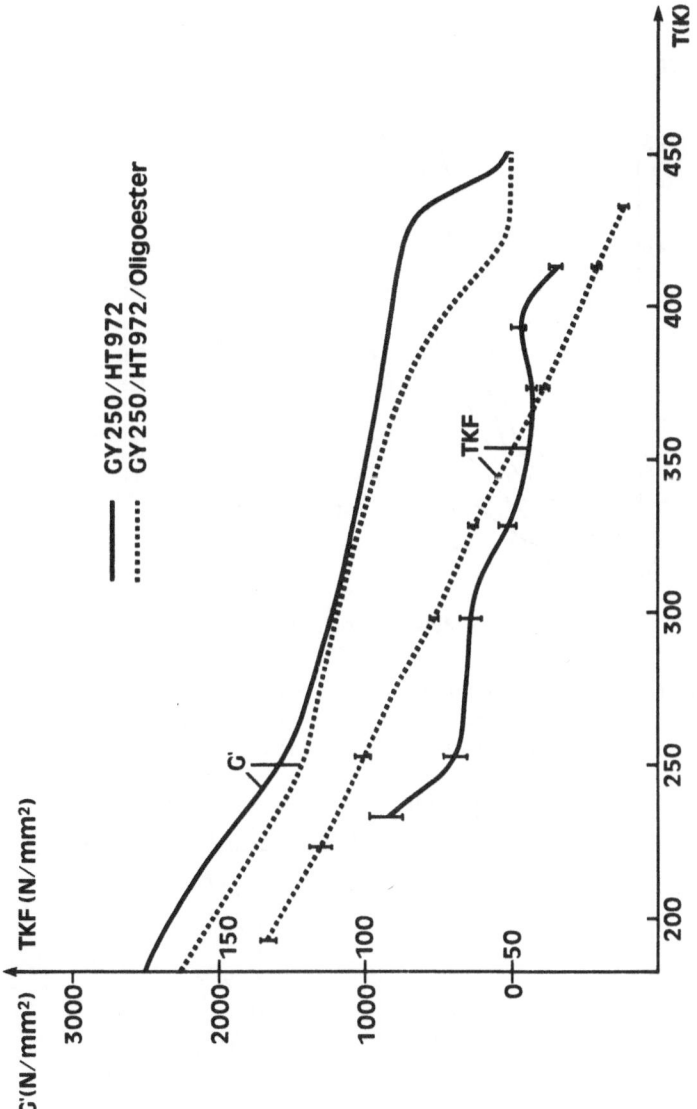

Fig. 25. Temperature dependence of shear modulus, G', and torsional adhesive
strength, TKF, as a function of flexibilization.

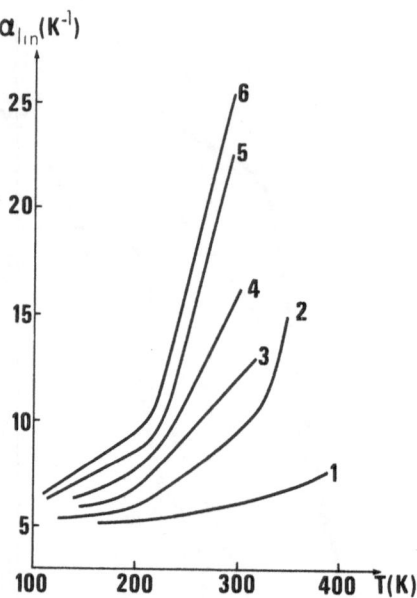

Fig. 26. Coefficient of linear thermal
          expansion, α_{lin}, as a function
          of flexibilizer content.

Curve	eq. HHPD	eq. HHPA	eq. sebacic acid neopentyl glycol (11:10)
1	1.0	1.0	–
2	1.05	1.0	0.05
3	1.1	1.0	0.1
4	1.2	1.0	0.2
5	1.5	1.0	0.5
6	2.0	1.0	1.0

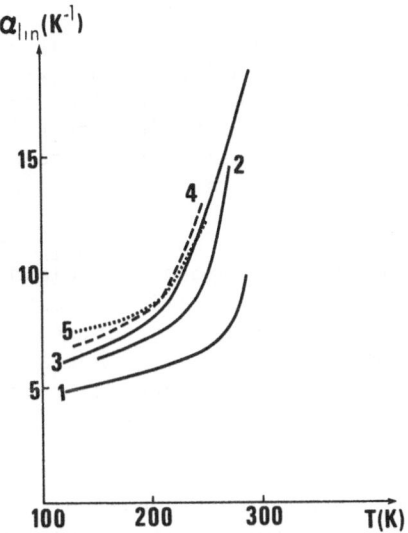

Fig. 27. Coefficient of linear thermal expansion,
α_{lin}, as a function of the chemical
structure of the flexibilizer segment in
a system comprising 1.5 equiv. Araldit
CY 175[R]/1.0 equiv. HHPA/0.5 equiv.
flexibilizer.

1 - succinic acid - neopentyl glycol (12:11)
2 - adipic acid - neopentyl glycol (11:10)
3 - sebacic acid - neopentyl glycol (11:10)
4 - Hycar CTBN[R]
5 - Empol 1024[R] - diethylene glycol (3:2)

Thermal Expansion

Not only the flexibility, but also the coefficient of thermal expansion, α, is of great significance in casting resins and adhesives. The coefficient of thermal expansion is usually much higher for polymers than for the parts to be encapsulated or for the glass fibre and fillers that are to be incorporated. Cooling, therefore, promotes extensive strain of the resin system in the boundary layer and may cause it to peel away from the substrate. The addition of a flexibilizer results in a coefficient of linear expansion that is much higher than that for the unflexibilized resin (Fig. 26), and it also causes a decrease in the torsional modulus. There is, consequently, an increase in the relaxation capability of the polymer and its ability to absorb internal stresses to some extent, at least above the level at which the α_2 relaxation begins.

The coefficient of linear expansion, α_{lin}, is closely correlated with the beginning of low-temperature relaxations. The values for α_{lin} are dependent on the flexibilizer content (Fig. 26). They are also determined by the chemical structure of the flexibilizer segment (Fig. 27). Unfortunately, α_{lin} increases with the chain length of the oligoester and, therefore, with the oligoester content. At low temperatures in particular, the Empol polyester gives rise to high α_{lin} values.

It is interesting to note that, despite the increase of the coefficient of thermal expansion by incorporation of flexibilizer into the epoxy resin system, the torsional adhesive strength increases to some extent with decreasing temperatures, as is shown in Fig. 25.

SUMMARY

The findings show that the introduction of aliphatic segments into the main chain, or, with the Empol-oligoester, also into the side chain of an epoxy resin system considerably enhances its flexibility and toughness at low temperatures. There is a close connection between the beginning of secondary relaxations and low-temperature properties which, because it correlates directly with the chemical structure of the molecules, makes feasible systematic study and improvement of the low-temperature properties of polymers.

ACKNOWLEDGMENTS

We would like to express our thanks to Prof. Dr. H. Batzer, head of the Research Department of the Plastics and Additives Division, Ciba-Geigy AG, for his generous support and valuable discussions, and to Dr. M. Fischer for mechanical investigations.

REFERENCES

1. H. Batzer, F. Lohse, and R. Schmid, Angew. Makromol. Chem. 29/30, 347 (1973).

2. F. Lohse and R. Schmid, Chimia 28, 576 (1974).

3. F. Lohse, Prog. Colloid. Polym. Sci. 64, 1 (1978).

4. C. Srna in: Kunststoff-Handbuch (R. Vieweg and L. Goerden, eds.), München (1973).

5. R.N. Haward, The Physics of Glassy Polymers, Applied Science Publishers Ltd., London (1973).

6. L.E. Nielsen, Mechanical Properties of Polymers and Composites, Marcel Dekker Inc., New York (1974).

7. N. Brown, J. Polym. Sci. Phys. 11, 2099 (1973).

8. A. Peterlin and H.G. Olf, J. Polym. Sci. Symp. 50, 243 (1975).

9. J.R. Kastelic and E. Baer, J. Macromol. Sci. Phys. B7, 679 (1973).

10. E. Plati and J.G. Williams, Polymer 16, 915 (1975).

11. L. Bohn and H. Oberst, Acustica 9, 191 (1959).

12. N.G. McCrum, B.E. Read, and G. Williams, Anelastic and Dielectric Effects in Polymeric Solids, John Wiley and Sons, London (1967).

13. M. Shen, Chem. Space Res. (1972), 319.

14. F. Lohse, R. Schmid, H. Batzer, and W. Fisch, Br. Polym. J. 1, 110 (1969).

15. J. Heijboer, J. Polym. Sci. C 16, 3755 (1968).

16. R.F. Boyer, Polym. Eng. Sci. 8, 161 (1968).

17. P.I. Vincent, Polymer 15, 111 (1974).

18. J.A. Manson, R.W. Herzberg, S.L. Kin, and M. Skibo, Polymer 16, 850 (1975).

19. J.M. Roe and R. Simha, Intern. J. Polym. Mater. 3, 193 (1974).

20. P.S. Wilson, S. Lee, and R.F. Boyer, Macromolecules 6, 914 (1973).

21. F.H. Müller, O. Heybey, and G. Knispel, Kolloid.-Z. Z. Polym.
     251, 932 (1973).

22. C.L. Choy, H. Huq, and D.E. Moody, Phys. Lett. 54A, 375 (1975).

23. Y.S. Papir, S. Kapur, C.E. Rogers, and E. Baer, J. Polym. Sci.
     A-2 10, 1305 (1972).

24. J. Heijboer, Ann. N. Y. Acad. Sci. (1977), p. 104.

25. T.F. Schatzki, J. Polym. Sci. 57, 496 (1962).

26. N. Hata, R. Yamanchi, and J. Kumanotani, J. Appl. Polym. Sci.
     17, 2173 (1973).

27. R. Schmid, Prog. Colloid. Polym. Sci. 64, 17 (1978).

28. W.D. Cook and O. Delatycki, J. Polym. Sci. Phys. 15, 1953
     (1977).

29. K.H. Illers, Makromol. Chem. 38, 168 (1960).

30. H. Jacobs and E. Jenckel, Makromol. Chem. 43, 132 (1961).

31. T. Kajama and W.J. MacKnight, Macromolecules 2, 254 (1969).

32. W. Pechhold, S. Blasenbrey, and S. Woerner, Kolloid-Z. 189,
     14 (1963).

33. U.T. Kreibich and R. Schmid, Prog. Coll. Polym. Sci. 62, 106
     (1977).

34. L.M. Soffer and R. Molho, J. Macromol. Sci.-Phys. B1, 709
     (1967).

35. M.B. Kasen, Cryogenics 15, 327 (1975).

36. M.B. Kasen, Cryogenics 15, 701 (1975).

37. R. Schmid, Kolloqium über Adhäsion, Leoben (1977).

MECHANICAL AND ELECTRICAL LOW TEMPERATURE

PROPERTIES OF HIGH POLYMERS

G. Hartwig

Institut für Technische Physik
Karlsruhe, West Germany

INTRODUCTION

High polymers have the disadvantage that they are too flexible
at high temperatures and too brittle at low temperatures.

What is the nature of brittleness? The chains of amorphous
polymers are entangled. At low temperatures most degrees of freedom are frozen; the chains, therefore, become stiff. This is the
situation where steric hindrance prevents chain movement relative
to one another. When an external load is applied, stress concentration occurs at some points. If no plastic flow is possible to
equalize the stress distribution within the material, crack initia-
tion and brittle fracture may occur.

STRESS-STRAIN BEHAVIOUR

Generally, the stress-strain behaviour of polymers can be
represented by a network of springs and dashpots which characterize
the elastic and damping elements. A very simple model is shown in
Fig. 1. The analogy between mechanical and electrical parameters
should be noted. Each element has an intrinsic dependence on temperature. At very low temperatures, the dashpots become more and
more frozen, thus causing the flow characteristics to deteriorate.
This is demonstrated in Fig. 2 for a typical epoxy resin (EP). EP
is a crosslinked system. For temperatures near the glass transition
temperature, the resin behaves viscoelastically and the stress-
strain relation is strongly dependent on the strain rate, $\dot{\varepsilon}$.
According to an intrinsic relaxation time, τ, a high or low strain
rate leaves the polymer little or much time, respectively, for

33

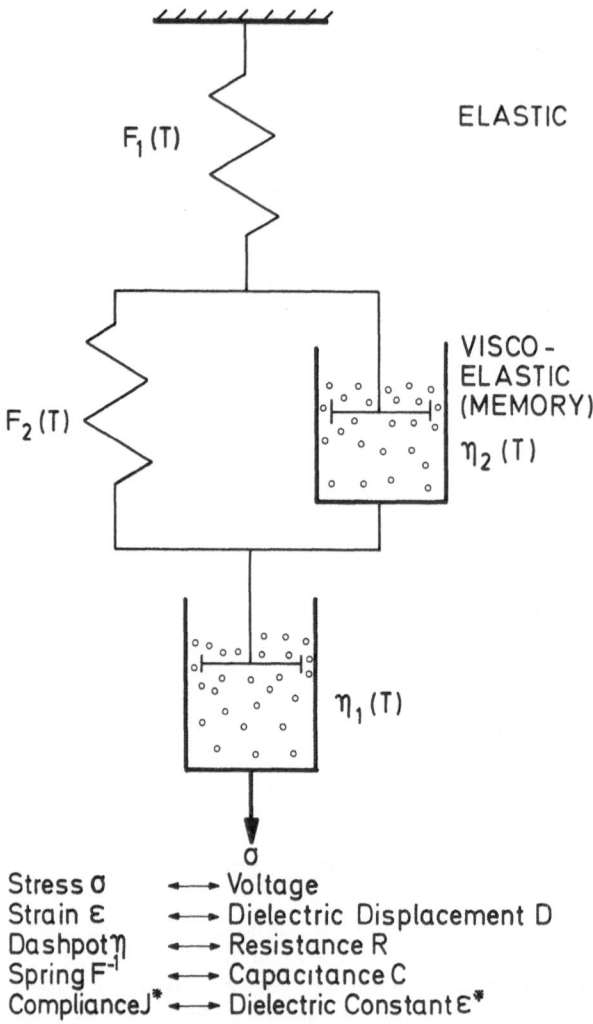

Stress σ ⟷ Voltage
Strain ε ⟷ Dielectric Displacement D
Dashpot η ⟷ Resistance R
Spring F⁻¹ ⟷ Capacitance C
Compliance J* ⟷ Dielectric Constant ε*

Fig. 1.   Schematic representation of viscoelastic properties.
Analogy between electrical and mechanical parameters.

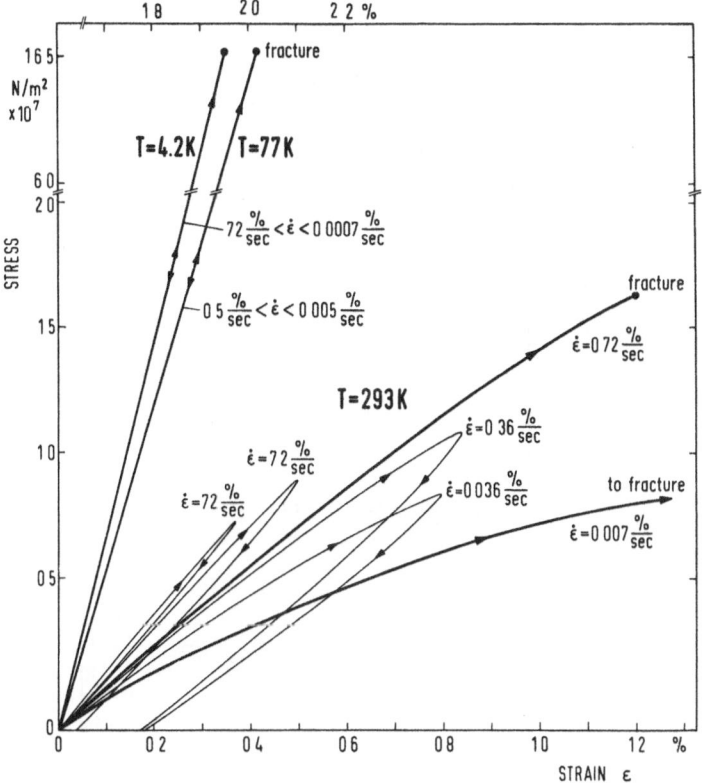

Fig. 2.   Stress-strain behaviour of epoxy resins at
           different strain rates, $\dot{\varepsilon}$, and temperatures, T.

plastic flow. The higher the strain rate, the more elastic and
stiff a material becomes. Usually, there are dissipative effects
connected with viscous flow, which manifest themselves in a hysteresis of loading and unloading curves.

At low temperatures, epoxy resins exhibit a linear elastic
behaviour up to the fracture point, no hysteresis effect, and no
strain rate dependence. The fracture strain is rather low, of the
order of 2% at 4.2 K (see Table I).

Table I. Mechanical Data at 4.2 K.

No.	Material	Nature at 293 K	Young's modulus, MN/m²	Fracture stress, MN/m²	Fracture strain, %	Poisson's ratio, ν
4	LMB 303/HT 907*	Flexible	7800	–	–	0.36
3	Cy 221/Hy 979*	Semiflexible	8100	179	2.1	0.37
2	My 740/Jeffamin D30*	Semiflexible	8170	200	2.4	0.37
5	X186-2476/Hy905*	Rigid	7650	142	1.9	0.37
	Polyethylene PE†	Partly crystalline, 60–80%	9730	175	4.4	
	Teflon†		8500			

*Resin/hardener
†Noncrosslinked polymer

Ductilization

What can be done to make high polymers ductile at very low temperatures? In polymer chains there are many vibrational, rotational, and stretching modes, which have different activation energies. The lower the activation energy, the lower the temperatures at which the viscous flow is possible. The probability for displacements of chain segments, W, is roughly proportional to exp. $[-(U-A)/kT]$, where U is the net activation energy (barrier), taking into account the steric hindrance; A is the applied external strain energy; and kT is the thermal energy of oscillating segments. For rotational modes, the activation energy without steric hindrance is relatively low. In Fig. 3, the rotation is shown around a C-C bond. Crankshaft motions are assumed to have especially low activation energies per oscillating segment.[1] Thus, long chains should exhibit some flexibility by rotational modes.

This assumption was tested for epoxy resins by insertion of cyclo-aliphatic and phenolic segments into the crosslinked network. The structure of these resins has been described in the preceding paper[2] and Table II. Depending on the segment length, a flexibilization down to LN_2 temperature was achieved. But at LHe temperature no plastic flow was found, at least not before fracture occurred. Some reasons for the behaviour may be that the rotational

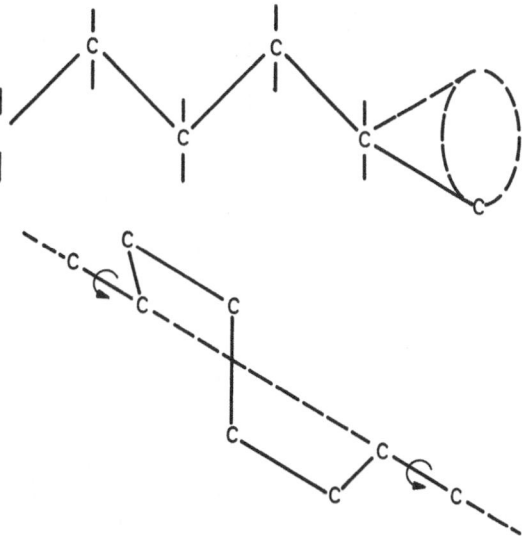

Fig. 3.   Schematic representation of rotational modes and crankshaft motion.

Table II

No.	Ingredients	Chemical classification	Nature at 293 K	Density, g/cm^3	Mean crosslink distance, S, nm
	Commercial systems (Ciba-Geigy, Basel)				
1	X 186/2476*	Glycidylesther	Rigid	1.26	≈ 150
5	Hy 905†	Phthalacid-anhydride			
3	Cy 221*	Bisphenol A + poly-propylenglycol	Semi-flexible	1.22	≈ 350
	Hy 979†	Aromatic + cyclo-aliphatic amines			
2	My 740*	Bisphenol A	Semi-flexible		≈ 300
	Jeffamin D230†				
1	Cy 221*	Bisphenol A + poly propylenglycol	Flexible	1.1	≈ 750
	Hy 956†	Aliphatic amines			
	Special Systems				
	My 970	Hardener: hexyhydrophtalacid anhydrid			~ 150
	Araldite B				~ 300
	A 6084				~ 500
	A 6097				~1000

* Resin
† Hardener

potential is increased by steric hindrance (a) in the vicinity of
crosslinks, (b) in segments with rigid side groups, and (c) in segments which have oxygen or nitrogen atoms or aromatic rings in the
chain backbone. When external tension is applied, stress concentration and microcracking may occur at these points. Another reason
may be that the segments between crosslinks are too short for
plastic flow. Even segment lengths of 10 to 15 nm showed no effect
at 4.2 K. But at room temperatures, those systems with very long
chains are like rubber and not useful. This is shown in Figs. 4
and 5, where Young's modulus is plotted versus temperature. Figure
4 shows a schematic presentation for resins that are flexible, semiflexible, and rigid at room temperatures. In Fig. 5, measured
values are plotted. It should be mentioned that, at helium temperature, all high polymers have roughly the same Young's modulus.

     Instead of crosslinking the segments, mechanical properties
can be improved by a partial crystalline polymer structure. As an
example, a system with long uniform hydrocarbon chains, polyethylene
(PE), is considered. The typical lamellar crystallite structure of
PE is shown in Fig. 6b. There are randomly distributed sandwich
structures of amorphous and lamellar layers. These lamellae can
also be a substructure of spherolites, as shown in Fig. 6c. A chain
length up to the order of μm, and a molecular weight up to 1 million
is possible. A typical stress-strain diagram of a high-molecular
PE (HMPE) is shown in Fig. 7.

1. A pronounced flow characteristic was found even at 4.2 K. A
    fracture strain of up to 0.5% was achieved at 4.2 K. This is
    a factor 2 larger than that for epoxy resins, which is an
    enormous gain. This material exhibits at least some toughness.

2. In contrast to the epoxy resins of Fig. 2, the stress-strain
    diagram of PE at 4.2 K is not at all linear. If the nonlinearity
    is connected with viscoelastic flow, a strain rate dependence
    is to be expected. Stress-strain measurements up to the yield
    point (between point 0 and A), with strain rates varying from
    ≈ 20 to $10^{-3}\%$ per second, were performed, and no dependence on
    $\dot{\varepsilon}$ was found. So, the stress-strain behaviour is elastic, but
    nonlinear. An explanation might be found in the superstructure
    of the crystalline and amorphous phases.

3. A small hysteresis between tensile loading and unloading curves
    was detected. The total strain is regained. Those processes
    are defined as anelastic ones. A temperature change occurs
    during deformation (adiabatic process). Thus, anelastic pro-
    cesses should depend on the strain rate, $\dot{\varepsilon}$, but this was not
    found in the experiment. This point cannot be explained.

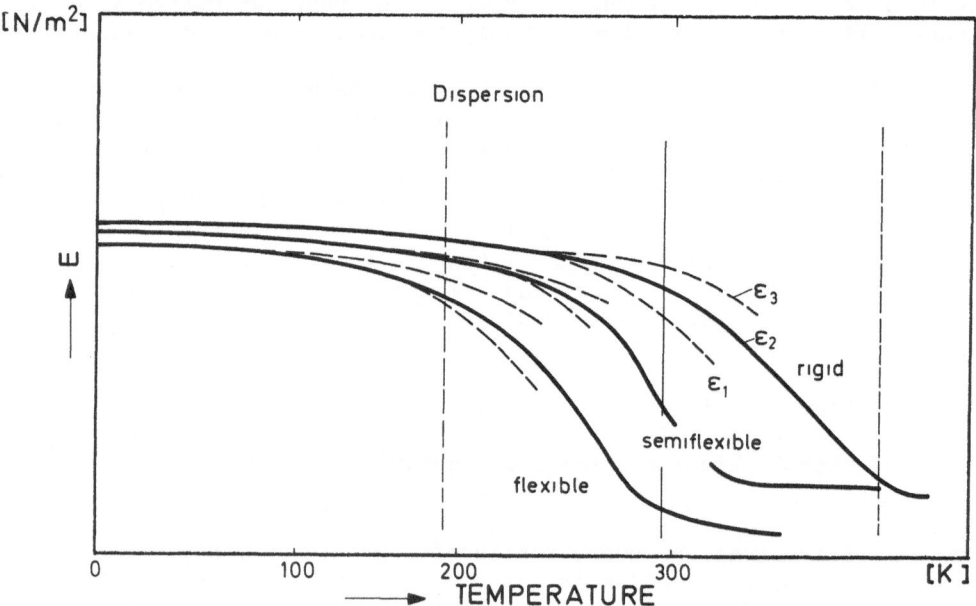

Fig. 4.   Schematic presentation of Young's modulus versus temperature.

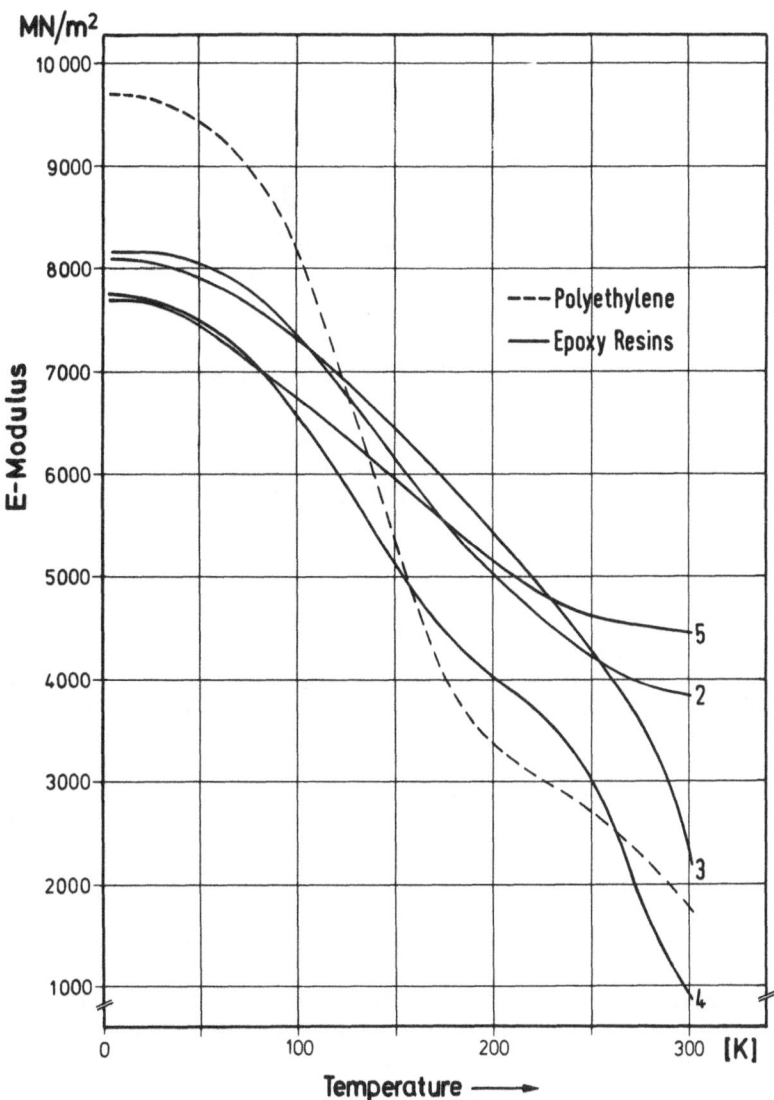

Fig. 5.   Young's modulus of several epoxy resins
versus temperature (numbers are defined
in Table I).

CRYSTALLINE POLYMERS

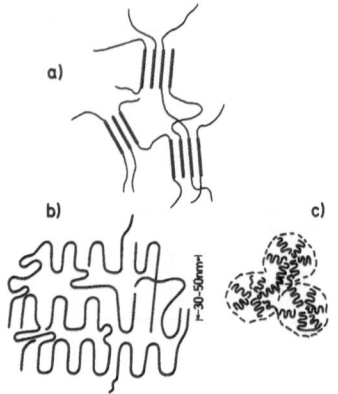

Fig. 6.   Crystalline structures:

a) Linear structure.
b) Lamellar structure.
c) Spherolitic structure.

Discussion

What are the reasons for plastic flow and nonlinear stress-stain behaviour? A well-known man, Norman Brown, wrote: "A most crystalline high polymer is much more complex than a most imperfect ionic or metallic crystal."[3] The structural parameters which might be of influence are (1) branching, (2) chain length (molecular weight), (3) crystallinity, and (4) additives (?).

The investigations are not yet finished, but these facts are presented for discussion:

1.   Branched PE ($\approx$ 35 branchings per 1000 C atoms; crystalline fraction $\approx$ 45%) showed no plastic flow and a nearly linear stress-strain behaviour. The branching is similar to the crosslinking of epoxies, which show a linear stress-strain relation at 4.2 K (see Fig. 8).

2.   The dependence on the chain length was studied by testing medium- and high-molecular linear PE, with molecular weights of some 10^5 and 10^6, respectively. Their flow characteristics and nonlinear stress-strain behaviour up to the yield point are identical (see Fig. 8).

3.   The crystalline fraction, C, is correlated with the chain length. High molecular polyethylene (HMPE) has a crystalline

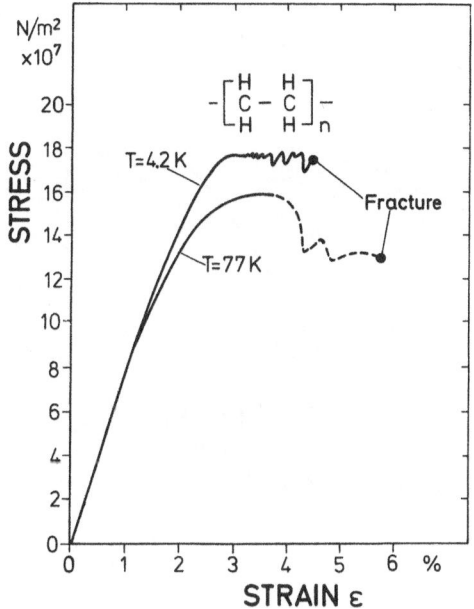

Fig. 7. Nonlinear stress-strain diagrams for
         high-molecular PE.

fraction of about 61%. For HMPE, it is very difficult to vary
the crystallinity drastically by annealing or quenching. Annealing of HMPE showed no effect. For quenched HMPE, a 54%
crystalline fraction was achieved, which yielded a lower modulus, as shown in Fig. 9.

Some explanation of plastic flow may be found in the amorphous
phase of PE. There are many chains that connect the crystalline
lamellae. If the chains are very long, they may be able to undergo
some plastic flow, e.g., by means of crankshaft processes. If they
are short, they can be pulled out or stuck into a crystalline lamel-
lae--a known process of low activation energy compared with the
mechanical binding energy of hydrocarbon chains. The crystallites

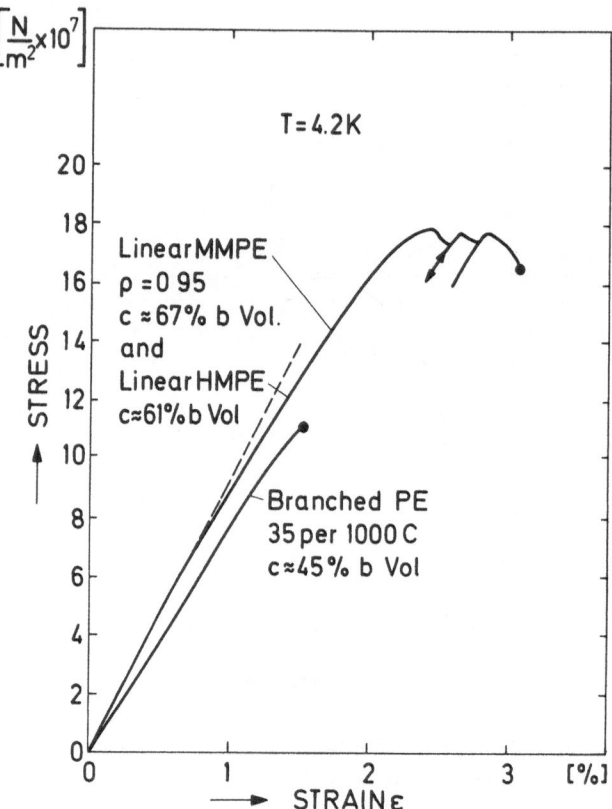

Fig. 8.   Stress-strain diagrams of different
           types of PE.

act as a soft connection of amorphous chains and may equalize local
stress concentrations.

This idea was applied to epoxy resins by Ciba-Geigy. Segments
which tend to crystallize were inserted into the network of epoxy
resins. The segments that are not exclusively hydrocarbon chains
have been described in the preceding paper.[2] Two different systems
containing crystalline fractions of 30 < C < 60% were tested. (A
quantitative analysis is not yet available.) No plastic flow was

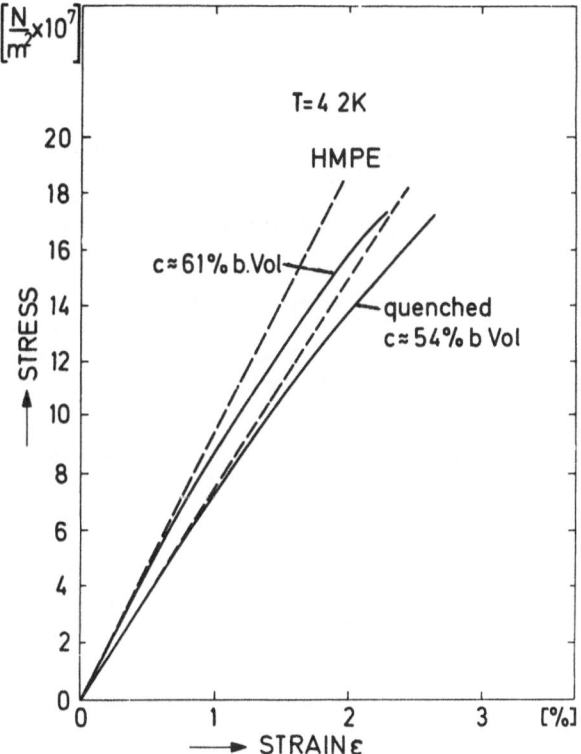

Fig. 9.   Stress-strain diagrams of PE with different
          crystalline fractions.

found under tension at 4.2 K.* The stress-strain behaviour was
rather linear up to fracture, even at strains where HMPE showed a
pronounced nonlinearity. The elastic properties of these partially
crystalline epoxy resins were similar to the amorphous ones. But
their inelastic properties had to undergo drastic modifications.
This will be discussed in the next paper.

---

*The reason may be found again in crosslinking or in the type of
 crystallites. In one system, there are Ester groups within the
 hydrocarbon chains; the latter tend to crystallize. The Ester
 groups could hinder a plastic flow.

Conclusions

As a conclusion of this section, a rough summary of the stress-strain behaviour can be given:

1. Low temperature flexibilization of epoxy resins by insertion of very long amorphous chains between the crosslink points is not useful, because those systems exhibit a poor mechanical strength at room temperature. The stress-strain behaviour under tension is linear, even at strains where HMPE shows a strongly marked nonlinearity at 4.2 K.

2. Partially crystalline epoxy resins considered in this paper have not yielded plastic flow at 4.2 K.

3. Medium- and high-molecular-weight linear PE shows pronounced plastic flow even at 4.2 K. The stress-strain behaviour below the yield point is nonlinear, but elastic. The fracture stress at 4.2 K is about 20 to 30% higher than is usually found for epoxy resins.

4. Branched PE shows no plastic flow and a more linear stress-strain behaviour than unbranched HMPE. It is more like crosslinked epoxies.

In Table I, some mechanical parameters of several crosslinked and noncrosslinked high polymers are listed. At 4.2 K, they have within 15% and 30% similar Young's modulus and fracture stress values.

PE has good low temperature (and electrical) properties, but it is not applicable as an impregnation material or as a matrix for long fibre composites. It might be applicable to short fibre composites, for example, in connection with whiskers.[4] Such a material could have a high mechanical strength and some toughness even at helium temperatures.

LOSSES

Mechanical Losses

High polymers behave--cum crano salis--elastically at very low temperatures. But no deformation of materials is free from internal friction and damping processes. By means of the sensitive torsion pendulum method, the mechanical loss was determined. In Fig. 10, $\tan \delta_m$ has been plotted versus temperature, where δ_m is the mechanical loss angle. These are the results:

1. At very low temperatures, many amorphous epoxy resins considered
    show roughly the same mechanical loss.

2. Partially crystalline PE has a lower mechanical loss at low
    temperatures than amorphous epoxy resins. At elevated tempera-
    tures, the situation is reversed.

3. Partially crystalline epoxy resins undergo less mechanical loss
    than amorphous ones. They are more like PE. The crystalline
    fraction has not yet been determined, but may well be similar
    to that of PE.

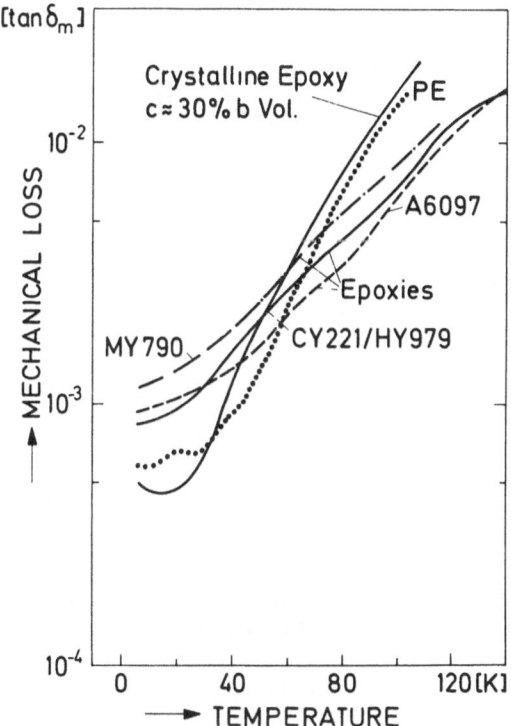

Fig. 10.   Mechanical loss versus temperature.

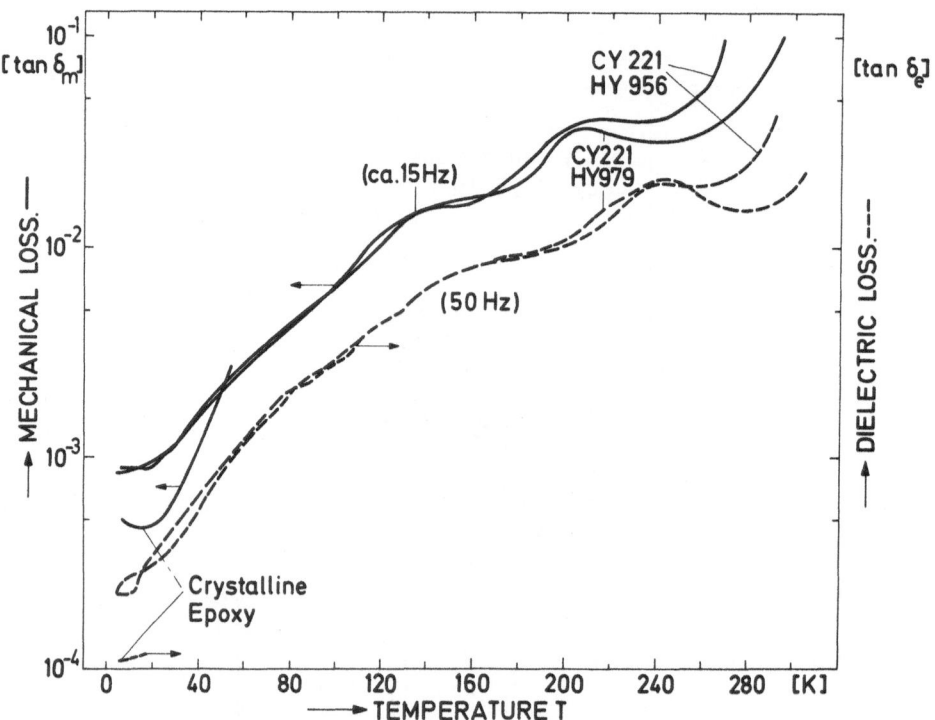

Fig. 11.   Dielectric and mechanical loss versus temperature.

DIELECTRIC LOSSES AT LOW FREQUENCIES

In both electric and mechanical cases, the loss tan, δ, is defined as the inelastic component normalized to the elastic component. In principle, dielectric and mechanical loss angles (δ) should be the same if the relaxation processes are the same and if both electric field and mechanical load are acting at the same dipole field. At least at low temperatures and low frequencies, the relaxation processes are the same if the same modes are activated. The activation can be different. For the dielectric losses, the net dipole moments are decisive. For many polymers with a certain regularity of structure, dipole moments are cancelled to some extent, thus yielding a lower dielectric loss. These materials are known as nonpolar ones. Polyethylene and Teflon are two examples.

Amorphous epoxy resins are polar polymers and have comparably higher losses. Some loss measurements on epoxy resins are plotted versus temperature in Fig. 11. It is interesting to note that the mechanical and dielectric losses are only different by a factor of roughly 2. Also, the dependence on temperature is rather similar. The dielectric and the mechanical parameters of Fig. 11 were determined at different frequencies. At low temperatures, this is a minor error. Mechanical measurements performed at a higher frequency, namely 50 Hz, would, at most yield lower values, more similar to the dielectric ones. Thus, for epoxy resins, the electrical and mechanical dipole forces are similar.

The mechanical losses of partly crystalline epoxy resins were found to be lower than those for amorphous ones. First measurements also show a similar situation for dielectric low-frequency losses. The value plotted in Fig. 11 is an upper limit and the true value is probably lower. Further measurements are in preparation. However, despite the fact that plastic flow has not yet been found, because of their good dielectric properties, crystalline epoxy resins seem to be of advantage.

ACKNOWLEDGMENTS

The assistance of Ciba-Geigy, of BASF, and Hoechst is greatly acknowledged. The author would like to thank Dr. Haberkorn, BASF, Ludwigshafen for performing the measurements and P. Raber, B. Vogeley and W. Weiss for their cooperation.

REFERENCES

1. N.G. McGrum, B.E. Read, and G. Williams, in: <u>Anelastic and Dielectric Effects in Polymeric Solids</u>, John Wiley & Sons, London, pp. 38, 141.

2. U.T. Kreibich, F. Lohse, and R. Schmid, in: Nonmetallic
    Materials and Composites at Low Temperatures, Plenum Press,
    New York (1979), p. 1.

3. N. Brown, in: Microplasticity, Interscience Publishers, John
    Wiley & Sons, p. 71.

4. H.J. Schladitz, Z. Metallkd. 59, 18 (1968).

SPECTROSCOPIC ANALYSIS OF HIGH POLYMERS

FOR USE AT LOW TEMPERATURES

W. F. X. Frank

Universität Ulm
Ulm, West Germany

INTRODUCTION

The application of high polymers as insulating materials at low temperatures, as is well known, is limited by their brittleness. Consequently, it is highly desirable to develop methods of checking quickly and simply if a certain polymer can be used at low temperatures or not. In this paper, we propose the use of spectroscopy as a testing procedure. From simple considerations of the molecular structure of polymers, it appears that spectroscopy in the low frequency range of lattice vibrations, i.e., the far infrared (FIR), or the low frequency Raman spectrum, is an appropriate tool for this purpose.

Spectroscopic testing methods are, aside from the costs for the spectrometer, inexpensive, need only small amounts of material, and are easy to evaluate, provided the assignment of the spectrum with respect to the desired sample properties is known.

Until now, no systematic work has been carried out on this application. The following explanations are an attempt to introduce spectroscopy into the search for suitable low temperature materials.

The first step is to find connections between spectral features and the macroscopic behaviour of the samples tested. Let us try to translate the problem into the language of spectroscopists: Low brittleness or a reasonable amount of ductility, disregarding any exact physical definitions of these quantities, means, from a molecular point of view, that a certain amount of energy applied to the sample; e.g., static stress, strain, or periodical

enforcement, will be quickly distributed over a broad variety of
molecular mobilities. At higher temperatures, this process is
assisted by the thermal excitation of atoms or atom groups, which
move as more or less coupled oscillators, and this helps to decrease
local stress. This may be illustrated by a simple picture. We
describe the potential energy of a row of atoms in a bulk polymer
as a function of their position. This is schematically plotted
in Fig. 1.

The single atom can (a) undergo vibrations of large amplitude,
which lead to hopping processes into free places in the neighbourhood (so-called displacement processes), and (b) vibrate around its
equilibrium position with small amplitudes. For the case of weak
intermolecular coupling, this leads to wide-range lattice waves,
which are very important for the energy distribution within the
material. We will discuss these two aspects under the special
point of view of their relevance at low temperatures.

MOLECULAR MOTION MECHANISMS IN POLYMERS

Relaxation Behaviour

A suitable property to describe the mechanical behaviour of a
polymer under time-dependent enforcement is the complex Young's
modulus. For a polymer, which, in general, is composed of crystalline and amorphous parts (for a comprehensive description of polymer

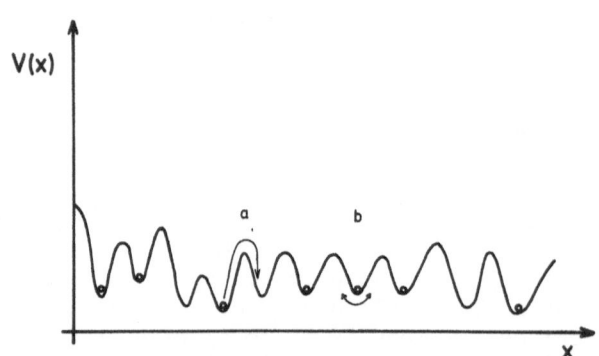

Fig. 1. Potential energy of atoms in a bulk,
         e.g., cross section of some polymer
         chains nearly well ordered. The arrows
         indicate possible motions.

morphology see Wunderlich),[1] the modulus is plotted schematically
in Fig. 2. The real part of the modulus, E', increases to higher
frequencies, ω, in a stepwise manner, and the single steps can be
assigned to molecular motions. In the loss modulus, the steps are
accompanied by sharp peaks. What we are most interested in here
are the relaxation times, τ, which are defined by

$$\omega\tau = 1 \qquad\qquad\qquad\qquad\qquad (1)$$

τ describes how fast the system responds to an external stress
(under the proposition of simple relaxation mechanisms). The energy
dissipated is given by the integral over the imaginary part E".
A complete treatment of the relaxation behaviour has been given by
Ferry,[2] Stavermann and Schwarzl,[3] and Müller.[4] As an example, we
mention the so-called γ process in polyethylene. The change of
location of a pair of CH_2 groups gives rise to a step in the E'
curve and a peak in the E" curve. This has been described extensively by Pechhold.[5]

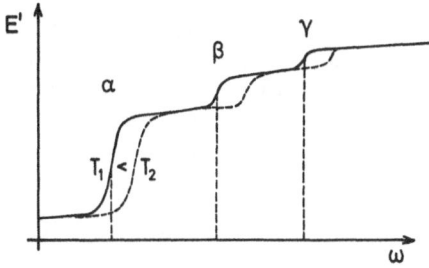

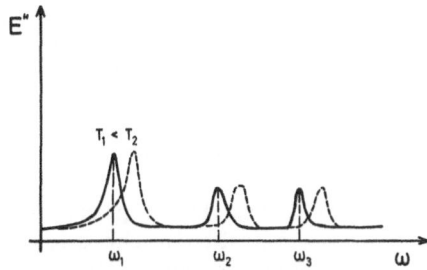

Fig. 2.   Real and imaginary part of Young's
          modulus as a function of the fre-
          quency of the acting force.

An important point is that all these processes are enhanced by heat energy, according to the well-known Arrhenius equation,

$$\tau = \tau_o \, e^{-U/RT} \tag{2}$$

This means the relaxation times of all processes increase drastically when T is reduced. The amount of increase is given here by the activation energy, U. This quantity can be determined by measuring the elastic constants as a function of frequency with variation of the temperatures.

For polyethylene (PE) this has been done by Pechhold et al.[6] The results are given in Fig. 3. The activation energy increases from right to left, and this means in this diagram that processes described by each straight line become more and more cooperative and more molecular groups are involved. For the processes with small groups, like the CH_2 motion, we find U and τ to be independent of temperature. For cooperative motion of bigger groups, this is no longer the case.

The question now is what happens when we extend the abscissa to the temperature region in which we are interested. This is plotted in Fig. 4, where, on this scale, Fig. 3 degenerates to a small field at the left. It is now obvious that processes like those just discussed are no longer observable, because the relaxation times have reached tremendous values (one year is equal to $\approx$ 30 Ms). If the activation energy is larger than 0.5 kcal/Mol or ca. 2 kJ/Mol, there is no chance of observing the molecular motion process at 4 K, and, therefore, the process cannot be used as an energy distributor.

In principle, all these molecular motions can be observed dielectrically, provided that the groups involved have a dipole moment and all the above remarks are valid for this type of measurement. Dielectric measurements at low temperatures had been carried out by several authors (Müller,[6] Phillips,[7] Sievers,[8] and Hunklinger[9]). Absorptions were observed from radio frequencies up to the very far infrared, but the origins of these processes are not completely clear. Abstract low-level systems have been discussed to reproduce the temperature dependence,[10] but a concrete molecular model is difficult to establish.

The energy absorption of a sample is given by the integral over the E" curve. From this, we obtain the idea that, for a material suitable for low temperature use, no sharp distribution function of relaxation times must exist. The broader this function, the more mechanisms exist to distribute mechanical energy over the entire sample and convert it into ordinary heat. In the region of 4 K, there is only a small possibility of location changes

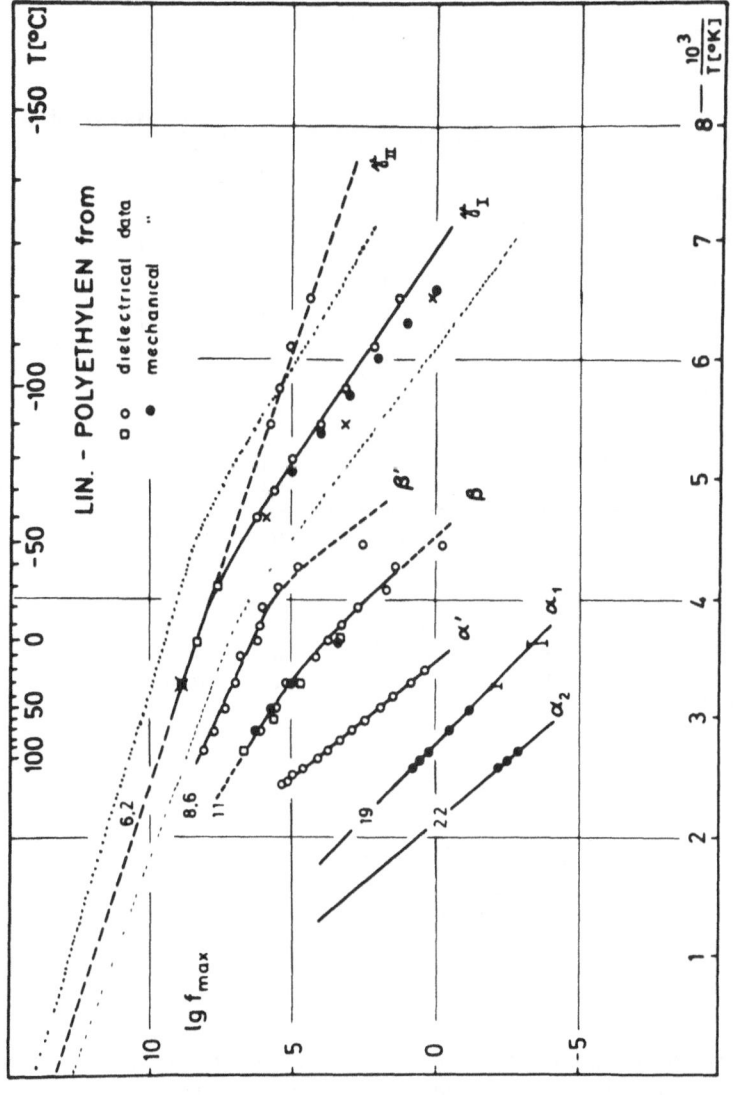

Fig. 3.   Activation diagram of polyethylene by Pechhold.[5] Each straight line belongs to a special molecular motion process. The number of molecular groups involved increases from right to left.

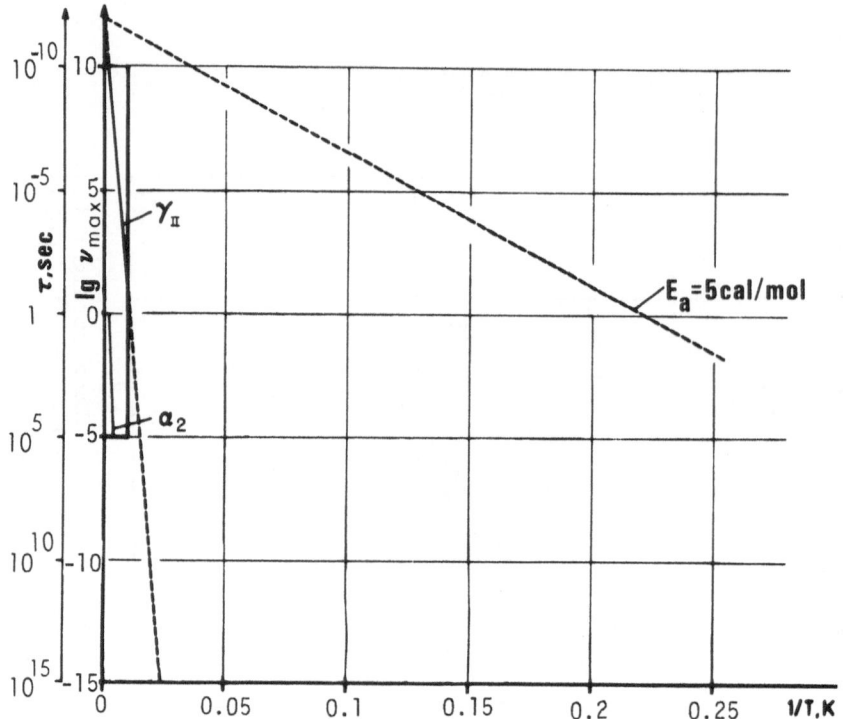

Fig. 4.   Analogue diagram to Fig. 3 with abscissa down to
          4 K. There is no chance to observe a displacement
          process at 4 K. The straight line with E_a = 5 cal/
          Mol ($\approx$ 22 J/Mol) is dielectrically measured by
          Müller.[6]

contributing to this process. We conclude that the sample should
definitely not be well ordered but "as amorphous as possible."
In the next section we will discuss this point further.

Elastic Waves in a Bulk Polymer

     If displacement processes are no longer possible at lower tem-
peratures, the distribution of mechanical energy in a bulk sample
is mainly provided by elastic waves. That means that the vibrations
of atoms are coupled within big arrays, where the atoms move with
fixed phase relations. These collective excitations are called
phonons.[10,11] Distributing the energy of a lattice vibration in

several degrees of freedom is now equivalent to the idea of scattering the phonons in the lattice inelastically. This process is schematically drawn in Fig. 5.

Our task of decreasing local stress in the material can now be reformulated as: how do we obtain a high scattering rate of phonons, especially at low temperatures?

We will estimate the order of magnitude of the wave length of the phonons under consideration. From the influence of T on the dominant phonon frequency, we obtain from Debye's equation

$$h\tilde{\nu}_{max} = (k_B/c)\theta_D \tag{3}$$

where h is Planck's constant; k_B is Boltzmann's constant; $\tilde{\nu}$ is frequency in "wavenumbers," cm^{-1}, i.e., ν/c; c is the velocity of light in cm/s; and θ_D is the Debye temperature.

For $T \geq \theta_D$, the mean frequency of vibrating atoms is equal to $\tilde{\nu}_{max}$.

For $T \leq \theta_D$, this frequency is reduced to values of the order of

$$\tilde{\nu} \sim \tilde{\nu}_{max}(T/\theta_D) \tag{4}$$

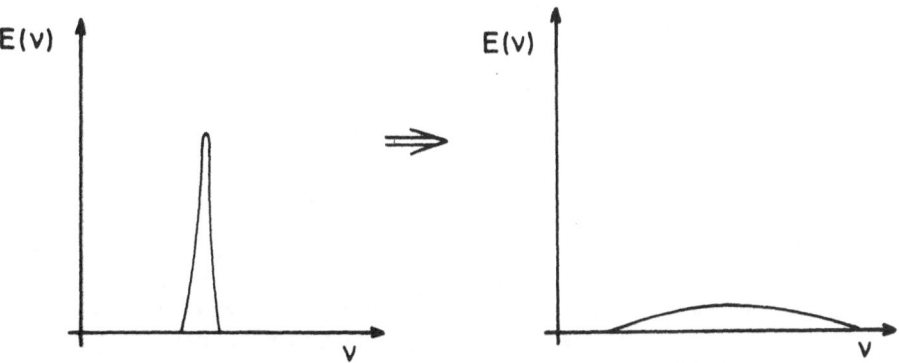

Fig. 5. A hypothetical sharp frequency distribution with large amplitude is smeared out over a broad frequency range.

This corresponds to a wavelength, λ, which for $T > \theta_D$ has the order of a lattice constant ($\lambda \approx a$) and increases with lowering temperature to

$$\lambda = a(\theta_D/T) \tag{5}$$

For our region of interest, λ reaches values of 100 or more lattice spacings.

The question of how these phonons are scattered is closely related to the problem of heat conduction[12] or the unidirectional transport of energy. The expression for this is

$$k = c_p \ell v/3 \tag{6}$$

where k is the heat conductivity, c_p is the specific heat, ℓ is the mean free path of the phonons providing the heat transfer, and v is the phonon group velocity. c_p shows the typical temperature dependence, in the simplest case according to Debye's T^3 law. So for a certain temperature, e.g., for liquid helium temperature, the scattering rate is proportional to $1/k$ and therefore to $1/\ell$.

Consequently, we have to look for scattering mechanisms that give high scattering rates, especially at low temperatures. The possible processes are calculated and listed by Erdmann.[13] For our purpose, the temperature behaviour of the scattering rates of different mechanisms is shown schematically in Fig. 6.

The phonon-phonon interaction is unimportant, because at low temperatures all vibrations are harmonic without regard to the degree of order (that means: the phonon gas is ideal and free of interaction). The scattering rate at point defects decreases proportionally with lowering temperatures. Only the scattering at boundaries gives a contribution in the opposite direction.

From this point of view, we see that a material like a partially crystallized polymer has a good chance of being suitable as an insulating material at low temperatures. The main reason for this is that these materials contain a large number of boundaries belonging to the small crystallites, so that a material like the Hostalen GUR* seems to be useful.[14] This polymer is composed of extremely long chains that tend to reduce the crystalli ation so that this material reaches only a 50% volume crystallinity, which is dispersed over a very large number of small crystallites. Fig. 7 shows a sketch of a real polymer built up as a two-phase

---

*Made by Hoechst AG, Frankfurt/Main, West Germany.

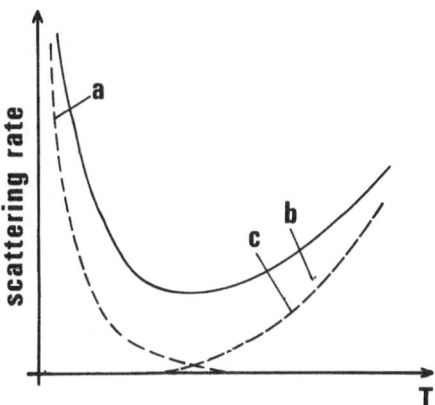

Fig. 6. Phonon scattering rate proportional to 1/k
for (a) scattering at boundaries, (b) scattering at point defects, (c) phonon-phonon
interaction. The full line is the resultant
of (a) - (c).

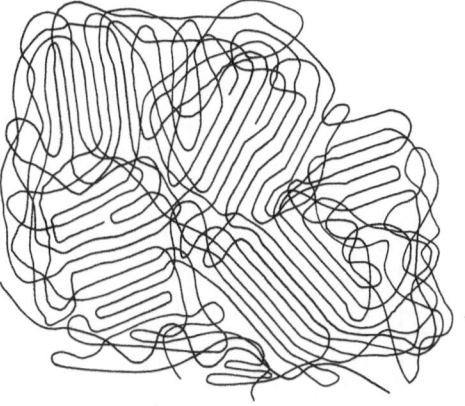

Fig. 7. Two-dimensional model of a partially
crystallized polymer. The structure of
the boundaries is not exactly known at
the present time.

system of crystallites and amorphous regions where the important role, for our purposes, is played by the boundaries between crystallites and amorphous parts.

The total energy of a solid can be considered as composed of contributions from all degrees of freedom that are excited at a certain temperature. From energy distribution for phonons (these are Bosons), we see that the vibrations of low frequency give the biggest contribution to the energy of the sample (see Fig. 8). So it is clear that, for our purposes, we have to study the low frequency part of the vibrational spectrum, which can be carried out in principle by far infrared or low frequency Raman spectroscopy.

Phonons in a Polymer

In the case of a polymer formed by molecules of high molecular weight, we consider what kinds of vibrations in the low frequency regions are possible and how can they be observed spectroscopically.

Let us start with the vibration of a single oscillator bound at a polymer chain backbone, e.g., on O atom at a C-C zigzag chain. Its frequency depends on the force constant and the mass involved:

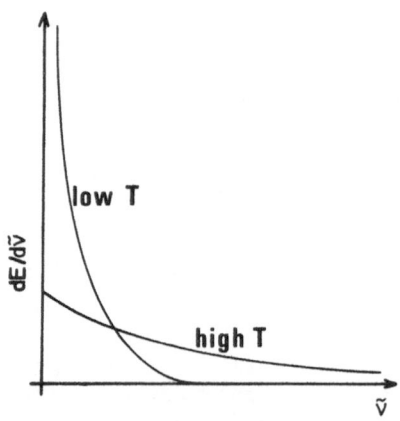

Fig. 8.   Energy distribution function of phonons: at low temperatures only the frequency vibrations give considerable contributions.

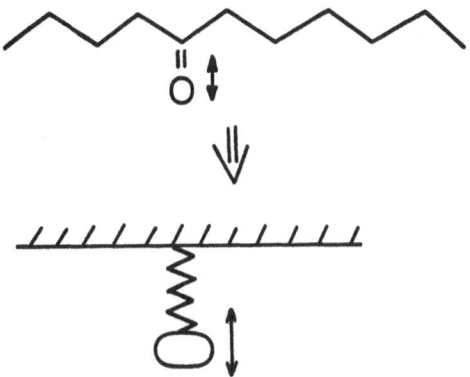

Fig. 9. Molecular oscillator and its abstract
model.

$$\tilde{\nu} \sim (f/m)^{1/2} \tag{7}$$

where f is the force constant and m is the mass of the vibrating
atom. In the example above, a sharp absorption band is observed
near 1800 cm^{-1}, which may be assigned as the eigenvibration of the
O atom against the C atom. Observation of this vibration as a
sharp absorption band in the IR spectrum is possible, because the
force constant in this case is quite different from those acting
in the neighbourhood and coupling to other vibrations can be disregarded. Considering an arbitrary oscillator, the frequency may
be lowered by increasing the vibrating masses or by weakening the
force constant. The first case, i.e., mass increase, does not
occur in a normal polymer, but the second occurs when we consider
van der Waal's interactions instead of main valence bonds. Figure
10 shows the relations in a well-ordered polymer. The forces in
chain direction are 10 to 100 times stronger than those in the
other directions, and these lead to absorptions in the mid-infrared.
What we are interested in here are the weak interactions between
nonbonded groups, the van der Waal's or intermolecular forces.
With the exception of H bridges, intermolecular forces are all of
the same order of magnitude, so localized vibrations are not to be
expected, but since the forces are weak, their influence extends
over long distances, which makes collective excitations (phonons)
possible. To get a feeling for what kind of phonons are possible
and how to observe them, we discuss briefly some special cases.

Let us simplify the macromolecular crystal lattice by a one-
dimensional row of atoms, a so-called linear chain (Fig. 10). The

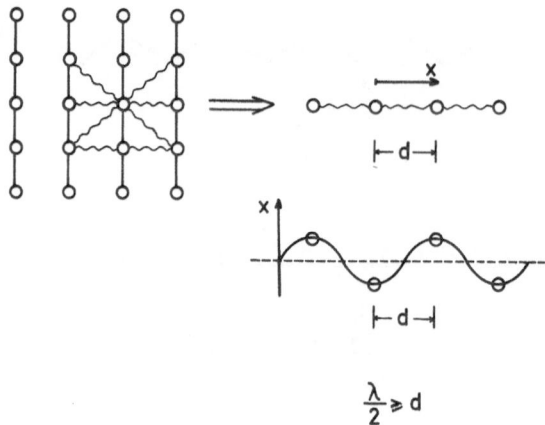

Fig. 10.   Idealized polymer lattice. Thick lines =
           main valence bonds, wavy lines = van der
           Waal's bonds. At the right side, the
           shortest wavelength in a one-dimensional
           chain is shown. It corresponds to twice
           the atomic distance.

possible frequencies depend on the wave vector, $2\pi/\lambda$, where the
shortest wavelength exists belonging to the highest frequency (see
Fig. 10).   Shorter waves would vibrate between the atoms, which
has no physical meaning. So we can restrict our considerations on
a range for the wave vector of $-\pi/d \leq k \leq + \pi/d$ (for the two propagation directions). This region is called the first Brillouin
zone, drawn in Fig. 11 (see any textbook of solid state physics,
e.g., Kittel).[10]

     The actual motion of the atoms is given by calculating the
phase difference of neighbouring elements by

$$(2\pi/\lambda)\Delta x = \Delta\phi \tag{8}$$

where Δx is the difference of the displacements from the equilibrium
position of two atoms. At $k = 0$, we have a motion in phase, which
means no vibration takes place, only a translation of the lattice
occurs. If we go slightly to right or left of $k = 0$, there are
small phase differences, $\Delta\phi$. These belong to long waves propagating
with the velocity of sound. Therefore, we call them acoustical
waves and the ω-k curve, an acoustical branch. At the maximum of
the branch, at $k = \pi/d$, we have $\lambda = 2d$, $\Delta x = d$ and, therefore,
$\Delta\phi = \pi$, which means, neighbouring atoms move in opposite directions,
and we have a standing wave with no energy transport.

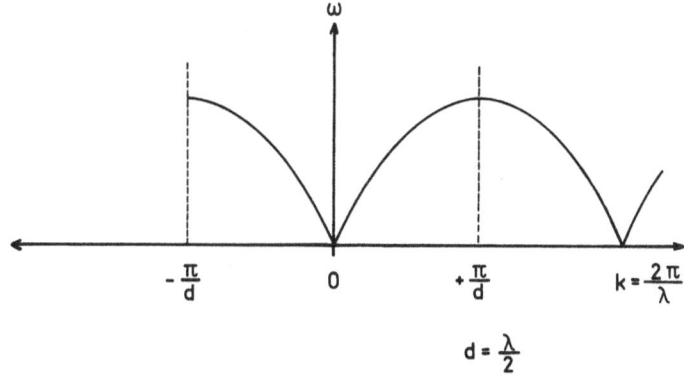

Fig. 11. Dispersion relation for the linear chain of
Fig. 10. The range $-\pi/d \leq k \leq + \pi/d$ is the
first Brillouin zone.

If we complicate this linear chain step by step, we can study,
in principle, all possible features of molecular vibrations in a
solid. First, let us assume that the row consists of two kinds of
atoms that are marked, but which have equal mass (Fig. 12a). The
translational symmetry of our lattice has now changed. The elemen-
tary cell is twice that of Fig. 10, and the corresponding Brillouin
zone has shrunk. The upper part of the branch is folded back, as
shown in Fig. 12a. If we now introduce different masses and different force constants, we obtain Fig. 12b. There is now a gap in the
frequency distribution, the width of which depends on the difference of the masses or force constants, respectively.

OBSERVABILITY OF VIBRATIONS BY IR SPECTROSCOPY

Ordered Structures

Not all vibrations that occur in a solid lead to absorption
of electromagnetic radiation. There are certain conditions, called
selection rules, which must be fulfilled to get an absorption of
light energy by molecular vibrations. In addition to the ordinary
condition for IR activity (existence of a transition moment, see
Krimm[15] and Hummel[16]), we have lattice vibrations for which no
periodicity in space is allowed, which means only frequencies with
phase difference $\Delta\phi = 0$ or infinite wavelength are observable. The
quantity that gives the density of vibrational states falling into
a certain interval, $\Delta\omega$, called the density of states $Z(\omega)$, plays an
important role, since it is proportional to $(d\omega/dk)^{-1}$. Its maxima
occur where the ω–k curve has a horizontal tangent. The rule is

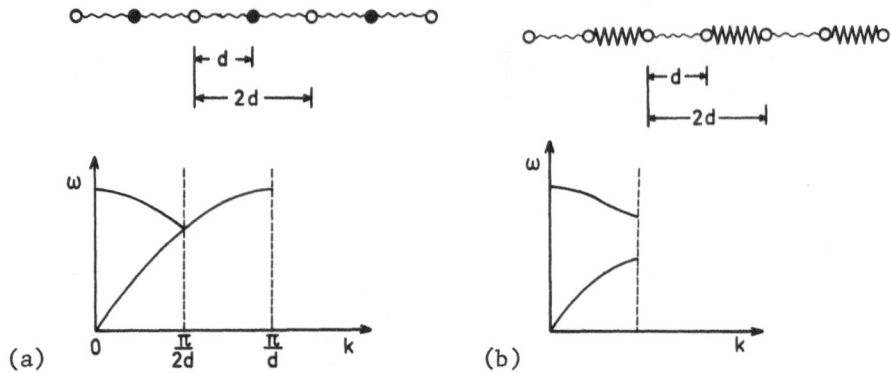

Fig. 12. (a) Formal change of the upper part of the acoustical
              branch in the dispersions relation of Fig. 11
              into an optical branch by lowering the symmetry.
          (b) Different masses and/or force constants generate
              a frequency gap between optical and acoustical
              branch.

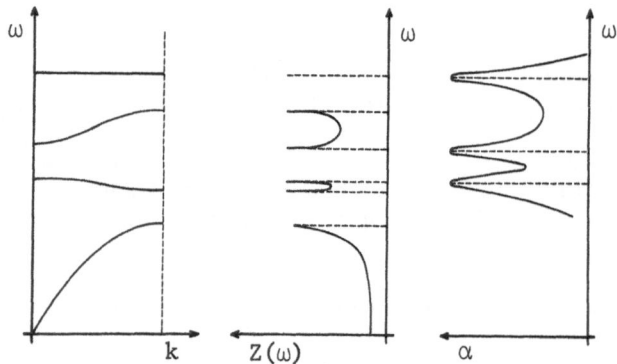

Fig. 13. Principal relationships between the dispersion
          relation, density of states, and absorption
          spectrum (from left to right).

that the density of states must have a sharp maximum (van Hove
singularity)[17] as a condition for observing an absorption peak.
These two conditions, namely, that only vibrations are to be
seen that (a) belong to a maximum in the density of states and
(b) are $k = 0$ vibrations, drastically restrict the possibility of
observing the lattice vibrations in an ordered structure. In
Fig. 13, the hypothetical spectrum generated by a relatively complicated dispersion relation is shown.

Disordered Structures

Let us discuss briefly which type of spectrum we have to expect from a sample that is highly disordered or amorphous. The
description of the absorption behaviour of a disordered structure
is not as simple as that of an ideal crystal. The methods for
evaluation are reported extensively by Bell.[18] However, we can
understand qualitatively the absorption spectrum of a disturbed
lattice on the basis of the principles developed for an undis-
turbed one. A common property of the spectra obtained from amorphous specimens is the broad background intensity, which grows
with increasing frequency (see next section). It has been determined experimentally that this intensity depends strongly on the
thermal treatment of the sample. With better annealing, the background intensity decreases, whereas absorption bands assigned to
be of crystalline origin are sharpened.

We propose a simple qualitative model to explain this behaviour.
Starting from an ideal crystal and its first Brillouin zone, the
basic idea is to conserve the translational symmetry when passing
into a disturbed structure. Dependent on the degree of order, we
have to choose n atoms instead of one or two to form a new "ele-
mentary cell." This is sketched in Fig. 14, where the first
Brillouin zone is drawn for an ideal one-dimensional lattice, as
in Fig. 11. We compare it to the first zone of a disturbed lattice. The minimum distance that conserves the translational symmetry is taken as a parameter characterizing the degree of disorder.
The variance of distances between the elements of the lattice leads
to different force constants caused by the anharmonic lattice potential. So the acoustic branch degenerates to a high number of
optical branches with corresponding maxima in the density of states.
We can assume that at least half of them are optically active. In
this sense, the disordered state is considered as a state of high
density of optically active modes, which gives rise to a continuous background absorption.

From these simple considerations, we can now give requirements
for the use of spectroscopic methods in testing polymer materials
for their applicability at low temperatures.

1. They must have a broad absorption background of large
   mean intensity which indicates a mainly disordered
   system.

2. There should be single additional absorption bands
   whose origins are small crystallites, which are embedded within an amorphous matrix. The crystallites
   should be as small as possible, because their boundaries must act as two-dimensional scattering centers,

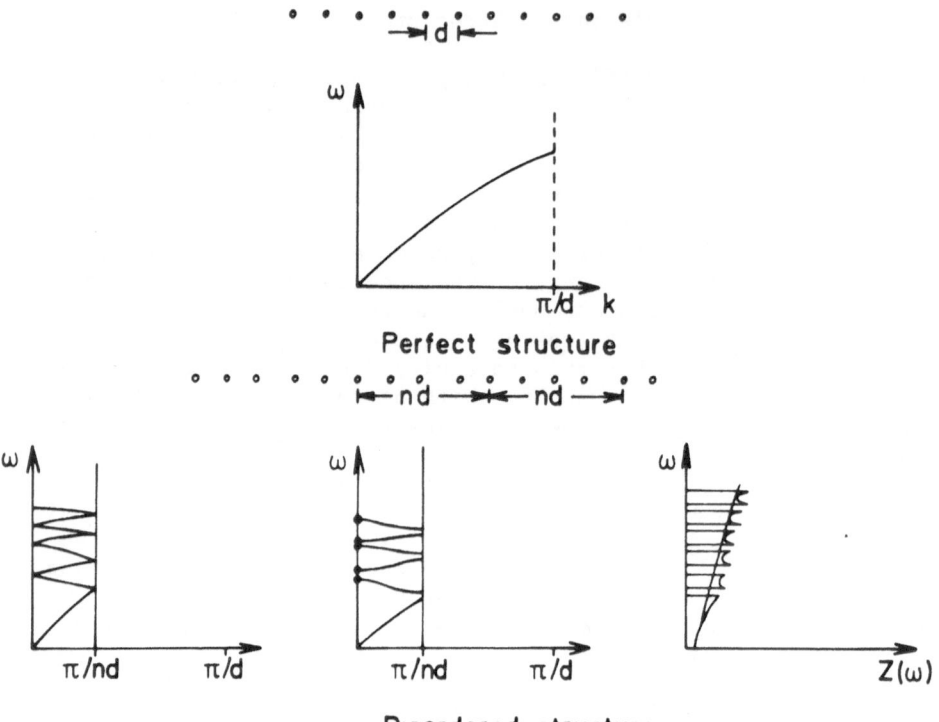

Fig. 14. Conservation of the translational symmetry in a dis-
          ordered structure by the introduction of a "supercell,"
          which contains a certain number of n atoms instead of
          one or two. The Brillouin zone shrinks, and the
          acoustic branch changes to many optical branches with
          many maxima in the density of states, increasing with
          higher frequencies.

which, especially at low temperatures, provide the
main part of the energy distribution.

EXPERIMENTAL RESULTS OF FAR INFRARED SPECTROSCOPY

OF SOME POLYMERS

     In the following, we report the results of some polymers which
are of possible interest for use as low temperature insulating
materials of general interest, because their far infrared behaviour
can be considered as typical for other polymeric materials. The
usefulness of the basic principles discussed in the previous section
will be illustrated, but it should be stressed that this is the

first attempt to relate spectroscopic features of polymers with their behaviour at low temperature, and, consequently, there remains a great need of a future systematic study of this problem.

The absorption spectra were measured with two Fourier spectrometers from Beckman-RIIC Company, models FIR 30 and FS 720. The spectral resolution was 3 cm^{-1}. The absolute frequencies could be determined with an accuracy of $\pm$ 0.25 cm^{-1}. Samples were cooled with a cryostat from Cryogenics Inc., Waltham, Massachusetts, U.S.A., model CT 20, from room temperature down to 12 K.

Polyethylene (PE)

This material with the simple chemical structure, $(CH_2)_n$, is obtainable with a large variety of physical properties. These properties depend mainly on the degree of crystallinity, W^c, which means the volume fraction of crystallized matter within a sample.

If v_c and v_a are the specific volume for the crystalline and amorphous phase, then we can calculate the actual specific volume by

$$v = v_c \ W^c + v_a(1 - W^c) \tag{9}$$

The possibility to define a degree of crystallinity depends on the validity of such a two-phase crystalline-amorphous model, which is, for practical purposes, a good approximation. However, this quantity says nothing about the distribution function of the crystallites. The same value of W^c can be realized by a few, large crystallites or by many small ones. If we remember the statements of the earlier discussion, we see that this point is of particular interest for the low temperature behaviour of the material: Material with many internal boundaries should be most suitable for low temperature applications.

The absorption spectrum of polyethylene in the far infrared is very simple: a broad background increasing to higher frequencies and only one sharp band at 73 cm^{-1}. Figure 15 shows an example for a high density polyethylene (HDPE). The background comes from the amorphous part of the material, which can be described in terms of the variable Brillouin zone. The absorption band itself, called the "73 cm^{-1} band," is generated by a lattice vibration of the orthorhombic lattice, which is shown in Fig. 16.

Group theoretical considerations[19,20] assign the vibration of of the two partial lattices (see Fig. 16) in the b-axis direction to the 73 cm^{-1} band. The vibration perpendicular to that is related to a band at 109 cm^{-1}, which at room temperature is not observable.

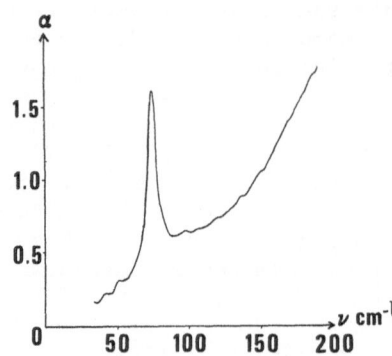

Fig. 15.   Absorption spectrum of polyethylene
           (HDPE, 6011L) at room temperature
           in the far infrared.

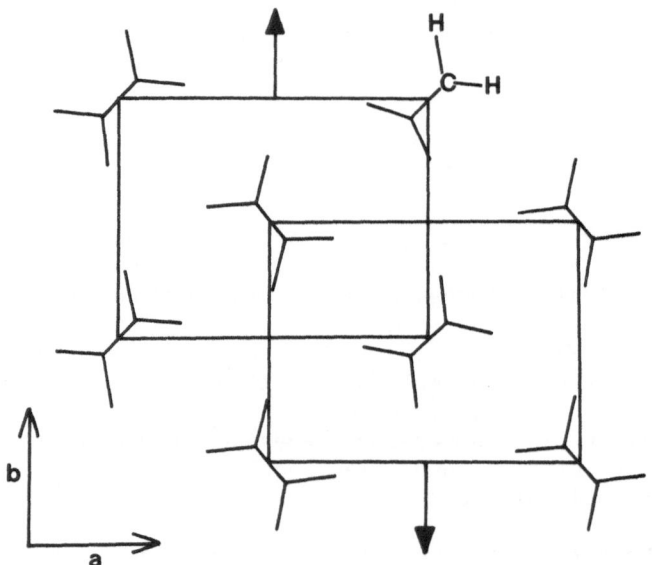

Fig. 16.   Cross section of the unit cell of polyethylene
           perpendicular to the chain axis.

It was first observed by Dean and Martin[21] at 2 K, in good agreement with the assignment given by Tasumi and Shimanoudu[19] and
Tasumi and Krimm.[20]   It seemed interesting to us to measure the
complete temperature dependence of PE from low temperatures up to
the melting point to get information about the temperature

dependence of the following quantities: frequency shift, maximum absorption intensity, integrated absorption, intensity, and the half-width of the observed bands.

We started a series of measurements, including linear, branched, and high-pressure crystallized material. The results on linear PE are published by Frank, Schmidt and Wulff.[22] Table I gives a survey of the materials and their treatment before they were measured in the spectrometer.

The absorption coefficient as a function of the wavenumber from 50 cm^{-1} to 120 cm^{-1} was measured for several temperatures in the interval from 14 K to the melting point, 412 K. The complete temperature dependence of the far infrared absorption of samples 1 and 2 is shown in Figs. 17a and 17b.

Several details can be read from these three-dimensional pictures. Here we will consider an interesting point in the diagram of the frequency shift vs. temperature. For the three samples, the $\tilde{\nu}$ - T diagrams are shown in Fig. 18.

There is a small but distinct bend in the linear slope of all three curves, and the temperature at which this bend occurs depends on the sample density. If we plot the density vs. the bend temperature, we obtain Fig. 19. The bend temperature for the ordinary 6011 L and 1804 H is close to the temperature at which Wunderlich[23] found a step in the heat capacity, a phenomenon which is assigned to the glass transition temperature (T_g). The question is now, what is the connection between an absorption process taking place within the crystalline regions only and the glass transition temperature, which is associated with mobilities in the amorphous parts of the polymer. We propose the following explanation:

Table I. Material Treatment Prior to Spectroscopic Measurement

| Name | Density, Mg/m^3 | W^c, % | Annealing | | |
			Time	Temp.	Pressure
1804 H*	0.920	43	3 h	150°C	1 bar
6011 L*	0.973	81	3 h	150°C	1 bar
6011 L*	0.991	94	45 h	226°C	5 Kbar

*Made by Badische Anilin- und Sodafabrik, Ludwigshafen, West Germany.

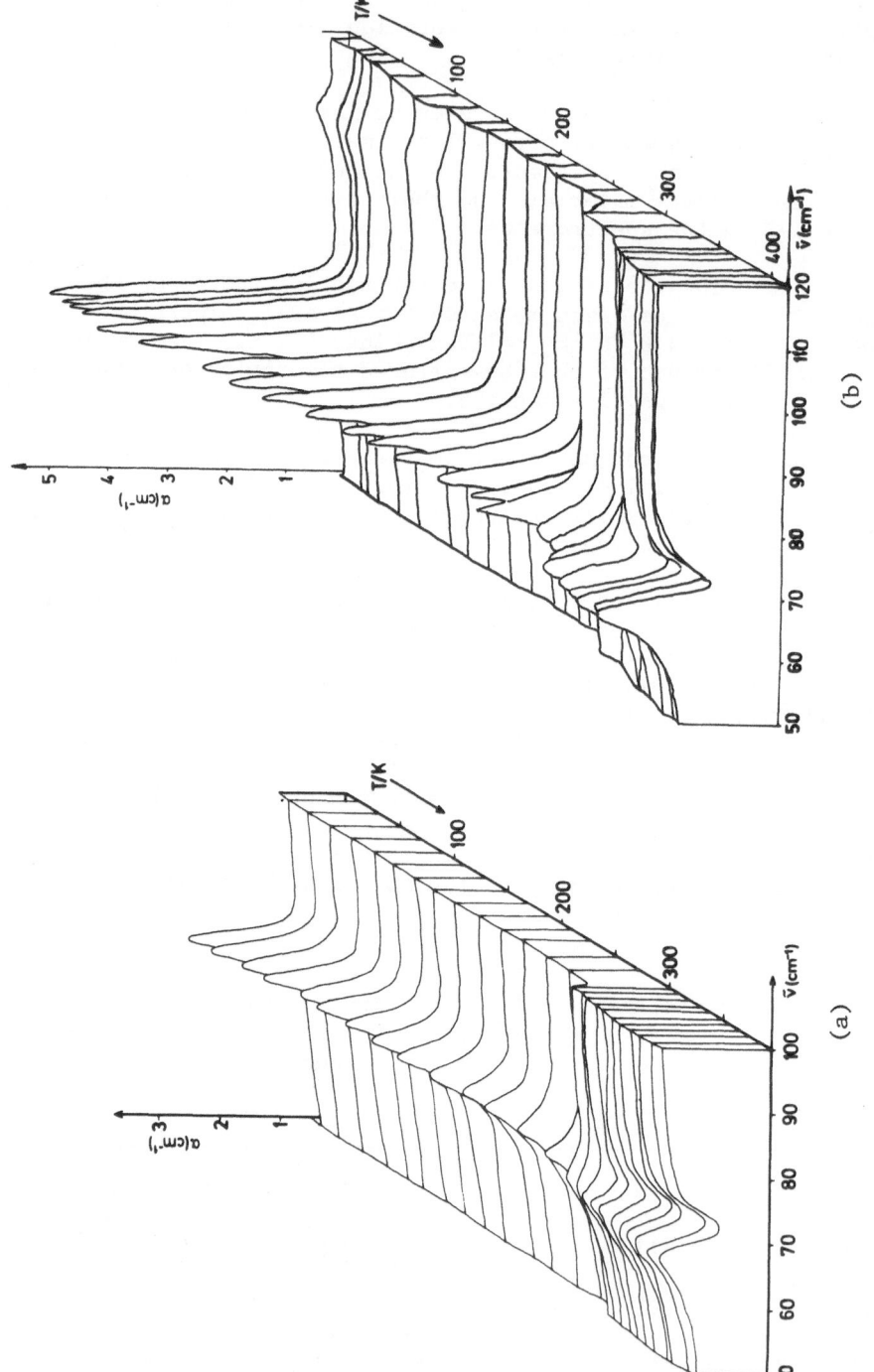

Fig. 17.   Temperature dependence of the absorption spectrum between 50 and
120 cm^{-1}.   (a) HDPE 6011 L; (b) LDPE 1804 H.

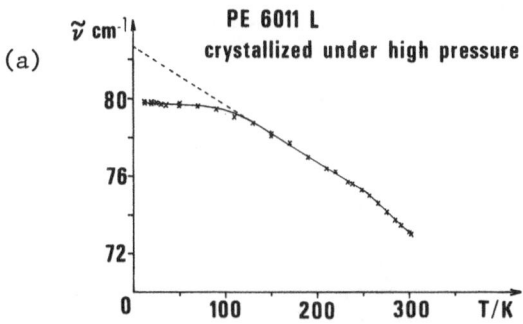

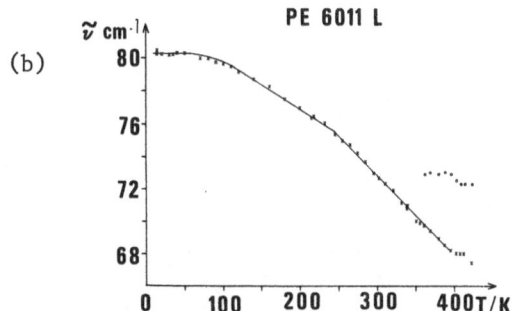

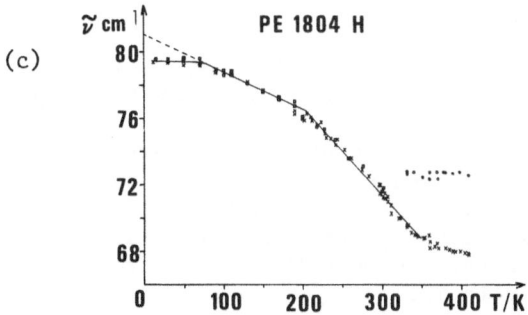

Fig. 18. Maximum absorption frequency as a function
of sample temperature. Sample preparation
is given in Table I.

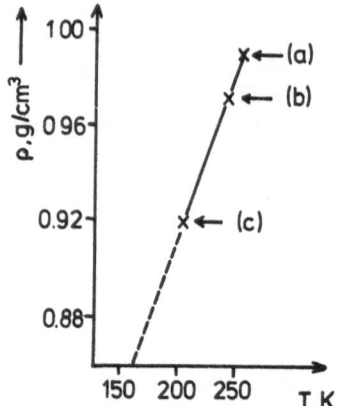

Fig. 19.   Bend temperature of the $\tilde{\nu}$ - T diagrams
of Fig. 18 in correlation with the
sample density.

(a) - high pressure cryst. 6011 L
(b) - 6011 L
(c) - 1804 H

Above T_g, chain defects like kinks can diffuse from the amorphous
parts into the crystallites, so that we have an equilibrium defect
concentration at each temperature. Below T_g, this defect concentration is a constant. This leads to the two different slopes
below and above T_g in Fig. 18. Below T_g, the influence of defects
is weaker, which results in a lower slope of the curve. The frequency shift is correlated to the thermal expansion of the corresponding lattice parameter, here the a axis. This is discussed in
more detail in Ref. 22.

    If we identify the bend temperature with the glass transition
temperature of the polymer, we obtain the following interpretation:

1. The glass transition depends on the degree of crystal-
    linity, since this leads to greater or less stress in
    the chains of the amorphous part of the polymer, which,
    in turn, influences the free mobility. The higher
    the crystallinity, the more the chains are stretched
    and the higher is T_g.

2. The extrapolation to the density of the amorphous
    material gives a T_g of 160 K (see Fig. 19) for the
    completely amorphous material, which is extremely
    difficult to obtain, since the crystallization speed
    is so high. This temperature is in good agreement

with Raman measurements on extremely fast quenched polyethylene made by Hendra.[24]

Poly(ethylene terephthalate) (PET)

PET is an example of a more complicated polymer, which seems to be typical for a partially crystallized material. It is a linear polymer with the monomeric unit

$$- \left[(CH_2)_2 - O - \overset{\overset{O}{\parallel}}{C} - \bigcirc - C - \overset{\overset{O}{\parallel}}{O} - \right]_n^{-}$$

The material can be obtained in several degrees of crystallinity, from fully X-ray amorphous up to ca. 60% crystallinity. The absorption spectrum in the far infrared is more complicated than that of PE (Fig. 20), but the elements of the spectra are the same: a broad background absorption generated by the amorphous substance and additional absorption bands arising from the crystalline regions. Here there are three bands instead of one at

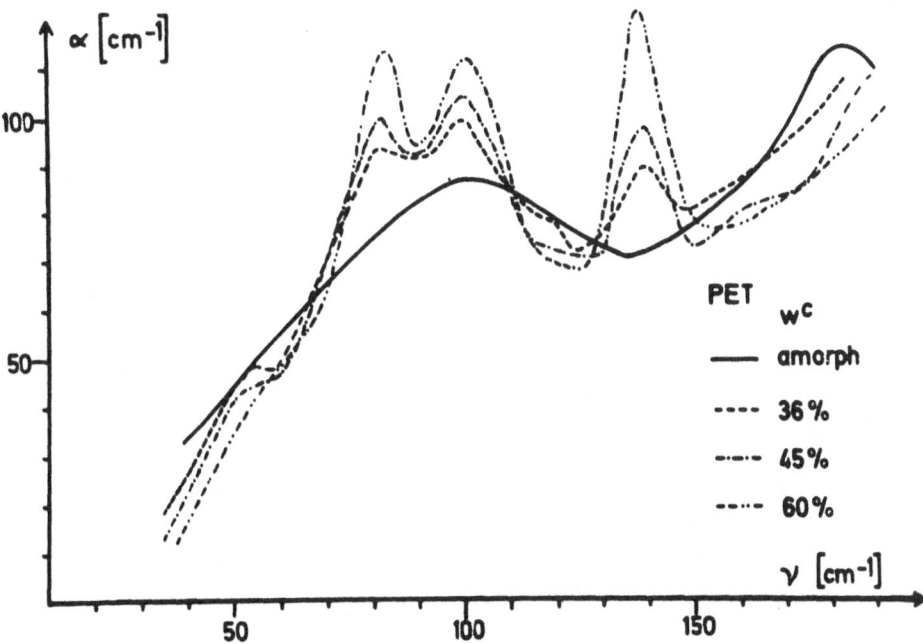

Fig. 20. Absorption spectra of PET for different degrees of crystallinity (Frank and Knaupp).[25]

PE. Their intensity grows with increasing degrees of crystallinity, as is seen in Fig. 20.

An assignment on the basis of a model of coupled torsional vibrations is given by Frank et al.,[25,26,27] where the benzene ring, the COO group, and the $(CH_2)_2$ group are considered as elements performing torsional vibrations. From our special point of view, we will discuss here only the connection of the spectral feature and the sample morphology. We consider the crystallization speed, v_{cryst}, of this material as an example for general polymer behaviour. It is sketched in Fig. 21. There are two temperatures where the crystallization speed becomes zero: at the melting point, T_m, and at the glass transition temperature, T_g. At T_m, the thermal energy is too high, so that any gain in energy by crystallization can be disregarded. At T_g, the thermal energy is too low, so that the chain molecules are no longer movable as a whole. Only short parts of the chains, called segments, can change their positions. The morphology, which we obtain by crystallizing the material in the region near T_g, is quite different from that near T_m.[28] Near T_m, the number of nuclei is low and crystallites that develop are large, whereas near T_g, the number of nuclei is high, and only small crystallites can grow; these soon obstruct one another, preventing further growth. These properties are now reflected in the far infrared spectra. A qualitative example is given in Fig. 22. We see that the absorption bands of samples annealed at higher temperatures are sharper and more intense than those treated at low

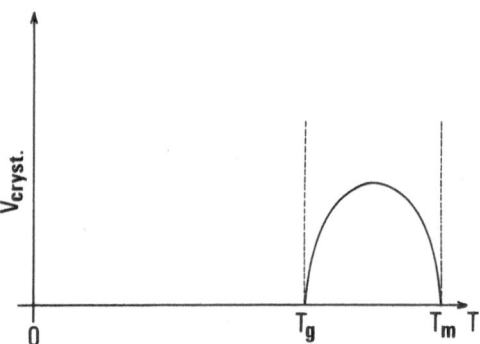

Fig. 21.   Schematic sketch of the crystallization
speed of a polymer. Below T_g, no crystallization occurs; above T_g, crystallization
speed undergoes a maximum.

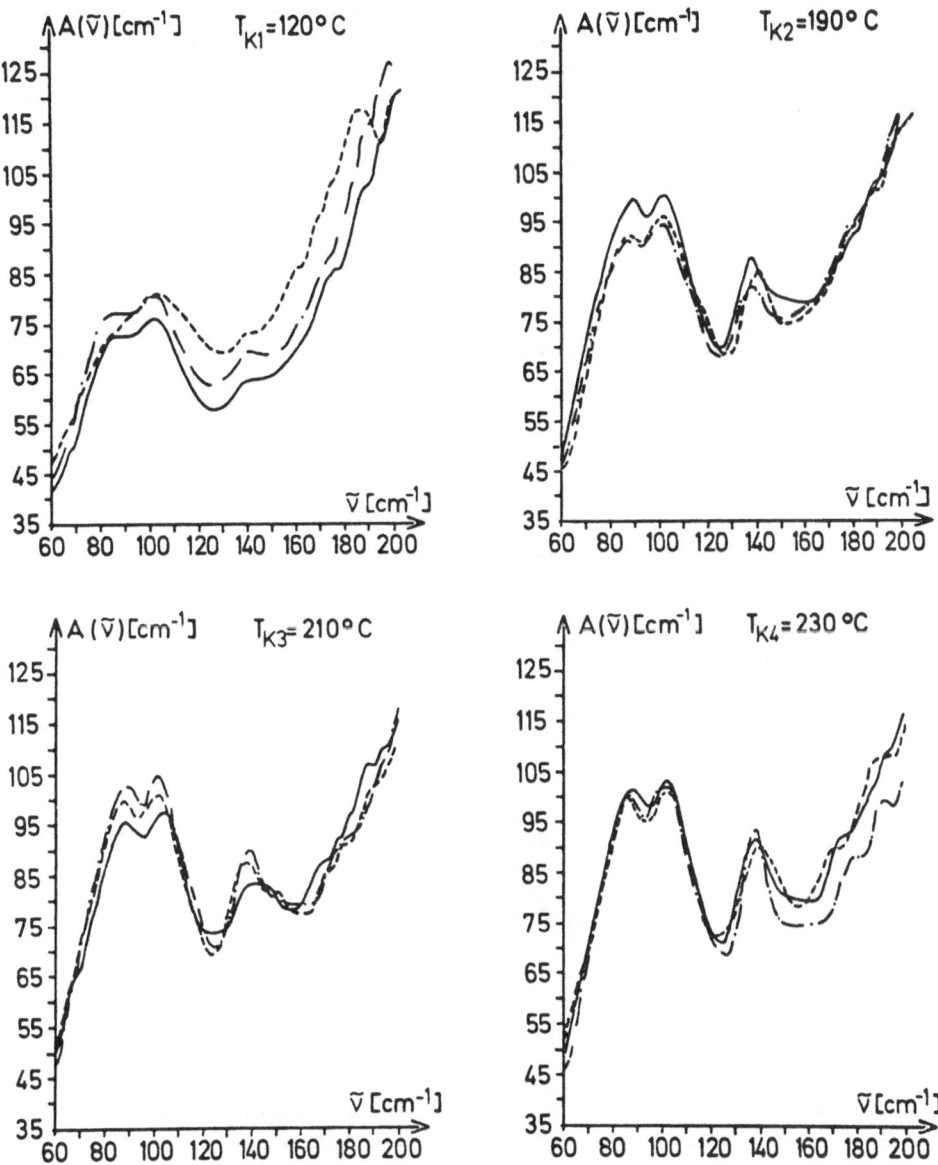

Fig. 22. Absorption spectra of PET samples crystallized at different temperatures, as indicated.

temperature. Figure 22 is a good demonstration of the fact that the degree of crystallinity, W^c, does not give sufficient information on the sample properties, because it contains no information on the size distribution function of the crystallites. On the other hand, these spectral features may be helpful in searching for suitable low temperature material. Annealing near T_g should produce a large number of crystallites with a big number of boundaries.

CONCLUSION

From this discussion, it is obvious that a definite decision on the suitability of a material with a special thermal treatment cannot be made from the results of spectroscopic measurements only. However, in combination with mechanical tests, spectroscopy in the far infrared is a quick guide to the sample morphology and its properties at low temperatures.

ACKNOWLEDGMENTS

The author is indebted to BASF, Ludwigshafen and Kalle, Niederlassung Hoechst AG, Frankfurt am Main for the samples of PE and PET.

The Deutsche Forschungsgemeinschaft provided financial support (projects FR 442/2-7).

REFERENCES

1. B. Wunderlich, Macromolecular Physics, Vol. I, Academic Press, New York and London (1973).

2. J.D. Ferry, Viscoelastic Properties of Polymers, John Wiley, New York-London (1961).

3. A.J. Staverman and F. Schwarzl, "Linear Deformation Behaviour of High Polymers," in: Die Physik der Hochpolymeren, Vol. IV, (H.A. Stuart, ed.), Berlin-Göttingen-Heidelberg (1956).

4. F.H. Müller, in: Das Relaxationsverhalten der Materie, Vol. 2, Marburger Diskussionstagung, Dr. D. Steinkopff, Darmstadt (1953).

5. W. Pechhold, Kolloid-Z. Z. Polym. 228, 1 (1968).

6. F.H. Müller, O. Heybey, and G. Knispel, Kolloid-Z. Z. Polym. 251, 932 (1973).

7. W.A. Phillips, Proc. Roy. Soc. Lond. A 319, 565 (1970).

8. K.K. Mon, Y.J. Chabal, and A.J. Sievers, Phys. Rev. Lett. 35, 1352 (1975).

9. S. Hunklinger, H. Sussner, and K. Dransfeld, "New Dynamic Aspects of Amorphous Dielectric Solids," in: Advances in Solid State Physics XVI, Pergamon-Vieweg (1976).

10. Ch. Kittel, Introduction to Solid State Physics, John Wiley, New York, London (1971).

11. P.M.A. Sherwood, Vibrational Spectroscopy of Solids, University Press, Cambridge, Massachusetts (1972).

12. H.M. Rosenberg, "The Behaviour of Materials at Low Temperatures," in: Advanced Cryogenics (C.A. Bailey, ed.), Plenum, London-New York (1971).

13. J.C. Erdmann, "Wärmeleitung in Kristallen," in: Lecture Notes in Physics 1, Springer, Berlin-Heidelberg-New York (1969).

14. G. Hartwig, Institut für Technische Physik, Karlsruhe, West Germany, private communication.

15. S. Krimm, Polym. Sci. 2, 51 (1960).

16. D.O. Hummel, Polymer Spectroscopy, Verlag Chemie, Weinheim/ Bergstrasse (1974).

17. L. van Hove, Phys. Rev. 89, 1189 (1953).

18. R.J. Bell, Rep. Prog. Phys. 35, 1315 (1972).

19. M. Tasumi and T. Shimanouchi, J. Chem. Phys. 43, 1245 (1965).

20. M. Tasumi and S. Krimm, J. Chem. Phys. 46, 755 (1967).

21. G.D. Dean and D.H. Martin, Chem. Phys. Lett. 1, 415 (1967).

22. W. Frank, H. Schmidt, and W. Wulff, J. Polym. Sci., Part C, Polym. Symp. 61, 317 (1977).

23. B. Wunderlich and H. Baur, Adv. Polym. Sci. 7, 283 (1970).

24. P.J. Hendra, to be published.

25. W. Frank and D. Knaupp, Ber. Bunsenges. Phys. Chem. 79, 1041
     (1975).

26. W. Frank, H. Fiedler, and W. Strohmeier, J. Appl. Polym. Sci.,
     in press.

27. W. Frank. W. Strohmeier, and M.L. Hallensleben, to be pub-
     lished in Polymer.

28. H.G. Zachmann, Adv. Polym. Sci. 3, 581 (1964).

CORRELATION BETWEEN VALENCE ENERGIES AND

LOW TEMPERATURE FLEXIBILITY

R. Schmid

Ciba-Geigy AG
Basel, Switzerland

Attention should be drawn to one particular aspect of the low temperature flexibility of polymers. Kreibich's contribution and recently determined data indicate an interesting connection between secondary valence forces, primary valence energies, and low temperature flexibility.

Figure 1 shows the tensile strength of an epoxy resin as a function of temperature being flexibilized in different ways. At first glance, the properties and temperatures involved would seem to have no bearing on low temperature flexibility. The unflexibilized epoxy (EP) has a glass transition temperature (T_g) of 145°C, and, as temperature drops from this level, tensile strength first increases rapidly, then slowly, and after attaining a certain value decreases. The same EP flexibilized with polypropylene glycol segments has a markedly lower T_g, and, because of this, the increase in tensile strength occurs at lower temperature. But here, too, tensile strength drops after attaining a certain value, the maximum value being achieved at somewhat lower temperature and higher stresses. Using a higher amount of adipic polyester segments, the increase in tensile strength occurs even at lower temperatures, and, with sebacic polyesters, the maximum tensile strength is not even attained at -80°C.

Apparently tensile strength corresponds to yield stress up to the maximum. At lower temperatures, brittle fracture occurs. In other words, ductility is given for as long as the secondary valence forces or the energy needed to induce slippage of adjacent chain segments are inferior to the strength of the main chains. We assume that the strength of the main chains increases with lower

79

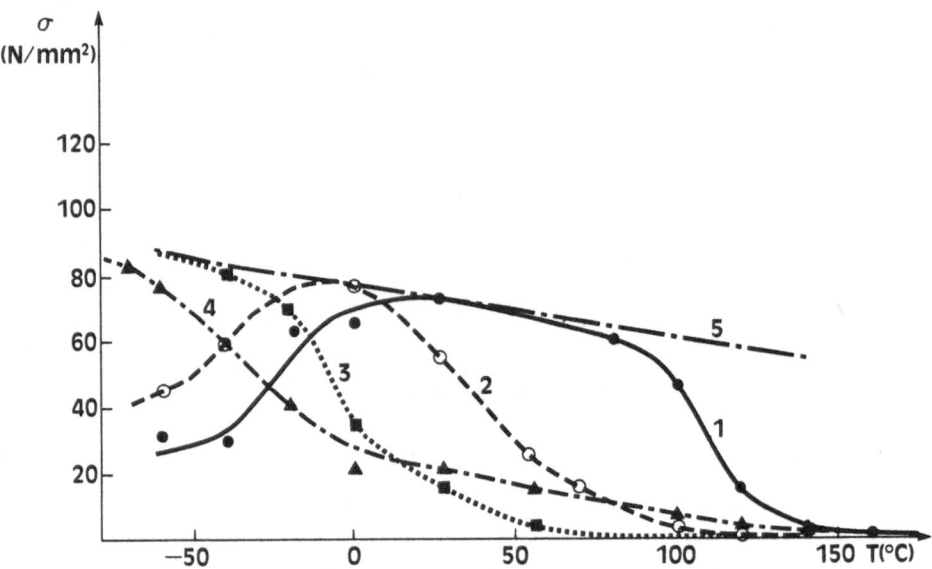

Fig. 1. Tensile strength as a function of temperature of an
         epoxide resin, flexibilized in different ways.

         1 – Unflexibilized resin (Bisphenol-A-diglycidylether
             cured with hexahydrophthalic anhydride).
         2 – The same as No. 1, but flexibilized with poly-
             propylene glycol segments.
         3 – The same as No. 1, but flexibilized with adipic
             acid polyester segments.
         4 – Cycloaliphatic resin, flexibilized with sebacic
             acid polyester segments.
         5 – Critical yield stress limit $\cong$ strength of the
             main chains.

temperatures, as shown by curve 5 of Fig. 1. Brittle fracture
occurs when the secondary valence forces attain values corresponding
to curve 5. This means that polymers that exhibit low polarity and
low yield stresses or a slow increase in the yield stress curve
should have good low temperature flexibility. And, in fact, polyethylene does have a particular flat-yield stress curve and low-
yield stress at -110°C. Nylon or flexibilized EP, which have poorer
flexibility at cryogenic temperatures, show much higher yield
stresses (Fig. 2).

     Intersetting low-polarity crosslinking polymers may be produced
by reacting crystalline acid polyesters with triepoxides (Fig. 3).

     The polyester segments also form crystallites in such a rela-
tively highly crosslinked polymer and impart adequate stiffness at

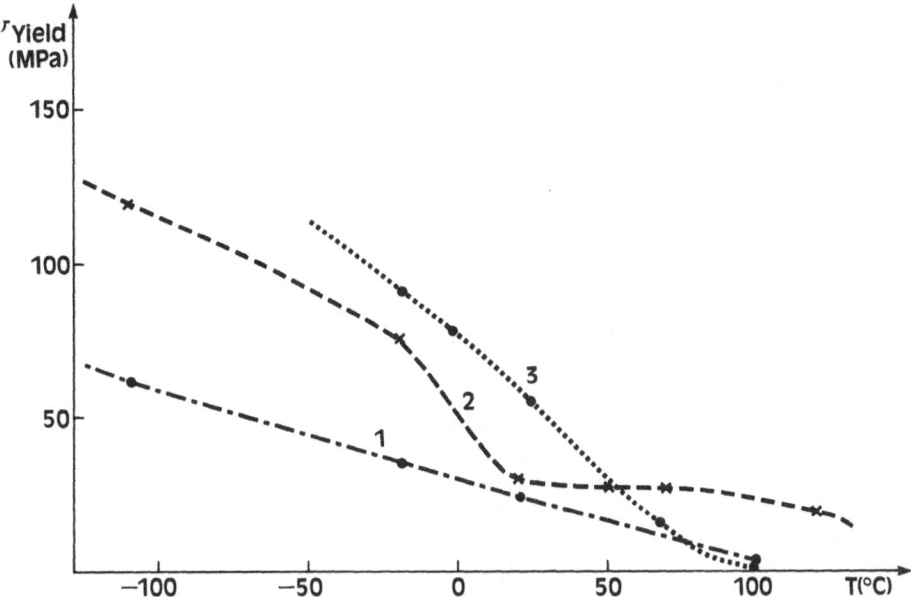

Fig. 2.   Yield-stress-temperature curves of 3 different polymers.

1 - Polyethylene (PE).
2 - Polyamide 6.6 (PA 6.6).
3 - Epoxy resin system (EP, system 2 of Fig. 1).

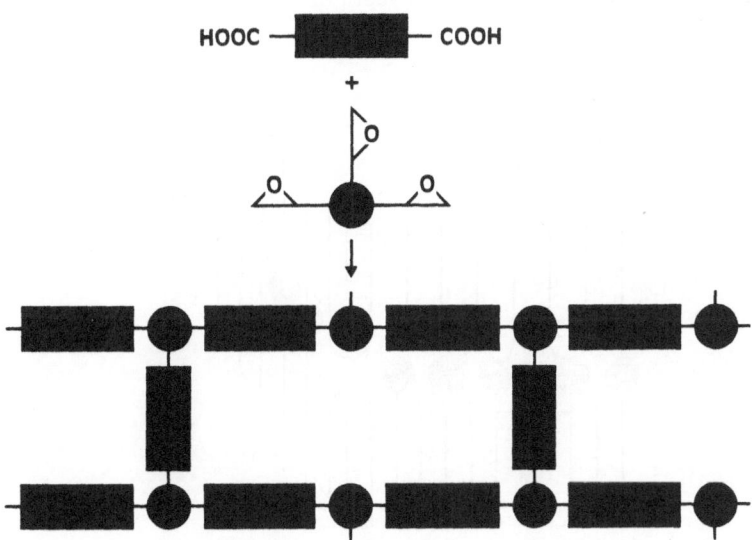

Fig. 3.   Schematic design of the formation of
           crystalline EP.

room temperature. These crystallites have a structure analogous
to that of PE, the x-ray spectrum showing similar dimensions. We
are dealing, therefore, with a structure comparable to that of PE,
except that the intermediate phase has a very different structure.
In Fig. 4, we see the folded polyester segments and the amorphous
interphase consisting of the EP component and the polyester remnants.

     Using noncrystallizing flexible segments, crosslinked, rubber-
like polymers can be produced in an analogous fashion. They have
the advantage that they are single phase without the relative polar
interphase (Fig. 5). Unfortunately, they are very soft at room
temperature.

     A comparison of the yield stress as a function of temperature
for different flexibilized epoxy resin systems is given in Fig. 6.
The crystalline EP exhibit lower yield stresses than the conventional flexibilized EP. The crystalline system, based on sebacic
acid even shows ductile behaviour at −80°C, whereas the adipic acid
system reaches its critical yield stress limit at −70°C, and brittle
failure at lower temperatures is found. Measurements of Hartwig[2]
also showed brittle fracture at cryogenic temperatures. But there
also seems to be no evidence of ductile behaviour in the case of
the sebacic acid polyester segments. The lowest yield stress at
−100°C is exhibited by the rubberlike polymer based on sebacic acid
polyester. Here too, the adipic acid polyester shows higher yield
stresses, but achieves ductile behaviour at −100°C. The more rapid
increase in yield stress is due to the absence of crystallinity.
These results show that good low temperature flexibility may be

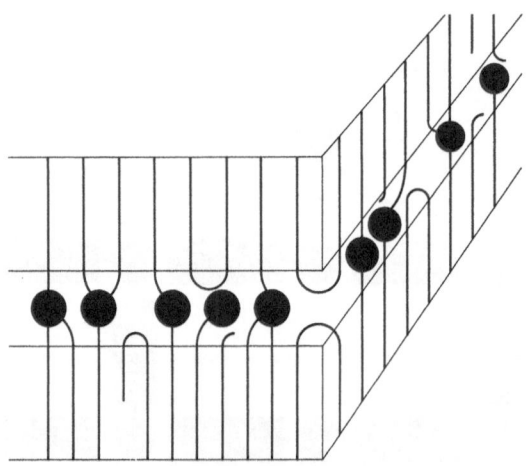

Fig. 4.   Schematic design of
a crosslinked EP.

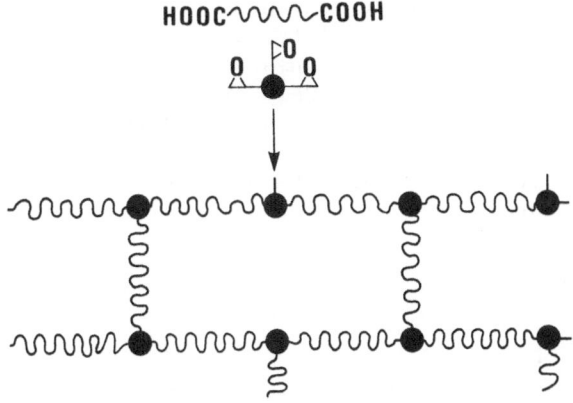

Fig. 5. Schematic design of the formation of rubberlike EP.

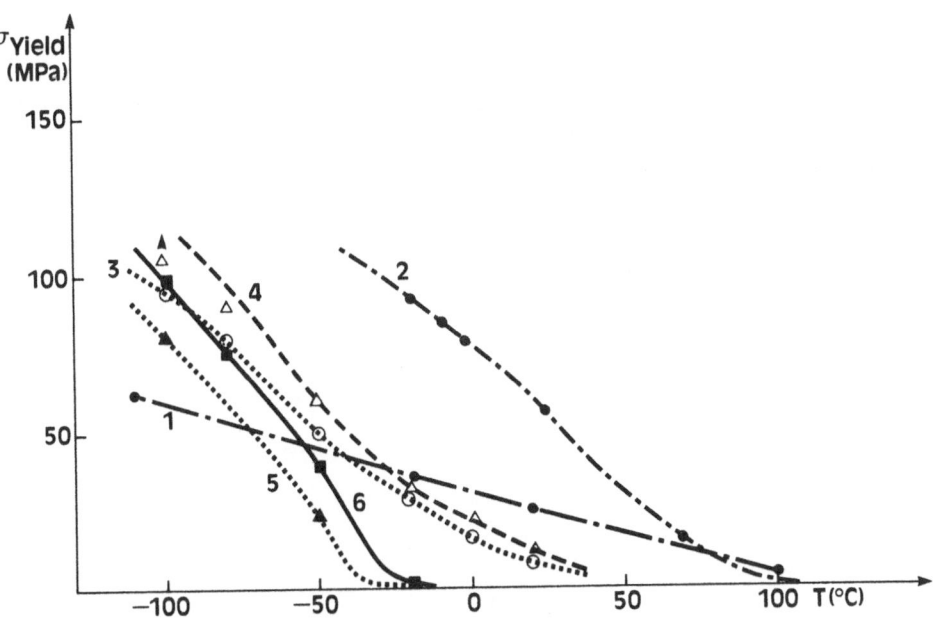

Fig. 6. Yield-stress-temperature curves of different
flexibilized EP's.

    1 - PE.
    2 - EP, flexibilized with propylene glycol.
    3 - Crystalline EP based on sebacic acid.
    4 - Crystalline EP based on adipic acid.
    5 - Rubberlike EP based on sebacic acid.
    6 - Rubberlike EP based on adipic acid.

achieved with longer nonpolar segments. They also show that crystallites tend to reduce flexibility, especially in the case of
crystalline EP. The amorphous zones comprising the epoxy compound
and the polyester remnants obviously are not flexible enough. Perhaps it might be possible to disperse the relatively stiff and polar
groups by more flexible segments to attain polymers that exhibit the
desired flexibility at cryogenic temperatures and adequate stiffness
at room temperature.

REFERENCES

1. U.T. Kreibich, in: Nonmetallic Materials and Composites at
    Low Temperatures, Plenum Press, New York (1979), p. 1.

2. G. Hartwig, in: Nonmetallic Materials and Composites at Low
    Temperatures, Plenum Press, New York (1979), p. 33.

THE DIELECTRIC LOSS OF POLYPROPYLENE FILMS AND

POLYPROPYLENE-POLYURETHANE LAMINATES AT CRYOGENIC TEMPERATURES

F. I. Mopsik, F. Khoury, S. J. Kryder, and L. H. Bolz

National Bureau of Standards
Washington, D.C., U.S.A.

INTRODUCTION

A summary is presented of measurements of the dielectric loss characteristics of six commercial polypropylene films, as well as laminates consisting of two or three polypropylene films bound together with a polyurethane.

This study was undertaken in collaboration with Brookhaven National Laboratory as part of an evaluation of the suitability of the aforementioned films and laminates for use, in tape form, as the electric insulation in high power superconducting ac trans- mission cables operated at 6 to 9 K. The superconductor in these cables is Nb_3Sn. Among the guidelines used for selecting polymer insulation suitable for such cables are: the dielectric constant should be less than 2.5, and the dielectric loss, tan δ, should preferably not exceed 20 x 10^{-6} at the cable operating temperature range and 60 Hz.[1] Although all the samples studied were suitable with respect to the criterion for the dielectric constant, signifi- cant differences in tan δ were observed in the temperature range of prime interest (4 to 10 K) among the polypropylene films, among the polypropylene-polyurethane laminates, and between the polypro- pylene films and the laminates. These differences are summarized and discussed briefly below, together with measurements of tan δ in the polypropylene films after they were annealed at 408 K in argon. The dielectric loss characteristics of all the samples were determined at temperatures from 4.2 to 323 K at both 100 Hz and 1 kHz.

SAMPLES

Six different commercial polypropylene films were examined. They are designated, alphabetically, PP-A to PP-F. The densities of these films, measured at 295.2 K using an ethyl alcohol/water density gradient column, are tabulated in Fig. 1, together with a summary of the results of a qualitative wide-angle x-ray diffraction study of the crystalline orientation(s) in the films using a flat plate camera. Three diffraction patterns were recorded for each film: with the x-ray beam parallel to the film normal (N), to the machine direction (M), and to the transverse direction (T). Except for the PP-C, all the films exhibited diffraction patterns consistent with the α-monoclinic crystal structure of isotactic polypropylene.[2] There were, however, distinct differences in crystalline orientation among the latter films, as indicated in Fig. 1. The PP-A, PP-B, and PP-F were "balanced" biaxially stretched films whose constituent crystalline regions consisted of two distinguishable populations of crystallites. One population exhibited preferential (040) planar orientation (b axis parallel to N), and the other exhibited preferential (110) planar orientation (<110>* reciprocal lattice vector parallel to N). These orientations are illustrated in the left-hand diagram in Fig. 1. The PP-D film exhibited two different orientations. The majority of the crystallites had their c axis parallel to (T), and a minor population was oriented with the a axis parallel to (T) (see the right-hand diagram in Fig. 1). The occurrence of all four different types of orientation depicted in Fig. 1 was detected in the PP-E film. Finally, no preferred crystalline orientation was observed in the case of the PP-C; in addition, the diffraction patterns corresponded to neither the α-monoclinic[2] structure nor the smectic[3] mesomorphous state of polypropylene, the latter being characteristic of the polymer when it is very rapidly cooled from the molten state. Evidence from the diffraction patterns, which were diffuse, indicated that the crystalline order in this film was intermediate between the above-mentioned mesomorphous state and the α-monoclinic structure. The relatively low density of this film is consistent with this observation. Upon annealing at 408 K, this film exhibited unoriented α-monoclinic x-ray diffraction patterns and a pronounced increase in density (see below).

Dielectric measurements were also made on all the films after they were annealed in argon at 408 K. The densities of the annealed films were also determined and found to be as follows: PP-A, -B, 0.907 g/ml; PP-C, 0.903 g/ml; PP-D, 0.910 g/ml; PP-E, 0.910 g/ml; PP-F, 0.908 g/ml. Comparison with the densities of the unannealed films given in Fig. 1 shows that the only substantial increase in density occurred in the case of the PP-C.

Details of the type and concentration of additives present in the various films are proprietary features which remain undivulged.

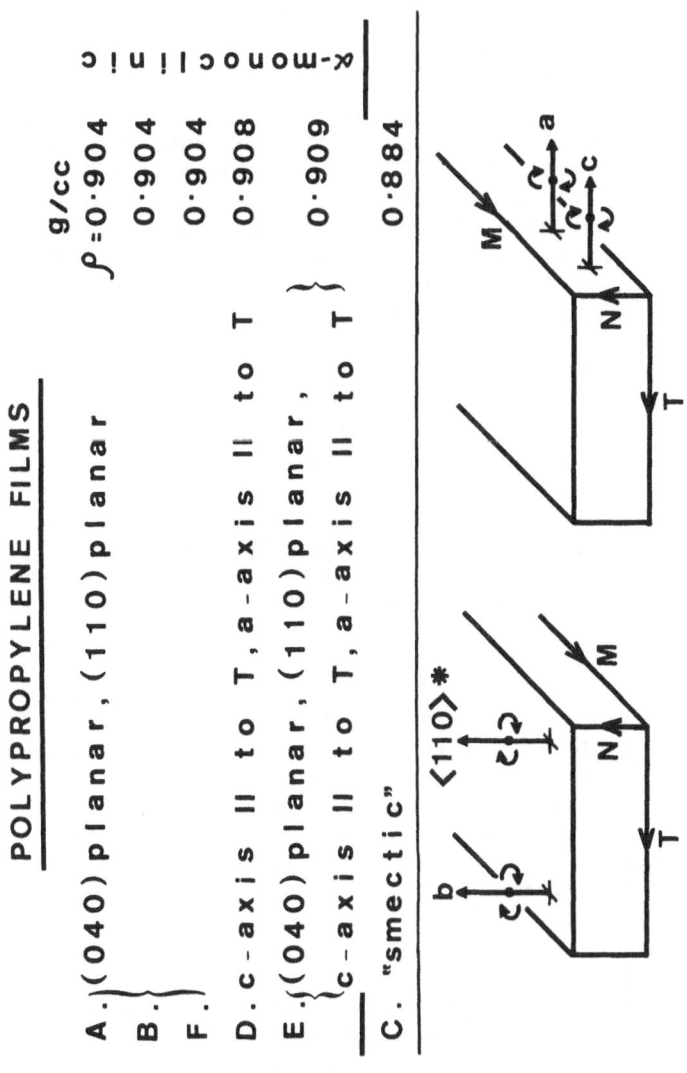

POLYPROPYLENE FILMS

	g/cc
A.) (040) planar, (110) planar	$\rho = 0.904$
B. }	0.904
F.	0.904
D. c-axis ‖ to T, a-axis ‖ to T	0.908
E. } (040) planar, (110) planar,	} 0.909
(c-axis ‖ to T, a-axis ‖ to T	
	α-monoclinic
C. "smectic"	0.884

Fig. 1. Densities, crystalline structures, and preferred crystalline orientations in the polypropylene films. The diagrams illustrate the tabulated preferred crystalline orientations in the films that exhibit the α-monoclinic structure of isotactic polypropylene.[2] The diagram on the left illustrates the (040) planar orientation (b axis parallel to N), and the (110) planar orientation (<110>* reciprocal lattice vector parallel to N). The diagram on the right illustrates the c-axis orientation parallel to T and the a-axis orientation parallel to T. See text.

The constituent polypropylene layers in all the commercially prepared polypropylene-polyurethane laminates were PP-F films. The single polyurethane layer in the PP-U-PP(B) and PP-U-PP(C) laminates contained a blue dye, as did the two polyurethane layers in the 3PP-2U(A) laminate. The two polyurethane layers in the 3PP-2U(B) laminate contained a violet dye.

DIELECTRIC MEASUREMENTS

The dielectric properties of the samples were measured with a three-terminal cell that could accommodate six specimens at a time.[4] Temperature control was better than 0.05 K at all temperatures from 4.2 to 323 K. The values of tan δ at 100 Hz and 1 kHz were determined by means of a transformer bridge using a substitution principle. In terms of tan δ, the bridge had a resolution of 1×10^{-6} or better.

RESULTS AND DISCUSSION

Polypropylene Films

The values of tan δ at 100 Hz from 4.2 K to 323 K for the unannealed and the annealed films are plotted in Fig. 2 and Fig. 3, respectively. The data for each film between 4 and 10 K are shown in detail in Fig. 4, in which the results of the measurements at 1 kHz are also included.

Confining ourselves to temperatures below 200 K and considering the data obtained from the unannealed films at 100 Hz, the following features may be noted. Three loss peaks, whose magnitudes vary widely among the films, are observed, namely, a broad peak at 125 to 150 K, an even broader peak centered at about 30 K, and a sharp peak centered at below 4.2 K. The occurrence of the latter peak is evidenced by the distinct increase in tan δ exhibited in some of the films as the temperature is lowered from 10 to 4.2 K (Fig. 4).

In general, the magnitudes of the peaks did not vary in a systematic manner that could be correlated with either differences in density or crystalline orientation among the unannealed films. Examination of Fig. 3 shows, however, that the overall effect of annealing is to bring the loss curves of the different films closer to one another. In particular, the peak at 30 K, which occurs with differing magnitude in four of the unannealed films, PP-C, -D, -E, and -F, is diminished after annealing.

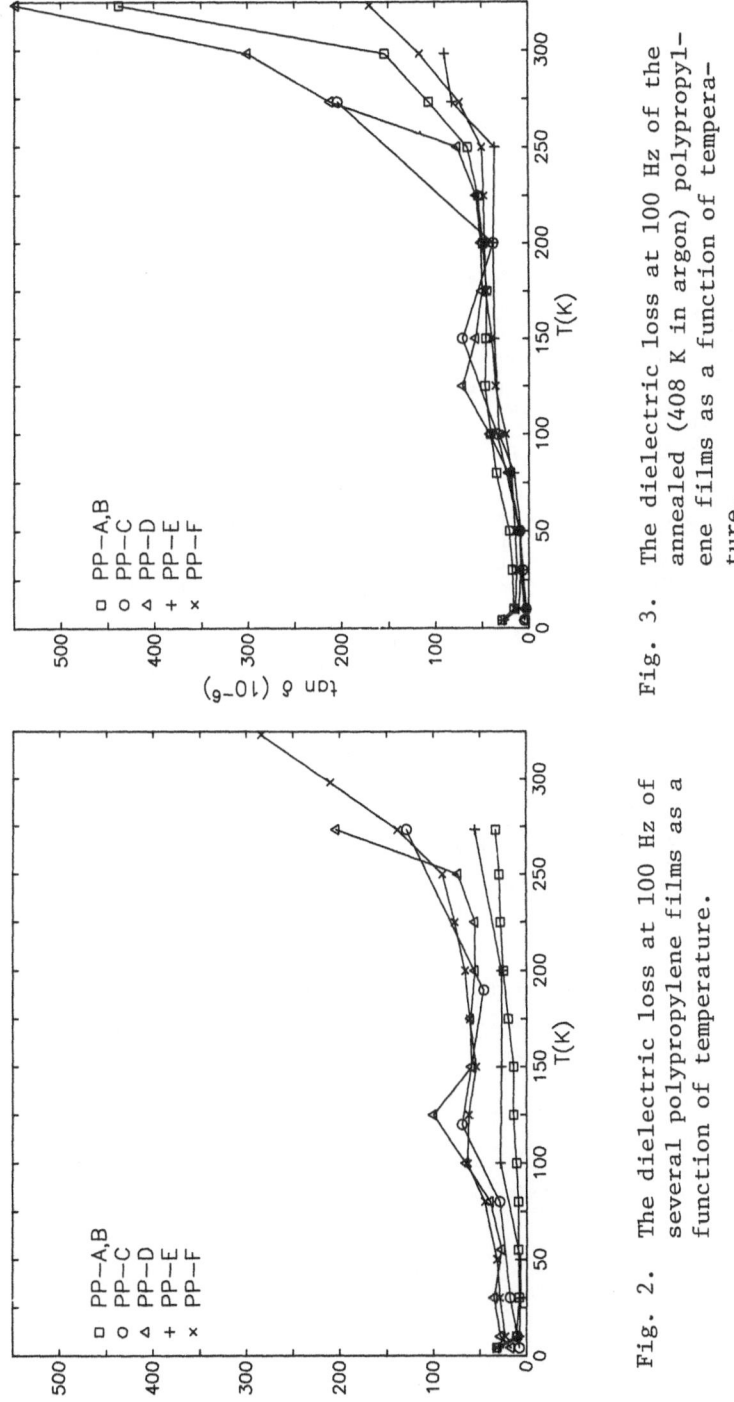

Fig. 2. The dielectric loss at 100 Hz of several polypropylene films as a function of temperature.

Fig. 3. The dielectric loss at 100 Hz of the annealed (408 K in argon) polypropylene films as a function of temperature.

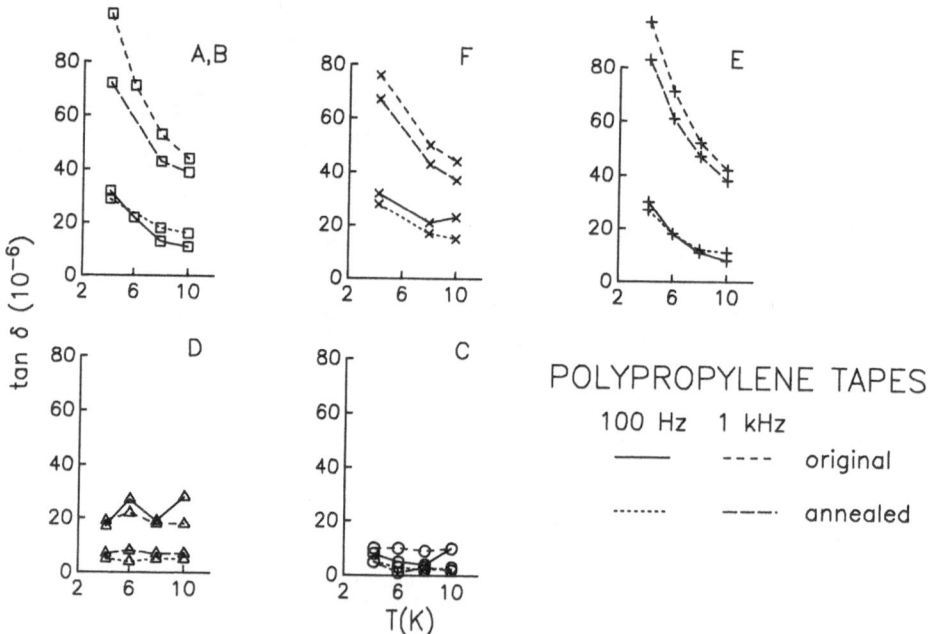

Fig. 4.   The dielectric loss at 100 Hz and 1 kHz of the unannealed
and annealed polypropylene films from 4.2 to 10 K shown on
expanded scale.

The origins of the loss peak at 125 to 150 K and that at 30 K are unknown at present. The former is probably related to a very broad loss peak in the range 123 to 223 K previously observed in dynamic mechanical measurements.[5] It is also interesting to note that a loss peak of unknown origin has been previously observed at 30 K in dynamic mechanical measurements.[6]

Four of the films, PP-A, -B, -E, and -F, exhibit the loss peak at below 4.2 K. A comparison of the data at 1 kHz and 100 Hz (Fig. 4) shows that the ratio of tan δ at these two frequencies is about 3, which is indicative of a loss peak that is very sharp for a polymer, but still broader than a Debye peak. By analogy with Thomas and King's[7] and our own studies on polyethylene[4] and a recent investigation of polypropylene by Gilchrist,[8] we believe that the loss peak at below 4.2 K in these four films is due to the presence of an antioxidant, whose contribution to the dielectric loss in the 4 to 10 K range was not eliminated by annealing.

It is of interest to consider the data in the 4 to 10 K range for the unannealed and annealed PP-D film, which exhibits no loss peak below 4.2 K. As can be seen in Fig. 4, annealing resulted in a considerable reduction in tan δ in the range 4 to 10 K from $\sim 20 \times 10^{-6}$ to $\sim 5 \times 10^{-6}$, the latter figure being very close to the lowest reported value of tan δ (3 or 4×10^{-6}) for polypropylene at 4.2 K.[9] In view of the absence of an antioxidant-related loss peak below 4.2 K in this film, this large reduction in tan δ by annealing indicates that there are factors other than antioxidant which can enhance tan δ in polypropylene in the range 4 to 10 K. The nature of these factors is not known at present. The following features may be noted, however. Among the films which we have studied, the 30 K peak was most pronounced in the unannealed PP-D (Fig. 2) and was suppressed upon annealing (Fig. 3). It appears that the higher loss exhibited at 4 to 10 K by the unannealed PP-D is associated with, and indeed corresponds to, the low temperature tail of the 30 K loss peak and that the reduction in loss (at 4 to 10 K) caused by annealing is a consequence of the suppression of that peak. Additional evidence that the dielectric loss in polypropylene at 4 to 10 K is influenced by the low temperature tail of the 30 K loss peak is indicated by the loss data at 100 Hz for the unannealed PP-F. The existence of a minimum in tan δ at 8 K (Fig. 4) may be attributed to a contribution to tan δ from that tail whose influence disappears with the suppression of the 30 K peak upon annealing.

In summary, the data described above indicated that the magnitude of the dielectric loss of polypropylene in the temperature range of interest for the superconducting cable application (6 to 9 K) is influenced by the relative contributions to tan δ resulting from the following two superposable features: (a) The presence of antioxidant, which gives rise to a loss peak below 4.2 K whose high temperature tail spans the range 4 to 10 K. (b) The manifestation

of a broad loss peak at 30 K (of as yet unknown origin) whose low
temperature tail also spans the 4 to 10 K range.

Polypropylene-Polyurethane Laminates

The results of the measurements of tan δ from 4.2 to 323 K at
100 Hz for all the laminates are plotted in Fig. 5. The data for
the PP-F film from which the laminates were made are also included
for the purpose of comparison. Confining ourselves to temperatures
below 200 K, the following features may be noted. All the laminates
exhibit a loss peak below 4.2 K, as evidenced by the upswing in tan
δ with decrease in temperature below 10 K. This peak may be attributed to the antioxidant in the polypropylene layers. In addition,
all the laminates, including the PP-U-PP(A) in which the polyurethane
layer contains no dye, exhibit a very broad and pronounced loss peak
centered at 150 K. This peak is a characteristic of the polyurethane
and, on the basis of what is known about the relative proportions of
polyurethane in the various laminates (there are uncertainties concerning the relative thickness of the polyurethane layers in them),
its magnitude appears to increase in accord with increasing polyurethane content.

The aspects of interest concerning the laminates are the relative contributions of the polypropylene and polyurethane layers to
tan δ in the region 4 to 10 K. Any changes that the PP-F layers
undergo due to heat or other effects associated with the laminating
process must be taken into account in this connection.

To examine these factors, some additional experiments were
carried out using the PP-U-PP(C). One of the polypropylene layers
was stripped from this laminate, leaving all the polyurethane
attached to the other layer. The stripped layer of polypropylene
was then folded on itself and the dielectric loss characteristics
of the resulting doubled-up layer (henceforth referred to as 0U;
see Fig. 6), were measured. The remaining polypropylene film with
the adhering polyurethane layer was doubled up to form a sandwich
consisting of two outer polypropylene layers and two inner polyurethane layers, thus effectively producing a laminate in which the
polyurethane is twice as thick as in the PP-U-PP(C). The dielectric
characteristics of this new laminate (henceforth referred to as
2U) were also measured.

The data for the 0U and 2U laminates are plotted in Fig. 6,
together with the results for the PP-F, the PP-F annealed at 408 K,
and the PP-U-PP(C) for comparison. The substantial increase in the
magnitude of the loss peak centered at 150 K in the 2U laminate,
compared with the PP-U-PP(C), confirms both the assignment of that
peak to the polyurethane and the attribution of variations in its
magnitude among the laminates (Fig. 5) to polyurethane content.

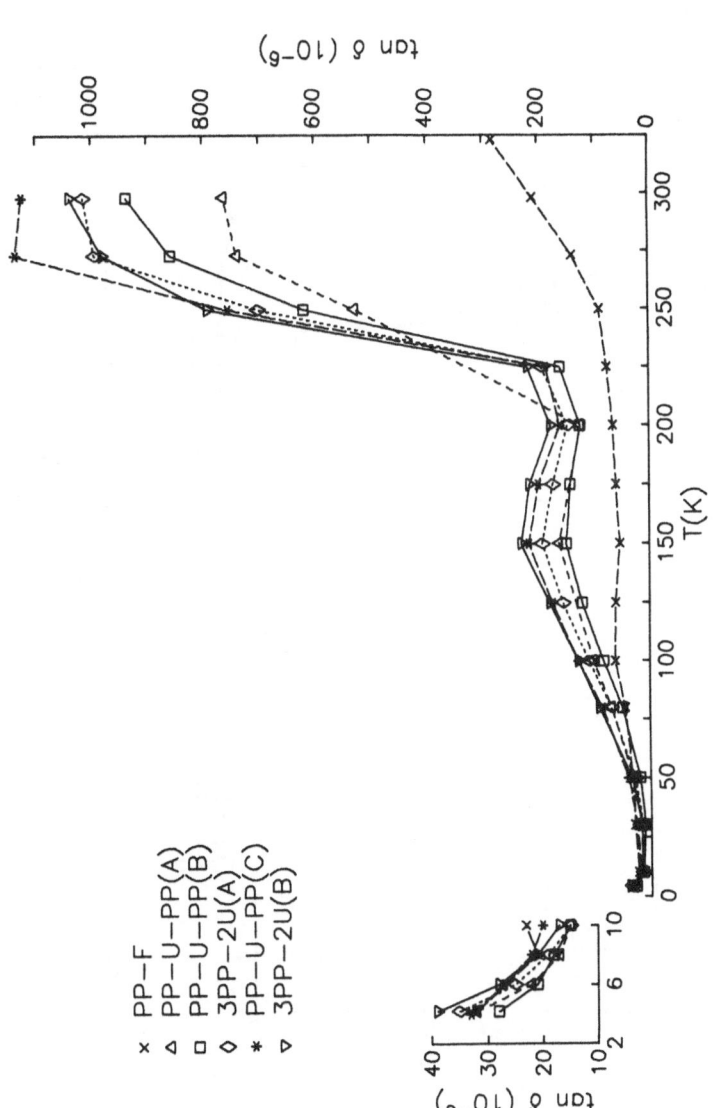

Fig. 5. The dielectric loss at 100 Hz of polypropylene–polyurethane laminates, as well as the PP–F from which they are made, as a function of temperature. The low temperature (4.2 to 10 K) data are shown on an expanded scale in the inset. See text for nomenclature.

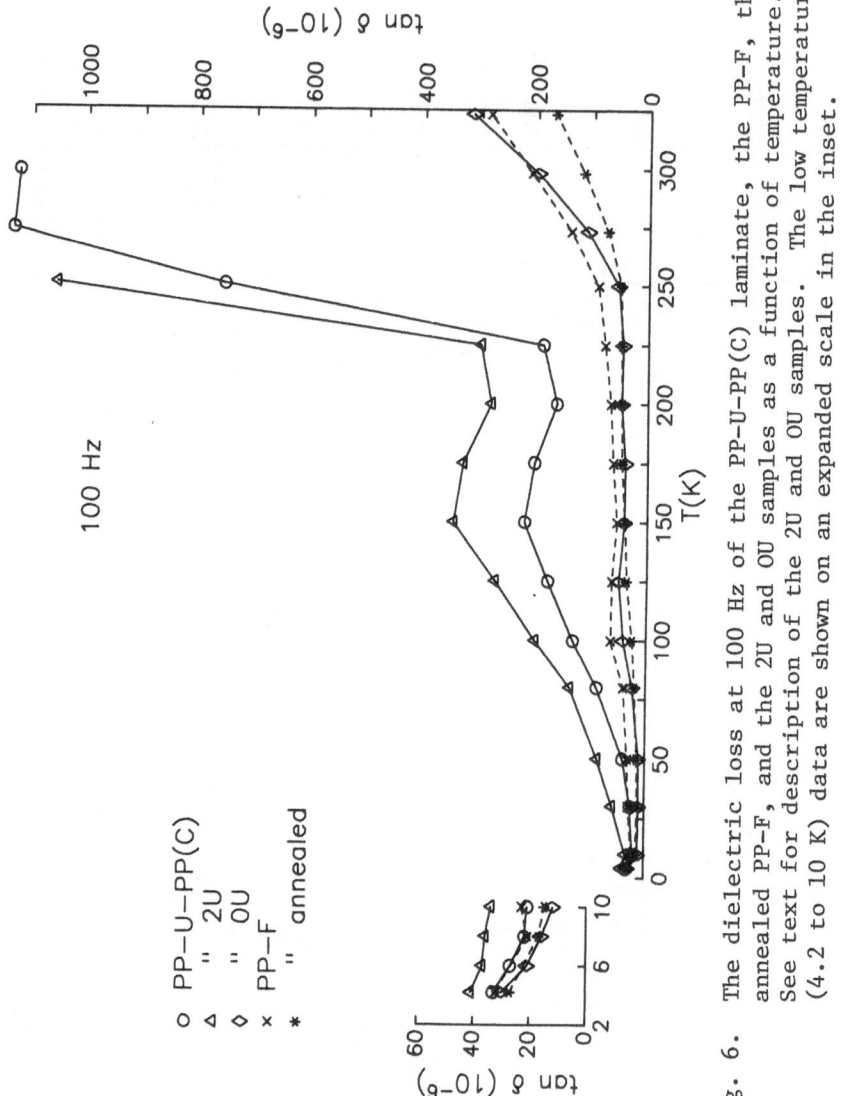

Fig. 6.   The dielectric loss at 100 Hz of the PP-U-PP(C) laminate, the PP-F, the annealed PP-F, and the 2U and OU samples as a function of temperature. See text for description of the 2U and OU samples. The low temperature (4.2 to 10 K) data are shown on an expanded scale in the inset.

The distinctly higher values of tan δ exhibited at 4 to 10 K (tan δ = 41 to 34 x 10^{-6}) by the 2U as compared to the PP–U–PP(C) (tan δ = 33 to 21 x 10^{-6}) clearly illustrates that the contribution of the polyurethane to tan δ in that temperature range is far from negligible. This is also confirmed by comparing the 2U data with the OU data (tan δ = 30 to 11 x 10^{-6}).

It is important to note that values of tan δ at 4 to 10 K for the OU sample coincide closely to those of the annealed PP–F. In both cases, the dielectric loss is lower than that for the un-annealed PP–F. This may be attributed to a suppression of the 30 K peak in the former two samples. In short, this illustrates that the polypropylene layers in the laminates undergo changes during the lamination process that must be taken into account in an analysis of the relative contributions of the polypropylene and polyurethane to tan δ.

In summary, the polypropylene-polyurethane laminates exhibited a broad loss peak centered at 150 K. This peak is due to the poly-urethane, and its magnitude varied according to the polyurethane content of the samples. In addition, the laminates exhibited a loss peak below 4.2 K, which can be attributed to antioxidant in the polypropylene layers. A more detailed study based on one lami-nate clearly showed that there is a significant contribution from the polyurethane to tan δ in the 4 to 10 K range. Direct evidence has been obtained that indicates that the dielectric loss of the polypropylene can undergo changes as a result of the laminating process. These changes, which can be compared with changes caused by annealing the polypropylene, must be taken into account in assessing the relative contributions of the polypropylene and poly-urethane layers to tan δ.

ACKNOWLEDGMENTS

This investigation was supported in part directly by the U.S. Department of Energy and in part by a subcontract from Brookhaven National Laboratories.

REFERENCES

1. E.B. Forsyth, A.J. McNerney, A.C. Muller, and S.J. Rigby, IEEE Trans. PAS 97, 734 (1978).

2. G. Natta and P. Corradini, Nuovo Cimento 15 (Supplement No. 1), 40 (1960).

3. G. Natta, M. Peraldo, and P. Corradini, <u>Stereoregular Polymers and Stereospecific Polymerization, Vol. 2</u> (G. Natta and F. Danusso, eds.), Pergamon Press (1967), p. 600.

4. F.I. Mopsik, S.J. Kryder, F. Khoury, J.P. Colson, and L.H. Bolz, in: Part II of ERDA Annual Report No. CONS/2062-1, U.S. Department of Energy, Washington, D.C. (1976).

5. E. Passaglia and G.M. Martin, <u>J. Res. NBS</u>, <u>68A</u>, 519 (1964).

6. J.M. Crissman, J.A. Sauer, and A.E. Woodward, <u>J. Polym. Sci. A3</u>, 5075 (1964).

7. R.A. Thomas and C.N. King, <u>Appl. Phys. Lett.</u> <u>26</u>, 406 (1975).

8. J. le G. Gilchrist, <u>Proceedings of the Sixth International Cryogenic Engineering Conference</u>, IPC Science and Technology Press, Guildford, Surrey, England (1976), p. 272.

9. P.S. Vincent, <u>Br. J. Appl. Phys. (J. Phys. D)</u>, Ser. 2, <u>2</u>, 699 (1969).

VARIATION OF DIELECTRIC MICROWAVE LOSSES IN POLYETHYLENE

AS THE RESULT OF DIFFERENT SAMPLE TREATMENTS

W. Meyer

Institut für Hochfrequenztechnik, Technische Universität
Braunschweig, West Germany

INTRODUCTION

During the past few years, there has been a growing need for
experimental data on the low temperature microwave dielectric
properties of insulators in connection with the development of
superconducting communication cables.[1] The feasibility of superconductive coaxial cables as a communication medium with an enor-
mous capacity may be increased by applying superconducting materials
with critical temperatures above 20 K using liquid hydrogen as a
coolant; therefore, the temperature region beyond 4.2 K is of
particular interest. With recently developed measurement equipment using superconducting microwave resonators made from Nb3Sn,[2]
we extended the accessible temperature range for precision measurements from 2.2 K up to 15 K, at frequencies between 100 MHz and
10 GHz. The absolute measurement accuracy is better than 10^{-7} for
the loss tangent, tan δ, being nearly one order of magnitude better
than with any other procedure in any other frequency region.

From the electrical point of view, the microwave region is of
special interest, too, because it exhibits the lowest dielectric
loss along the frequency scale.[3] Therefore, and because of the
high measurement accuracy, valuable insight can be gained into
molecular physics. Especially, the dielectric low temperature
behavior depends very critically upon the molecular structure
(e.g., defects and impurities incorporated into the guest lattice),
as has been shown by our measurements on various polyethylene
samples treated in different ways.

EXPERIMENTAL RESULTS

We continued with the investigations on the commercial grade
polyethylene Lupolen 4261 A,[4] which turned out to be the lowest loss
material measured so far.[5] This material contains uncommonly small
concentrations of chemical admixtures and antioxidants of only 1.4
percent. The chemical features of these admixtures are not known;
in fact, the general uncompleteness of the polymer's specifications
make a direct comparison of the vast literature on isolation data
very unsatisfying. Therefore, modifications of the same material
seem to be the only way to locate the structural units responsible
for the respective loss profile.

The different samples (cylindrical specimens, 6 mm in diameter,
60 mm long) were treated as compiled in Table I. They were soxhlet
methanol extracted for different periods and annealed for different
periods under different atmospheres. The respective frequency and
temperature loss profiles are shown in Figs. 1, 2, and 3.

DISCUSSION

The untreated material shows unique features, as it exhibits
a nearly perfect Debye relaxation centered around 400 MHz at 4.2 K.[6]
This nearly ideal behavior is very surprising because, in general,
at low as well as at high temperatures, the structural variations
of amorphous and partly crystalline materials tend to broaden the
loss-frequency curves.

Generally, various loss shapes occur at cryogenic temperatures.[1]
The loss might reflect the room temperature curve, as is the case
with polytetrafluoroethylene (PTFE) and polystyrene (PS), or might

Table I. List of Sample Treatments, Medium Density
Polyethylene Lupolen 4261 A (BASF).

No.	Treatment
1	Untreated
2	Annealed for 72 h at 110°C under nitrogen N_2
3	20 h soxhlet methanol extracted
4	20 h soxhlet, 24 h annealed at 80°C under nitrogen
5	20 h soxhlet, 24 h annealed at 110°C under nitrogen
6	20 h soxhlet, 72 h annealed at 110°C under nitrogen
7	20 h soxhlet, 24 h annealed at 80°C under air
8	72 h soxhlet methanol extracted
9	Untreated, best sample

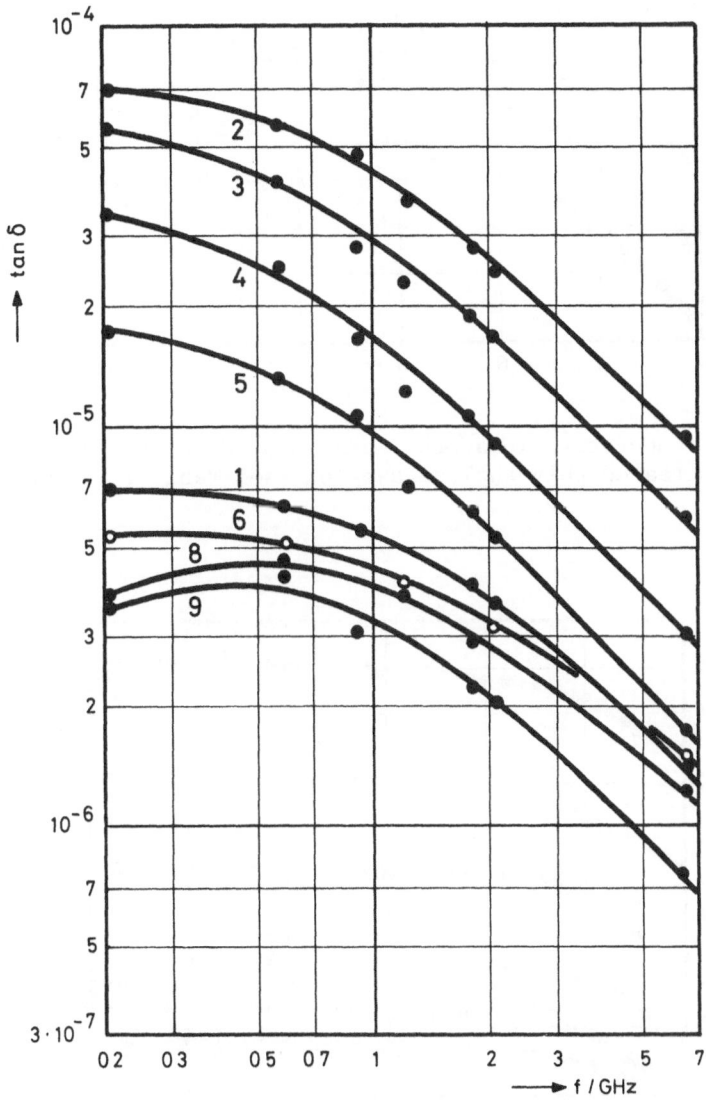

Fig. 1. Frequency loss profile at 4.2 K of different treated
polyethylene samples; see Table I for notation.

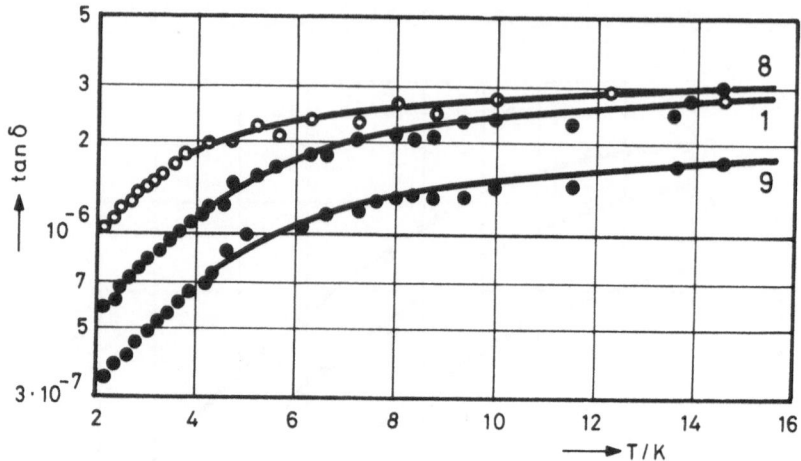

Fig. 2.   Temperature loss profile at 6.5 GHz of different
          treated polyethylene samples; see Table I for notation.

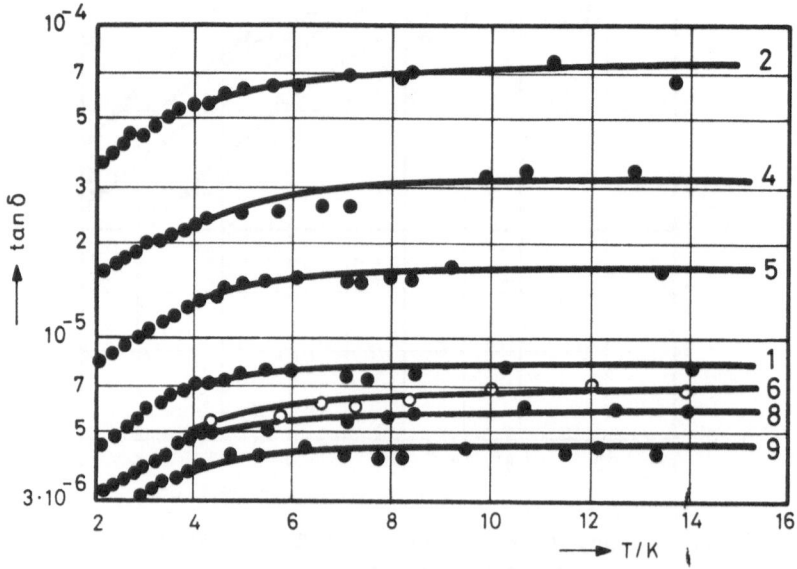

Fig. 3.   Temperature loss profiles at 0.57 GHz of different
          treated polyethylene samples; see Table I for notation.

differ considerably [high density polyethylene (HDPE)]. It might
exceed the room temperature values (HDPE) or stay below it [poly-
propylene (PP), PTFE, PS]. Also, various temperature profiles
were measured: whereas in the GHz region all samples showed increasing loss with increasing temperature, at lower frequencies
different tendencies were observed.[4] Furthermore, new saturable
absorptions occur in the microwave range, which increase when lowering the temperature and decreasing the incident electromagnetic
power.[7] The same low-energy excitations (LEE's) might be responsible for additional resonances with different temperature dependences appearing at mm wave frequencies (50 GHz) in glasses and
polymers.[8]

The physical nature of the LEE's as well as the relaxational
centres is essentially unknown; whereas low-frequency peaks are
tentatively assigned to hydroperoxide groups connected to the polymer backbone, the high-frequency loss strongly depends on the poly-
merisation conditions, degree of crystallization, oxidation, etc.,
and is due to phonon-assisted tunneling of hydrogen atoms.[9] A
complete understanding of the absorption data affords the knowledge of the frequency and temperature dependences of the matrix
elements, which couple the electromagnetic waves to the vibrational
modes, as well as the phonon density of states at a certain temperature. Both depend upon the content of (dipolar) impurities and the
degree of crystallization; both are modified by our annealing treatment, but not independently. Therefore, we can only conclude in a
merely qualitative manner that extraction and annealing procedures
do not diminish the microwave loss in the investigated medium density polyethylene. On the contrary, methanol extraction may affect
the loss tangent considerably (sample 3 in Fig. 2), whereas annealing subsequently lowers tan δ, which corroborates the results of
the low-frequency peak in polyethylene.[10] Interpretation of the
measurement results is further complicated by the finding that
samples prepared from the same moulded compact differed in their
loss magnitude up to 100 percent (samples 1 and 9). Very small
loss tangents (10^{-6}) are not stable in time either: degradations
up to 20 percent were measured at 7 GHz in different samples after
one year storage in humid atmosphere under darkness. Therefore,
the presently available data did not lead to a complete understanding of cryogenic losses in polyethylene, in contrast to silica
glasses where tunneling OH^- dipoles give rise to the microwave
absorption.[11]

REFERENCES

1. W. Meyer, IEEE Trans. Commun. COM-26, 449 (1978).

2. W. Meyer, paper presented at the 7th European Microwave Con-
    ference, Copenhagen, Denmark, September 5-8, 1977.

3. W. Meyer, in: <u>Proceedings 6th International Cryogenic Engineer-ing Conference</u>, IPC Science and Technology Press, Guildford Surrey, England (1976), p. 367.

4. Lupolen 4261A, BASF, Ludwigshafen, West Germany.

5. W. Meyer, paper presented at the International Cryogenic Materials Conference, Boulder, Colorado, U.S.A., August 2-5, 1977.

6. W. Meyer, <u>Solid State Commun.</u> <u>22</u>, 285 (1977).

7. M. v. Schickfuss, Thesis, CNRS, Grenoble, France (1977).

8. K.K. Mon et al., <u>Phys. Rev.</u> <u>35</u>, 155 (1975).

9. G. Frossati and J. le G. Gilchrist, <u>J. Phys. C: Solid State Phys.</u> <u>10</u>, L509 (1977).

10. J. le G. Gilchrist, in: <u>Proceedings 6th International Cryo-genic Engineering Conference</u>, IPC Science and Technology Press, Guildford, Surrey, England (1976), p. 372.

11. G. Frossati et al., <u>J. Phys C: Solid State Phys.</u> <u>10</u>, L515 (1977).

DIELECTRIC LOSS SPECTRA OF POLYETHYLENES

J. le G. Gilchrist

Centre de Recherches sur les Très Basses Températures
Grenoble, France

INTRODUCTION

The polymer tape to be used in the construction of an ac super-conducting power transmission cable should have a dielectric loss factor no greater than about 10^{-5}. The polymers with the most desirable mechanical properties, such as polyesters, polyimides, and polycarbonates, all have losses several times too large even at low electric stresses, so any possibility of reducing these losses would be worth investigating. Also under investigation are ways of increasing the mechanical strength of low-loss polymers, such as polyethylene. Pure unoxidised polyethylenes, either without electrodes or pressed between metallic plates, usually have dielectric loss factors below 10^{-5} from 10 Hz to 10 GHz and below 20 K.[1,2] It also appears that they may be modified by copolymerizing the ethylene with carbon monoxide or by chlorinating them to form a rubber without substantial loss factor enhancement,[3,4] even though these procedures result in the formation of carbonyl and chloride substituent groups, which greatly increase the room temperature losses.[5] It is likely that the introduction of dipolar cross-linking groups would also be permissible, and certainly also tert-butyl or other hydrocarbon side branches[6] and all textural modifications. In fact, substantially increased low temperature loss factors appear to result only from certain very specific causes, but because these are the oxidation, accidental or deliberate, of the polyethylene[7,8] and the addition of certain antioxidants,[9] they require investigation. The present paper reports low voltage dielectric studies of (a) deliberately oxidised polyethylenes, (b) polyethylenes prepared with various antioxidants, and (c) various "lossy" polymers and glasses.

OXIDISED POLYETHYLENES

Polyethylenes are liable to oxidation both during processing and prolonged room temperature storage. For our investigations we deliberately oxidised samples by exposing them to air in darkness for periods 1 to 100 h at 100 to 160°C. Such treatment resulted in the appearance of three dielectric relaxations with loss peaks well resolved in the frequency/temperature spectrum (Table I) and curves rather broader than Debye curves. A newly discovered peak comes closest to the part of the spectrum of interest for power transmission (50 to 60 Hz, 6 to 15 K), followed by the Vincett[1]-Phillips[7] peak, while Carson's[8] is furthest away. Various other characteristics are listed in Table II, from which it is apparent that the Vincett-Phillips relaxation is the most menacing, since

Table I.   Approximate frequencies in Hz where the three peaks appear when high density polyethylenes are oxidised and dielectric loss is measured at fixed temperatures. When peaks are observed with low density polyethylenes, the frequences are 2 to 3 times higher. Columns H refer to the normal peaks and columns D to the shifted peaks, which were observed after exposing the samples to labile deuterium, usually in alcohol C_2H_5OD. The shifts were reversed by subsequent exposure to C_2H_5OH or other labile protons. An asterisk denotes that the observations were made with oxidised poly(C_2D_4).

Temperature, K	Carson		Vincett-Phillips		New peak	
	H	D	H	D	H	D
.02		120				
.1		450	120			
.5		3 k	450			
2	1.8 M	11 k	1.5 k			
4	3.6 M	24 k	3.6 k			
10			50 k	20*		
20					55,24*	
50					3.3 k*	40*

Table II. Various Characteristics of the Three Loss Peaks of High Density and Low Density Polyethylenes.

	Carson	Vincett-Phillips	New peak
Maximum peak with poly(C_2H_4)	$\sim 10^{-3}$	$\sim 10^{-3}$	$\sim 3 \times 10^{-5}$
Temperature variation	Very little, 1 – 5 K	$\propto T^{-1}$, for T $\gtrsim$ 5 K	None, 17 – 50 K
Occurs during oxidation	Late	Early	Late
Maximum peak with poly (C_2D_4)	$\sim 10^{-3}$	$\sim 10^{-4}$	$\sim 10^{-3}$
Effect of...			
Vacuum anneal above 100°C	Fairly stable	Diminishes rapidly	Diminishes slowly
Cold roll	Slightly enhanced	Diminishes drastically	Diminishes
Subsequent anneal below 100°C.		Partly restored	Mainly restored
Dipolar group attached by...	$SOCl_2$, $(CH_3CO)_2O$	Cu^{2+}, Fe^{3+}, HI	KOH, $(CH_3CO)_2O$
Assignment, group	Alcohol hydroxyl $C\diagup O\diagdown H$	Hydroperoxide $C\diagup O\diagdown O\diagup H$	Probably another O–H
Movement	Rotation about C–O axis	Rotation about O–O axis	

it grows most rapidly during the early stages of oxidation, whereas, except for the case of the oxidised perdeutero polymer poly(C_2D_4), the maximum observed value of the new peak is much less than that of the other two. We have argued that the Vincett-Phillips relaxation is attributable to hydroperoxide groups[3] and investigated its

incidence on the 50 to 60 Hz to loss factor.[4] The large isotope
shifts (Table I) show clearly that all three relaxations are associated with the tunneling motions of labile hydrogen atoms. These
are most probably bonded in each case to oxygens. In general, it
may be said that the physical principles of these phenomena are
well understood,[7] and that the identification of the atomic groups
or radicals is well on the way to so being. We now enquire to
what extent the same principles apply and similar atomic groups
are active elsewhere.

POLYETHYLENES WITH ANTIOXIDANTS

The presence of certain antioxidants in polyethylene causes
the appearance of a loss curve with a peak occupying a similar
position in the spectrum as the Vincett-Phillips peak, but much
broader.[9,10] It is also distinguishable by the fact that a
methanol extraction, which would leave the Vincett-Phillips peak
unaltered, causes it to diminish[9] and eventually to disappear.
Thomas and King cited two substituted phenol antioxidants, one of
which caused a loss peak, the other apparently did not. We have
examined a series of low-density polyethylene samples each prepared
with 0.1% of a different substituted phenol antioxidant. As indicated in Table III, each caused a loss peak, the only exception
being the diffusive and volatile butylated hydroxytoluene (BHT).
Room temperature rolling caused some diminution of these peaks,
but the effect was less striking than in the case of Vincett-
Phillips. The occurrence does not seem to be confined to the 2,6
di-tert-butyl phenols nor to polyethylene as host, since a stabilized polypropylene showed a similar peak;[4] however, a polystyrene
with ethyl 330 did not.[10] We also examined low-density polyethylenes with 0.07 % of DPPD (diphenyl-N,N'-paraphenylene diamine)
having 3×10^{-3} M concentrations of the bis N,N diethyl dithiocarbamates NiDEC and ZnDEC. None of these nonphenol antioxidants
caused a peak, but we did not check for possible loss by diffusion
or evaporation.

The results call for two comments. Firstly, it seems reasonable to suggest that the occurrence is systematic for substituted
phenols in polyethylenes, but that the BHT may have been inadvert-
ently lost during moulding or room temperature storage, and also
partly the others with molecular weights below 500. Alternatively,
if they were not lost by evaporation, they may have migrated to
positions within the samples where they were dielectrically inactive,
although not necessarily ineffective, as antioxidants. Secondly,
it is not clear whether the phenomenon should be attributed to
hydroperoxides of the host (these can coexist with phenol antioxidants) or to the phenol groups of the antioxidant molecules, or
as Yano et al. suggest,[10] to some specific interaction between the
polymer and the antioxidant. Unless host hydroperoxides are

Table III. Dielectric Loss Peaks Observed at Several kHz of
Polyethylene Prepared with Various Antioxidants at
4.2 K.

2,6 di-tert-butyl phenol antioxidant	Other phenol	Other antioxidant	Molecular weight	D, $\mu m^2/s$ at 50°C	Dielectric loss peak, μ rad
Irganox 1010			1177		91
Ethyl 330 (ionox 330)			774		150[9]
Irganox 1035			642		155
	Topanol CA		554	0.2	68
	Nonox WSP		420	0.05	38
	Santonox		374	0.35	33
		ZnDEC	361		0
		NiDEC	355		0
		D.P.P.D.	260		0
BHT			222	1.15	0[9,10]

responsible, it is curious that, whatever the phenol, the peak
always occurs near the spectral position of the Vincett-Phillips
peak, and that it occurs with polyethylene and polypropylene
(which tend to form hydroperoxides) but not polystyrene (which
does not). If host hydroperoxides are responsible, then the effect
of the methanol extraction must be explained. Both to help answer
this question and as a likely means of stabilising polyethylene
effectively without dielectric loss enhancement, the dithiocarbamates
look promising and should be studied further. Their action is quite
different from the phenol and amine antioxidants, which act by
terminating the chain reactions caused by free radicals. The
dithiocarbomates prevent new chain reactions developing by causing
hydroperoxides to decompose without formation of free radicals.
Therefore, unlike the others, they do not tend to coexist with an
appreciable hydroperoxide concentration.

LOSSY POLYMERS AND GLASSES

Figure 1 shows the dielectric loss spectra of several typical "dipolar" polymer samples together with three glasses and an oxidised polyethylene with Vincett-Phillips and Carson peaks. Other "dipolar" polymers mostly have loss levels of the order of 10^{-4}. Two questions arise: are O-H groups involved and can the various spectra be understood with the same physical model (of atomic tunneling) as the oxidised polyethylene, but assuming a wide distribution of relaxation rates? For the silica glasses, the answer to both questions seems to be a qualified yes, but for the other dielectrics, no, or only slightly. The loss level of a silica glass depends strongly on its hydroxyl content, which suggests that hydrogen atoms may be tunneling as in oxidised polyethylene. There is a difficulty, though, in that very similar tunneling motions also seem to occur in hydroxyl-free silica where their dielectric effect is very weak, but they dominate the thermal and acoustic properties.[11] The hydroxyl groups then may merely act as transducers, making these motions dielectrically active.

Figures 2 and 3 illustrate one feature of the physical model as it applies to oxidised polyethylene and to hydrated silica glass. Figure 2 shows the dielectric susceptibility increment associated with the Vincett-Phillips effect in a lightly oxidised polyethylene.[12] Above about 90 mK, the increment is mainly related to the relaxation, but below 90 mK, the relaxation intensity falls rapidly while the susceptibility increment remains roughly constant and then is attributable mainly to a predicted high-frequency paraelectric resonance. The susceptibility determined at an intermediate frequency, f, exhibits a minimum as a function of temperature. By applying the same notions to silica, but assuming a wide distribution of relaxation rates and of tunnel-splitting energies (corresponding to 90 mK for the polyethylene), the susceptibility minimum is expected to occur at a temperature proportional to $f^{1/3}$, as it mainly does in fact (Fig. 3). Another more striking validation of the physical model for hydrated silica was the observation of electric dipole echoes.[16]

Regarding the other dielectrics in Fig. 1, it seems extremely unlikely that O-H groups are systematically involved, and perhaps not at all. We tested this idea by dissolving a polycarbonate sample in orthodichlorobenzene and adding thionyl chloride ($SOCl_2$) to remove any O-H groups that may accidentally have been present and then reprecipitating and remoulding it. The loss factor remained, as before, around the level of 6×10^{-5}. The flat loss spectrum is apparently characteristic of all partly or wholly amorphous materials in which the corresponding crystal's unit cell or the monomer unit (unlike the SiO_4 tetrahedron of silica) has a dipole moment. The relatively low loss of the polyoxymethylene (POM) is probably related to its rather high crystallinity. The tunneling model is

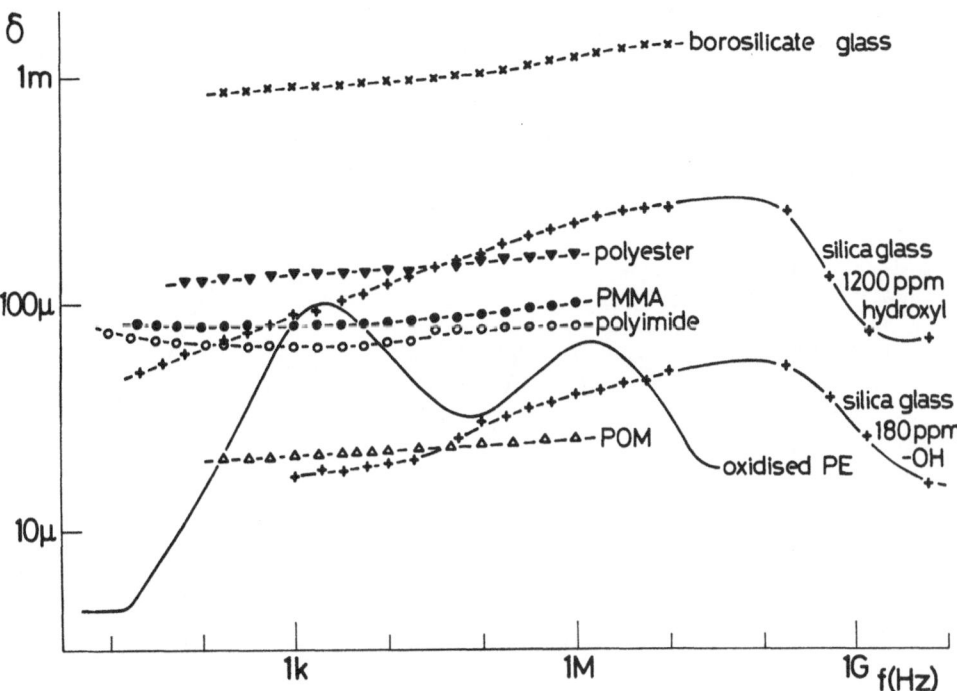

Fig. 1. Various loss spectra of dielectrics at 2.0 K (the polyester,
PMMA, and polyoxymethylene) and 4.2 (the others).

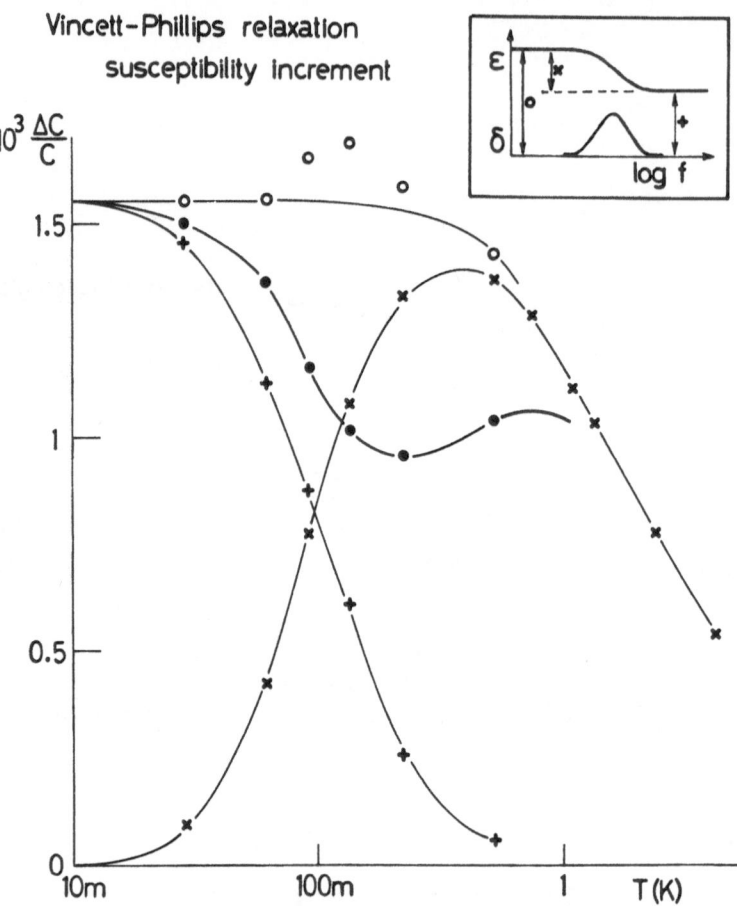

Fig. 2.   Temperature variations of a capacitance with a lightly
          oxidised polyethylene dielectric, exhibiting the Vincett-
          Phillips relaxation only.
              +....high frequency capacitance (experimental),
              x....increment due to the relaxation (deduced from the
                   area under the loss curve),
              o....low frequency capacitance (sum of + and x),
              o....capacitance at 220 Hz (experimental).

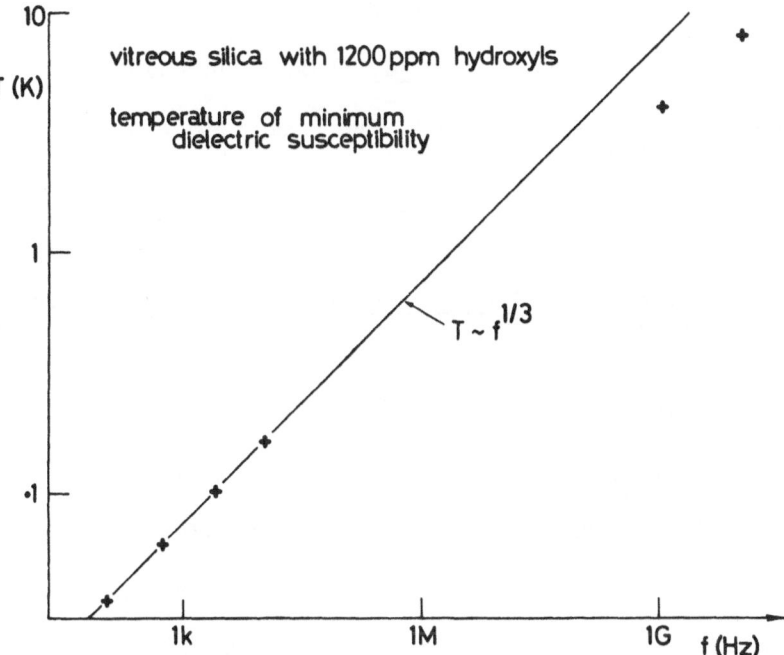

Fig. 3.   Temperature at which hydrated vitreous silica
          exhibited a dielectric susceptibility minimum
          as a function of measuring frequency. Data
          collected from Refs. 13 to 15.

not adequate to explain all the behaviour of the borosilicate
glass,[17] but a polyimide sample seemed to have a genuine susceptibility minimum like the hydrated silica.[18]

CONCLUSION

There appears little prospect of producing a noncrystalline
polymer with a dipolar monomer unit and a dielectric loss factor
below 10^{-5}. Hydrocarbon polymers with optimised mechanical properties and suitable antioxidants look much more promising.

REFERENCES

1. P.S. Vincett, J. Phys. D: Appl. Phys. 2, 699 (1969).

2. W. Meyer, ICEC 6 Proceedings, IPC Press, Guildford, England (1976), p. 367.

3. J.le G. Gilchrist, J. Polymer Sci.: Polymer Phys. Ed. (in press).

4. J.le G. Gilchrist, ICEC 6 Proceedings, IPC Press, Guildford, England (1976), p. 372.

5. C.R. Ashcraft and R.H. Boyd, J. Polymer Sci.: Polymer Phys. Ed. 14, 2153 (1976).

6. G.A. Mortimer, J. Appl. Polymer Sci. 15, 1231 (1971).

7. W.A. Phillips, Proc. Roy. Soc. A 319, 565 (1970).

8. R.A.J. Carson, Proc. Roy. Soc. A 332, 255 (1973).

9. R.A. Thomas and C.N. King, Appl. Phys. Lett. 26, 406 (1975).

10. O. Yano, T. Kamoshida, S. Sekiyama, and Y. Wada, J. Polymer Sci.: Polymer Phys. Ed. 16, 679 (1978).

11. S. Hunklinger, Spring Meeting of the Solid State Division of the German Physical Society, Friedr. Vieweg u. Sohn, Braunschweig, Germany (1977), p. 1.

12. G. Frossati and J.le G. Gilchrist, J. Phys. C: Solid State Phys. 10, L509 (1977).

13. M. von Schickfus, S. Hunklinger, and L. Piché, Phys. Rev. Lett. 35, 876 (1975).

14. M. von Schickfus and S. Hunklinger, J. Phys. C: Solid State Phys. 9, L439 (1976).

15. G. Frossati, J.le G. Gilchrist, J.C. Lasjaunias, and W. Meyer, J. Phys. C : Solid State Phys. 10, L515 (1977).

16. L. Bernard, L. Piché, G. Schumacher, J. Joffrin, and J. Graebner, J. Physique Lett. 39, L126 (1978).

17. G. Frossati, R. Maynard, R. Rammal, and D. Thoulouze, J. Physique Lett. 38, L153 (1977).

18. G. Frossati, private communication.

FLASHOVER BEHAVIOR OF SPACERS AT LOW TEMPERATURES

E. Telser

Anstalt für Tieftemperaturforschung
Graz, Austria

INTRODUCTION

The ever increasing demand for electrical energy requires the consideration of new kinds of devices in the future for the generation, transmission, and application of electrical energy, including the utilization of superconductivity。

As is known, superconductivity is the phenomenon that the resistance of certain metals and alloys drops entirely to zero under dc load and almost to zero under ac load at low magnetic field strengths and sufficiently low temperatures. The required low temperatures can be attained only by helium。

Knowledge of the flashover voltage of the necessary insulators in helium at low temperatures in addition to the breakdown voltage will be important. Flashovers can occur wherever surfaces of solid insulators are parallel to the electric field lines or their components.

Therefore, special low-temperature studies of the flashover behavior of solid insulation materials at various helium states were conducted in connection with superconducting cable development at the Anstalt für Tieftemperaturforschung (ATF) der Technischen Universität Graz (Institute for Low Temperature Research of the Technical University, Graz, Austria).[1,2,3] The materials tested were polyethylene, teflon, polyamide (PAS-60), polypropylene, paper-polyester, paper-epoxy, plexiglass, and porcelain.

TEST EQUIPMENT

A test device was designed, as indicated schematically in
Fig. 1. The device consists of two concentric tubes of stainless
steel (V 2 A) constituting the two cylindrical electrodes. Both
tubes were polished and rounded at their ends to prevent excessive
increase in the electric field. The tubes were of sufficient
length to attain uniform field distribution in the space between
the electrodes. In all tests, an inner tube of 24 mm diameter
was used. As an outer electrode, two tubes of different diameter
(32 mm, 40 mm) were used to permit tests at various electrode
distances. The annular samples of insulation material were placed
between the electrodes, and the whole test device was arranged
vertically in a cryostat filled with supercritical helium. Pressure
and temperature of the helium could be varied during the tests. The
high voltage was led to the inner electrode through a proper feed-
through in the cryostat. The voltage could be controlled in fine

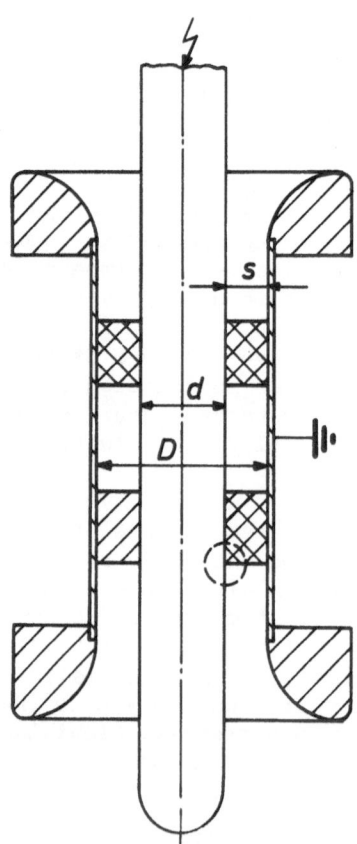

Fig. 1. Test device.

steps. The high voltage stressing the insulators was measured by
means of a calibrated voltage divider and an oscilloscope. The
ground current of the sample was measured by the voltage drop at
a low-valued resistor in the shielded ground wire.

TEST RESULTS

Annular samples of various materials were investigated at an
electrode distance of S = 4 mm (see Fig. 1). The density of the
helium remained nearly constant for each test and was always above
100 kg/mm ^{3}. Pressure and, minimally, also the temperature increased
some during the test due to the heat leaking into the cryostat from
outside by conduction and radiation. For the duration of the taking
of the measuring points, the helium pressure was kept constant. At
first the voltage was gradually raised until the first flashover
of short duration appeared and the measured voltage was noted as
"minimal value." Subsequently, the voltage was raised further
until continuous flashover occurred and this value was noted as
"flashover voltage." The flashover voltage as well as the minimal
value increased only insignificantly with increasing helium pressure, which may be due to the almost constant density (Fig. 2).

At an electrode distance of S = 4 mm and a constant pressure of
2.5 bar (0.25 MPa), increase of the flashover distance (by cutting
a groove into the side wall of the insulators, with increasing depth
at each test) yielded only an insignificant increase of the flashover

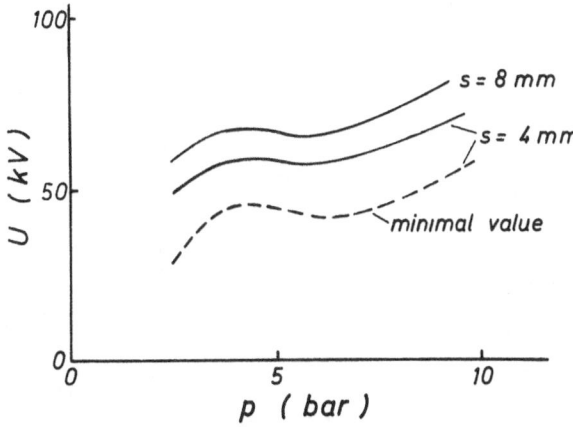

Fig. 2.   Flashover voltage vs. pressure for high-density
          polyethylene.

voltage (see Fig. 3). However, the minimal values were significantly higher and closer to the flashover voltage than those of
samples with smooth surfaces.

By increasing the electrode distance to 8 mm and, with this,
decreasing the electric field strength, higher flashover voltages
could be attained, however, not to the same degree as the electric
field strength decreased (see Fig. 1).

Generally, it can be said that at low temperatures the flashover voltages of various insulation materials are markedly below
those of the breakdown voltage, as in the conventional electrotechnology, and that increase of the flashover distance has less
effect on flashover voltage than the increase of insulator thickness has on breakdown voltage.

Since in superconductive equipment, temperature increase of
short duration is unavoidable in the case of a failure, the effect
of the helium temperature on flashover voltage was investigated,
too.

As shown in Fig. 4, the flashover voltage declines with
increasing temperature, and this effect is greater the greater the
electrode distance is. This is probably due to the decrease of
the density of the helium with increasing temperature at constant
pressure, and with this, a greater mean-free path.

Because of the lower minimal values in all tests, compared
with the measured mean values of the flashover voltage, it appeared
to make sense to observe also the ground current of the sample. It
has already been shown that before the occurrence of the continuous

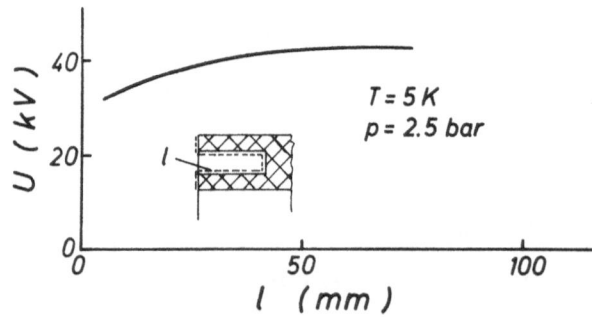

Fig. 3.   Flashover voltage vs. depth of groove for high-
          density polyethylene.

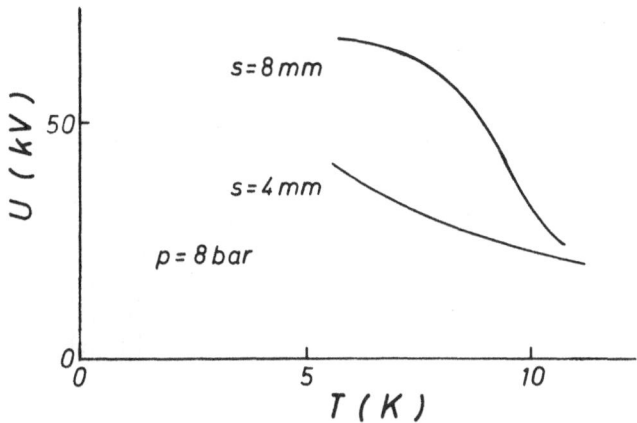

Fig. 4.   Flashover voltage vs. helium temperature for
          high-density polyethylene.

flashover current, impulses of short duration appear as soon as
certain voltage is reached. The number of these current impulses
increased with increasing voltage, while the shape remained almost
the same (Fig. 5). It was interesting that at all test samples,
the flashover always appeared when the number of current impulses
per voltage half wave reached a certain value, namely, about 30.
The impulse of the ground current appeared at unpolished surfaces
of the insulators, which were at lower voltages than the carefully
polished surfaces.

Since the tested insulator materials have different dielectric
constants, ε_r, the flashover values of the samples are entered in
dependence upon ε_r in Fig. 6. The measured values lie within the
dashed lines and show a certain deviation. However, it can be seen
that the flashover voltage decreases with increasing ε_r. The cause
for this is presumably ions eliminated from the helium; the ions'
adherence to the surface of the insulation material is stronger
the higher the ε_r value of the insulation material, leading to
flashover by field distortions.

Figure 7 shows the dependence of the flashover voltage upon
the product of the dielectric constant and the tangent of the loss
angle (ε_r, tan δ). The dielectric loss in the insulation material
and the dielectric loss at its boundary layers with the helium
each are proportional to this product. These losses lead to heating of the helium and a decrease of its density, and, in further
sequence, to flashover. These measured values also show deviation
and lie within the dashed lines. But here, too, decrease of the

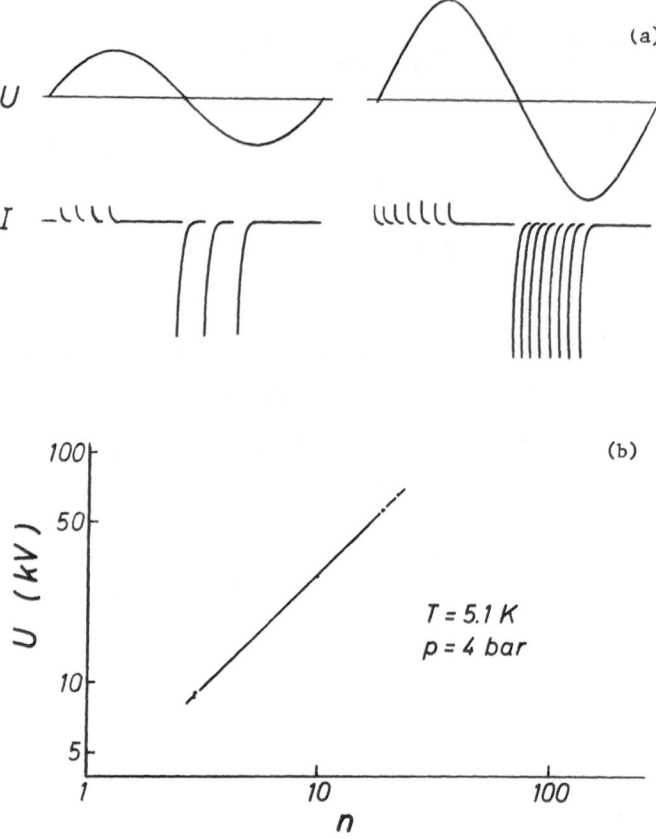

Fig. 5. Flashover current pulse shapes and numbers (polyethylene).

flashover voltage with increasing value of this product can be clearly seen.

    Those spots where three media meet,[4] namely, the electrode, the insulation material, and the helium, are indicated in Fig. 1 by a circle in dashed lines. Those spots appeared to be particularly critical. Special consideration was given to improvements to relieve those spots of the electric stress as far as possible. Because of a simpler test arrangement, these experiments were performed at first in liquid nitrogen. The particularly proper insulators were tested in helium. It was found that a conductive layer on the inner surface of the insulator contacting the inner electrode resulted in significant improvement of the flashover behavior. The reason for this is probably that the insulator and

inner electrode are in contact only at some points and otherwise are
separated by small gaps. The helium in these gaps is particularly
stressed because of its lower dielectric constant, ε_r, compared with
that of the insulation material. By exceeding the dielectric
strength of the helium in the gaps, predischarges appear. It comes
to a temperature increase of the helium in the gaps and to a decrease of its density, which facilitates the occurrence of flashover.

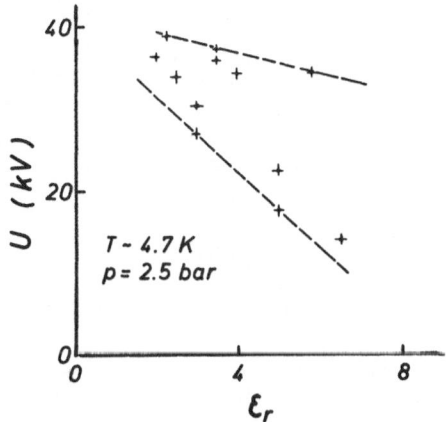

Fig. 6.   Flashover voltage vs. dielectric constant
          for various materials.

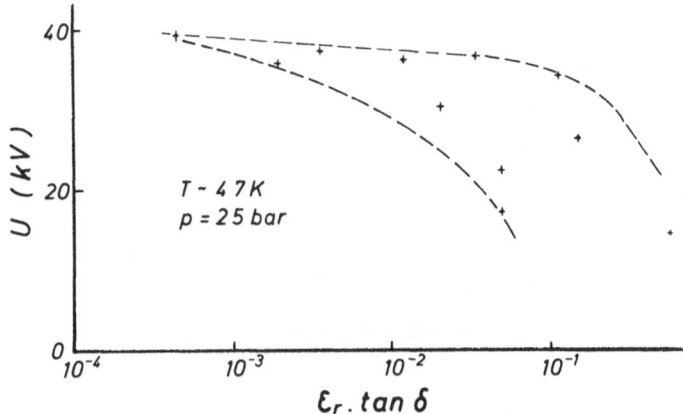

Fig. 7.   Flashover voltage vs. dielectric constant times
          loss tangent for various materials.

Improvements of flashover voltage were also achieved by suitable shaping of the annular insulators, especially at the mentioned spots.

Figure 8 shows a spacer at which both methods were applied. It is coated inside with a conductive layer. Furthermore, copper spirals are inserted into grooves of the spacer for control of the electric field. By these measures, relief of the critical spots mentioned above could be attained. During the experiments, a special bushing of Pertinax with a capacity of 1.6 nF was used. Under helium at 8.5 K/5 bar (8.5 K/0.5 MPa) with this spacer, field strengths of 7.5 MV/m could be attained using a sinusoidal ac voltage of 50 cps. Also, measurements with a standard shock wave of 1/50 μs were performed. Values of the impulse field strength of 16.5 MV/m at the inner electrode were attained, which are expected roughly also in other reports.[5]

SUMMARY

Investigations of the flashover voltage of various solid insulation materials in helium at low temperatures near zero have shown that, in general, their flashover voltage is markedly lower than their breakdown voltage; this is also the case in conventional electrotechnology. The flashover voltage is dependent upon the material of the insulator and the state of the surrounding helium, particularly its density. Increasing the flashover distance yields only small increase of the flashover voltage.

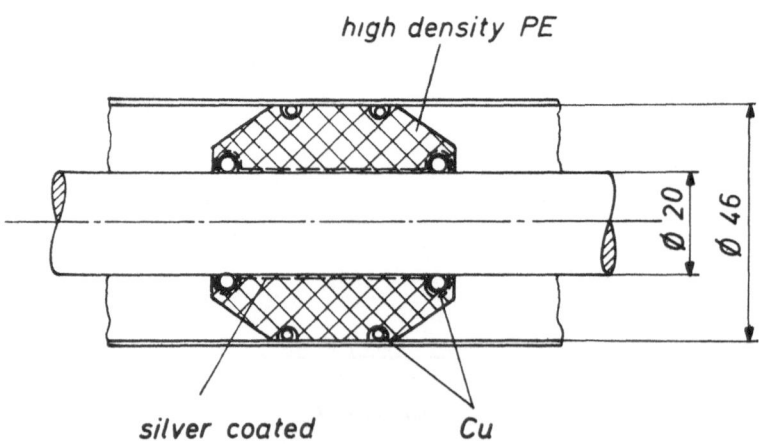

Fig. 8.   Spacer

By conductive coatings of the surface contacting the inner electrode, inserting copper spirals into the insulator for electric field control, and proper shaping of the insulator, higher flashover voltage can be attained. Experiments in this direction are not yet finished. Hope exists that further improvements can be achieved.

Basically, the results found could also be applied to the design of high-voltage bushings in low temperature devices, especially where limited space is available.

REFERENCES

1. P. Klaudy, Elektrotechnik und Maschinenbau 89(3), 93 (1972).

2. J. Gerhold, "Durchschlagsfestigkeit von Helium bei tiefen Temperaturen," Diss. Technische Hochschule, Graz, Austria (1970).

3. R. Wimmershoff, "Durchschlagsfestigkeit von gewickelten Folien-isolationen in Helium bei tiefen Temperaturen," Diss. Technische Hochschule, Graz, Austria (1973).

4. P. Weiss, Rotationssymmetrische Zweistoffdielektrika, Diss. Technische Universität, Munchen, 1972.

5. R.J. Meats, Cryogenics, 17, 77 (1971).

FRACTURE PROPERTIES OF EPOXY RESINS

AT LOW TEMPERATURES

B. Kneifel

Institut für Experimentelle Kernphysik
Karlsruhe, West Germany

INTRODUCTION

There are many applications for epoxy resins in cryogenic devices. One of these is the impregnation of superconducting magnets to prevent movement of the conductor and to insure the accuracy of the field. Therefore, in cryogenic engineering, it is very important to know the low-temperature properties of the materials. Examples are: dilatation, moduli of elasticity, and shear. In this paper, the stress intensity factor, the fracture toughness, and the surface energy of four different types of resins will be discussed. Knowledge of the surface energy and of the stress intensity factor of a resin shows how well a resin will resist cracking under certain stresses. In this way, one can make the choice from a variety of resins.

SPECIMEN

Measurements were performed with CT (compact-tension) specimens. The crack-starter slot was a chevron type notch.[1] The CT samples have different sizes, as seen in Fig. 1. Different specimen dimensions were tested to be sure that geometrical effects of the absolute sample size are insignificant.

TEST MACHINE

The apparatus was a hand-driven spindle-lever-transmission machine. The principles of this machine are shown in Fig. 2.

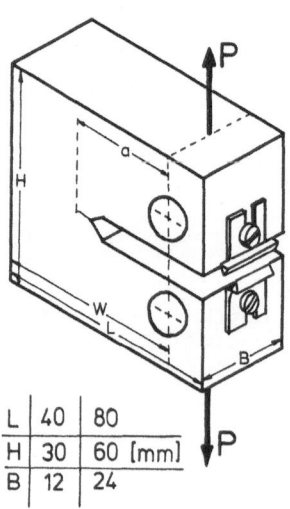

L	40	80
H	30	60 [mm]
B	12	24

Fig. 1.   Compact-tension
          (CT) specimen.

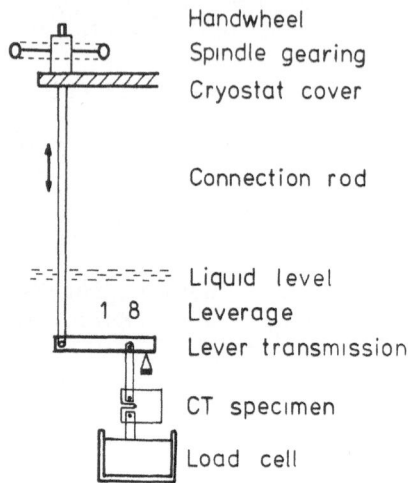

Fig. 2.   Load machine

The crucial point here is that the strain energy must be stored
within the specimen and not in the machine. This was achieved by
using a 10 Mp (98 kN) machine loaded to only 1.5% of the maximum
load. The force is measured by a load cell and the deflection is
determined by a displacement gage[1] and a knife construction
(Fig. 1).

METHOD

        Figure 3 shows a schematic load-deflection curve. As one can
see, at low temperatures the epoxy resins follow Hook's law until
fracture. To get good results, one must have initial cracks for
defined starting conditions at the same temperature at which crack
growth is measured. This is a great problem for the experimenter.
It was solved by the use of a short load shock. Otherwise, pre-
cracking has to be achieved by fatigue, which is the method
recommended by the American Society for Testing and Materials
(ASTM).[1]

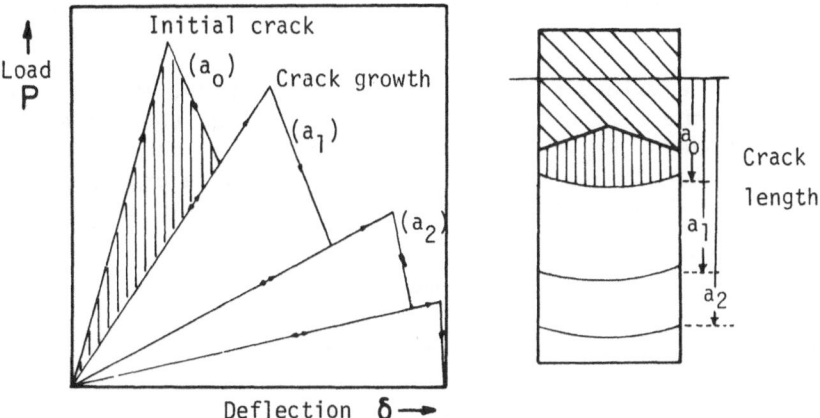

Fig. 3.   Schematic load-deflection curve and fracture surface.

THEORY

When will a crack growth take place at all? According to
Griffith,[2,3] the condition is:

$$-\frac{\partial U}{\partial A} \geq \gamma \tag{1}$$

which means that the change of elastic energy (U) stored in the
specimen divided by the change of fracture area (A) must be greater
than the effective surface energy (γ). The second derivative
($\delta^2 U / \delta A^2$) is a measure of whether the crack will stop within the
specimen or not. If this value is negative, a point may be attained
during crack propagation where ($\delta U / \delta A$) becomes less than γ. Under
this condition, external work must be done to keep the crack moving.
For load measurements, one has the load cell at hand, but in a
closed cryostat determining the crack length is a problem. The
compliance method[4] may be used. In this case, inelastic processes
are helpful, because they give rise to the formation of plastic
zones at the crack tip.

During the dynamic crack growth, local heat generation takes
place at the crack tip. If the temperature is high enough,* a
plastic zone is formed at the tip (Fig. 3) and its traces are the
so-called "arrest lines" which appear on the fracture surface after

_____

*It was found for several glasses that the temperature at the
crack tip is higher than 1500 K.[5]

a crack has stopped. This can be used to determine the crack
length.

Knowing the values of crack length, deflection, and load, one
can determine the stress intensity factor, K_I, and the plane-strain
fracture toughness, K_{Ic}. Also, the surface energy can be calculated, but as pointed out, plastification prevents an exact determination, since the energy dissipated has to be taken into account.

DEFINITIONS

The definitions of K_I, K_{Ic}, and γ are:

1. K_I is a measure of the stress-field intensity near
    the tip of an ideal crack in a linear elastic med-
    ium deformed such that the crack faces are displaced
    apart, normal to the crack plane (that means Mode 1).
    K_I is directly proportional to the applied load and
    depends on the ratio of the specimen dimensions.[1]

2. K_{Ic} is called the plane-strain fracture toughness.
    The measurement of K_{Ic} is based on the lowest load
    at which significant measurable extension of the
    crack occurs.[1] K_{Ic} should be a material constant.

3. The specific surface energy (γ) is determined by
    the total work done. This is given by the area under
    the load-deflection curve divided by the sum of the
    fracture area created on both sides of the sample
    wings.

MATERIALS AND RESULTS

The four materials[6] measured are: MY 740 - Jeffamin D-230,
rigid at room temperature (RT); CY 221 - HY 979, semiflexible at
RT; CY 221 - HY 956, flexible at RT; and LMB 234(X183/2476) -
HY 905, semiflexible at RT.

The results of the measurements (Tables I and II) show that
one cannot measure every specimen at each temperature. This is
due to the rigidity as a function of temperature for the different
epoxy resins. The number of specimens tested are given in parentheses. The fracture toughness, K_{Ic}, (Table I) is much better at
77 K than at 5 K. This shows that the notion that there are no
great differences in the material properties at the lower temperatures is incorrect.

One interpretation is that the notch effect is reduced by
plastic deformation at 77 K with respect to that at 5 K. At room

Table I.   Fracture Toughness

	K_{Ic}, kp $\times$ cm$^{-3/2}$ *		
	293 K	77 K	5 K
MY 740 Jeffamin D-230	103 (8)†	255 (4)	149 (4)‡
	100 (7) (Rigid)	251 (4)	148 (5)
CY 221-HY 979	(Semi-flexible)	187 (6)	135 (11)
CY 221-HY 956	(Flexible)		120 (10)
X183/2476 HY 905	(Semi-flexible)		139 (5)

* 1 kp $\times$ cm$^{-3/2}$ = 9.806 kPa m$^{1/2}$.

† The numbers in parentheses are the number of specimens tested.

‡ CT sample large.

temperature, the plastic volume is even greater, but the low K_{Ic} value can be explained by the breaking stress, which is two to five times less than at 77 K and 5 K. The surface energy results (Table II) derived from the load-deflection curve have a temperature dependence that is similar to that of K_{Ic}. It should be noted, however, that these results are uncertain, because the plastic deformation energy has not been identified and subtracted. The problem is that plastic energy is a function of the crack velocity, which is related to the growth of the crack length and, of course, to the temperature rise. One would like to determine accurately the specific surface energy. But this depends on how well one can estimate the crack tip temperature and the plastic volume. Better results on surface energy call for some further research.

Table II. Surface Energy

	γ, J/m²		
	293 K	77 K	5 K
MY 740 Jeffamin	199 (6)*	281 (4)	208 (4)†
D-230	202 (4)	290 (4)	210 (5)
CY 221- HY 979		270 (6)	161 (14)
CY 221- HY 959			106 (9)
X183/2476 HY 905			146 (6)

* The numbers in parentheses are the number of specimens tested.

† CT sample large.

COMMENTS

Some comments should be added in conclusion: The aim is to develop a new sample for measuring K_{Ic} in a simpler way. Consequently, it would be helpful if the determination of the K_I and K_{Ic} were independent of the crack length. In this way, the crack should always grow at the same rate over the whole sample. This could be done by increasing the stiffness of the specimen in the direction of crack propagation by expanding either the thickness or the height as a function of the crack length. One can try to solve these problems by finite element calculations, but for samples with varying thicknesses, a three-dimensional computer program is required. A two-dimensional computer program (STRUDL from MIT)[7] fits the results quite well for short crack lengths of normal CT samples. The agreement is within 2 to 4%. For longer crack lengths, the calculated displacements are too small by 5 to 15%.

REFERENCES

1. "Proposed Method of Test for Plane-Strain Fracture Toughness of Metallic Materials," Book of ASTM Standards, Part 31, ASTM, Philadelphia, Pennsylvania, U.S.A. (1969), p. 1099.

2. A.A. Griffith, Philos. Trans. R. Soc. Lond. A 221, 163 (1920).

3. A.A. Griffith, Proc. 1st Int. Congr. Appl. Mech., Delft, Netherlands (1924), p. 55.

4. K. Heckel, Einführung in die tech. Anwendung der Bruchmechanik, Carl Hanser Verlag, München, West Germany, p. 28; or G.R. Irwin, Fracture Testing of High-Strength Materials under Conditions Appropriate for Stress Analysis, Report No. 5486, Naval Research Laboratories, Washington, D.C., U.S.A. (1960).

5. K. Schönert, Universität Karlsruhe, Inst. f. mech. Verfahrenstechnik, private communication.

6. Materials from Ciba-Geigy AG, Basel, Switzerland.

7. D. Roos (ed.), ICES System-General Description, Report No. R 67-49, Dept. of Civil Engineering, M.I.T., Cambridge, Massachusetts, U.S.A. (1967).

INTERESTING LOW TEMPERATURE THERMAL AND MECHANICAL

PROPERTIES OF A PARTICULAR POWDER-FILLED POLYIMIDE

G. Claudet, F. Disdier, and M. Locatelli

Centre d'études nucléaires de Grenoble
Grenoble, France

INTRODUCTION

The use of insulating composite materials in cryotechnology can be considered in two separate ways, according to whether the filling of the resin is achieved with a fiber or a powder filler.

Fiber reinforcement (glass or carbon) increases the mechanical resistance strongly and decreases the thermal expansion, whereas thermal conductivity tends to increase. Powder filling does not affect the mechanical properties very much but reduces both thermal expansion and thermal conductivity at low temperatures. These two possibilities can be used to advantage for specific problems.

The feasibility of powder-filled resins has been considered in project studies concerning a tokamak machine proposed for the investigation of plasma physics by the EURATOM - CEA association. The project "TORE II SUPRA" is described elsewhere.[1]

A toroidal niobium titanium superconducting winding cooled to 1.8 K is projected. The winding is made with 24 torus coils, each of them enclosed in a strong stainless steel casing that provides the mechanical resistance and the screening of the winding against fast external field disturbances. In view of the relatively high heat input from field variations, these strong casings are cooled separately by a 4.5 K supercritical helium auxiliary circuit. Owing to the transients, they may reach temperatures of 11 or 12 K.

Insulating spacers, prestressed at room temperature, are placed between the coils at 1.8 K and the strong casing at 4.5 K.

131

Normal working conditions include compressive stress levels of about 100 MN/m^2. The thermal expansion of these spacers must be similar to that of the stainless steel to maintain the prestress in spite of the cooling. The principal heat load on the 1.8 K superfluid helium refrigerator is due to the thermal conduction between the coils and the casing. This means that a material must be used that has a thermal conductivity as low as possible in the temperature range 1.5 K to 12 K.

A few results obtained from the literature or from experiments are reported to define a material satisfying the three following conditions:

1. Ultimate compressive strength more than 300 MN/m^2 at 4.2 K.

2. Integral thermal expansion between 3 to 4×10^{-3} in the range 300 to 4.2 K.

3. Thermal conductivity less than 10^{-4} W/cm K at 4.2 K (10 times smaller than commercially available epoxy fiber glass).

CHOICE OF COMPONENTS

Resin

In Table I are given the three essential properties of the resins that were considered. The thermal expansion of the epoxies and nylon is too large. Nylon is, in addition, the most compressible. Aramid, Vespel, and Kerimid are very similar, but supply and casting considerations led us to choose Kerimid, which is especially designed for the construction of composites. Kerimid is commercially available in powder form and therefore is very suitable for homogeneous mixing with other powders.

Filler

In this case, the most important criterion to be considered for the powder design, is the influence of the powder on thermal conductivity. Previous studies have shown that the thermal conductivity of a powder-filled composite at low temperature can reach smaller values than that of the resin for both dielectric[7] or metallic[5,8,9] fillers.

The main results, confirmed by investigation of phonon diffusion by extended defects[10,11] are:

Table I. Mechanical and Thermal Properties of Some Selected Resins

	Temperature	Epoxies	Polyamides		Polyimides	
			Nylon (DPN)	Aramid KS 105 (DPN)	Vespel SP4 (DPN)	Kerimid (Rh.P.)
Ultimate compressive strength, MN/m²	300 K	100–170*	50–90*			
	4.2 K			380	370	> 590
Thermal expansion, $\frac{\Delta\ell}{\ell} \times 10^3$	300–77	10†	13†	6.5	7.7‡	7.4
	300–42	11†	14.5†		9‡	
Thermal conductivity, W/Cm K	4.2 K	$5\text{–}7 \times 10^{-4}$§	1.25×10^{-4}†	3.5×10^{-4}†	1.1×10^{-4}#	3.75×10^{-4}

* Guide de l'utilisateur de pièces moulées en plastiques.[2]

† Scott.[3]

‡ Perrot, Blin, Bouriot, and Gilquin.[4]

§ Hartwig.[5]

Van de Voorde.[6]

1. The decrease of thermal conductivity is due to phonon
    scattering at the boundaries between the two media
    (powder and matrix).

2. Phonon scattering depends on the acoustic mismatch
    between the two media. These phenomena are accen-
    tuated if sound velocities and densities are very
    different. Diamond could give the best efficiency,
    but alumina is better than glass and is more readily
    available.

3. The temperature below which the thermal conductivity
    falls depends on the size of the powder. This
    temperature reaches about 10 K if the powder size
    is decreased to 1 μm.

4. Below 10 K, thermal conductivity is smaller if the
    filling factor is higher.

5. A saturation effect appears if a higher filling
    factor is used when the filler size is smaller.

For these reasons, alumina powder is used, which is commercially available in a large range of convenient particle sizes.

SAMPLE PREPARATION

Samples are prepared by casting under a pressure of 20 MPa
(200 bars) at a temperature of 220°C.

To look at the influence of the particle size of the filler,
two different alumina powders, called A_1 and A_2, were used whose
main characteristics are given in Table II. Note the large difference of particle size and specific surface area of the two powders.

In the nomenclature used, samples are called Kx Ay, where
x and y represent the mass concentration in percent for each
component, K corresponds to Kerimid, and A to alumina powder.

In Table III, the values of the measured and calculated
specific mass are given for different samples. Calculated values
were obtained by taking densities of 1.3 for Kerimid and 4 for
alumina. The A_1 filler gave a homogeneous composite. The A_2
powder induced a mass defect, which increased with the concentration. This was probably due to the fact that the resin could not
perfectly fill the agglomerates.

RESULTS

Comparison between the two kinds of alumina and several concentrations of the filler was carried out from compressive
strength, thermal dilatation, and thermal conductivity measurements.

Compressive Strength

The ultimate compressive strength given in Table IV was
measured with cylindrical samples, the length and diameter of which
were 10 mm. The limitations of the apparatus prevented rupture
with A_1-filled composites. With A_2 alumina, the ultimate compressive

Table II. Characteristics of Alumina Powders

Alumina designation	Crystalline system	Particle size, μm	Size of Aggregates		Specific surface, m^2/g
			Average value, μm	Maximal value, μm	
A_1	hexagonal α	1.5	2	4	1
A_2	cubic γ	0.01 to 0.02	0.5	3	100

Table III. Specific Mass of the Composites

Symbol of the composites	$K44A_156$	$K35A_165$	$K43A_257$	$K35A_265$
Calculated value, g/cm^3	2.09	2.32	2.11	2.32
Measured value, g/cm^3	2.06	2.29	1.67	1.62

strength falls dangerously for filling factor higher than about
60%. All the measurements performed at 4.2 K show elastic
behavior up to the ultimate strength.

Thermal Dilatation

In Table IV are given the results of some 30 mm long samples
measured between room temperature and 77 or 4.2 K. In comparison
with the pure resin, type A_1 alumina gives a reduction of thermal
expansion in good agreement with the results obtained by Hartwig[5]
for epoxy resins. A thermal expansion comparable with that of
metal can be obtained. At the same mass concentration, the effect
is larger with A_2 powder.

Thermal Conductivity

Samples for measurement were parallelepipeds (5 x 5 x 30 mm).
Results are listed in Table IV and shown in Fig. 1. They were
obtained with the double flux method, which permits the elimination of first order systematic errors owing to the parasitic heat
leak.[12] Measurements below 7 K were taken with superconducting
heating leads to reduce the heat leak. The temperature difference
was measured with a gold/iron - Chromel thermocouple, which consti-
tutes a thermal shunt. This must be taken into account to calculate the sample thermal conductivity. At values near 10^{-5} W/cm K
at 2 K, for instance, the calculated correction amounts to nearly
30%.

The results obtained on samples containg A_1 alumina are in
good agreement with those of previous studies given in the earlier
references. The thermal conductivity of these composites is less
than that of the pure resin below about 9 K, with a saturation
effect at concentrations near 56 wt.%. The results are very different for A_2 alumina. The thermal conductivity becomes lower than
that of the pure resin below about 20 K, and the saturation is not
yet observed at 65 wt.%. This difference is attributed to some
porosity, which is suggested by the appreciable differences between
the calculated and measured densities (see Table III). Near 4 K,
thermal conductivity is reduced with respect to the resin value by
a factor of 2 for A_1 alumina and more than 10 for A_2 alumina.
Values such as 10^{-5} W/cm K can be obtained, giving an improvement
of more than 30 times, compared with current materials, such as
epoxy fiberglass.

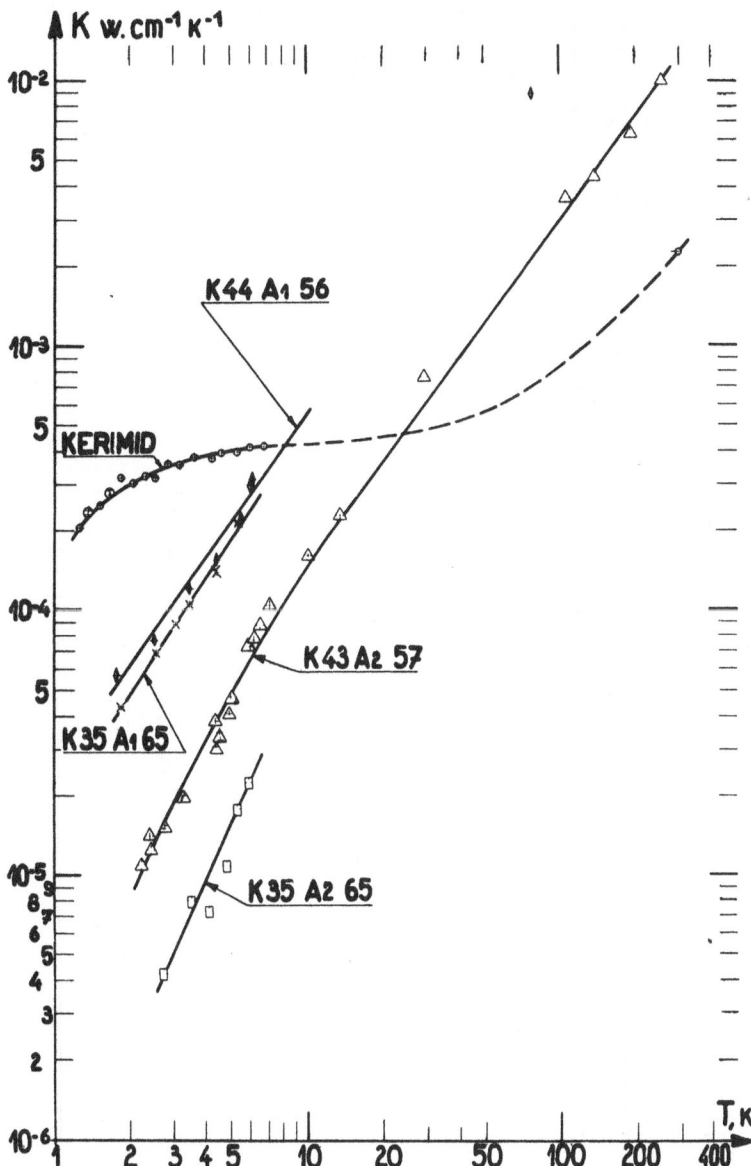

Fig. 1. Thermal conductivity of some polyimide–alumina.

Table IV. Some Selected Mechanical and Thermal Properties of the Composites

	Temperature	Kerimid	K43A$_2$57	K35A$_2$65	K30A$_2$70	K44A$_1$56	K35A$_1$65
Ultimate compressive strength, MN/m^2	300 K		240	150	120		370
	4.2 K	>590	410	300	240		>690
Thermal expansion $\frac{\Delta \ell}{\ell} \times 10^3$	300–77	7.4	3.8	3		4.8	3.42
	300–4.2					5.5	3.95
Thermal conductivity at 4.2 K, W/Cm K		3.8×10^{-4}	3.9×10^{-5}	1.7×10^{-5}		1.55×10^{-4}	1.4×10^{-4}

CONCLUSIONS

These preliminary results show that polyimide-alumina composites combine together some properties that are very useful in cryogenic technology. In the specific case of the TORE II project, A_1 alumina does not give a sufficiently low value of thermal conductivity; on the other hand, A_2 alumina seems to be limited by mechanical considerations. An intermediate powder quality could be the best compromise.

In the near future, industrial production problems will be studied, and a new characterization program will be prepared including some bending, creep, and fatigue measurements.

ACKNOWLEDGMENTS

Thanks are due to the Rhone Poulenc Company for preparing the samples in their laboratories and also to our colleagues at CEA who performed the measurements.

REFERENCES

1. R. Aymar et al., "Conceptual Design of a Superconducting Tokamak: TORE II SUPRA," paper presented at the Applied Superconductivity Conference, Pittsburgh, Pennsylvania (1978) and at the Symposium on Fusion Technology, Padova (1978).

2. Guide de l'utilisateur de pièces moulées en plastiques, 3rd Edition SNMP, 65 rue de Prony, 75017 Paris (1916).

3. R.B. Scott, Cryogenic Engineering, Van Nostrand, New York (1959).

4. A. Perrot, M. Blin, P. Bouriot, and J. Gilquin, "Propriétés mecaniques et thermiques des matériaux organiques aux températures cryogéniques," CERN 77-8 ISR-BOM (Jan. 1977).

5. G. Hartwig, in: Advances in Cryogenic Engineering, Vol. 24, Plenum Press, New York (1978), p. 17.

6. M. Van de Voorde, "Los Temperature Irradiation Effects on Materials and Components for Superconducting Magnets for High Energy Physics Applications," CERN 77-03 ISR Division (July, 1977).

7. K.W. Garrett and H.M. Rosenberg, J. Appl. Phys. D 7, 1247 (1974).

8. F.F.T. de Araujo and H.M. Rosenberg, J. Phys. D. 9, 665 (1976).

9. C. Schmide, Cryogenics 15, 17 (1975).

10. K. Guckelsberger and K. Neumaier, J. Phys. Chem. Solids 36, 1353 (1975).

11. M. Locatelli, Journal de physique C7 37(12), 322 (1976).

12. D. Howling, E. Mendoza, and J. Zimmerman, Proc. R. Soc. London, Ser. A. 229, 86 (1955).

RADIATION EFFECTS ON INSULATORS FOR SUPERCONDUCTING MAGNETS

C. J. Long, R. H. Kernohan, and R. R. Coltman, Jr.

Oak Ridge National Laboratory
Oak Ridge, Tennessee, U.S.A.

INTRODUCTION

A power-producing thermonuclear plasma produces large amounts of gamma and neutron radiation. Materials for fusion reactors must be chosen with this radiation in mind, taking into account the fact that materials in different components experience different fluxes. Magnetic-confinement fusion devices are expected to use large superconducting magnets. Although these are protected from the plasma radiation by both the power-absorption blanket and a shield whose principal purpose is magnet protection, radiation at the magnets must still be addressed during design. There are a number of limitations on radiation tolerance, depending on the particular design. Most of the potential problems relate to changes in material properties.

For assessing radiation resistance of the magnet, there are four functions for which we must choose materials. These are structural support, superconductor, stabilizer, and electrical insulation. The structural materials commonly used in current designs are austenitic stainless steel and high-strength aluminum alloys;[1] neither type is expected to experience significant degradation at the radiation levels other magnetic components can tolerate. If organic-matrix composites are used structurally, as has been suggested for tokamak poloidal-field coils, then their radiation stability must be considered.

The superconductor is expected to be either Nb-Ti alloy or the compound Nb_3Sn. Particularly for the Nb_3Sn, whose electrical properties are sensitive to the long-range order of the lattice,

radiation effects have been observed[2] at neutron fluences comparable
to those expected in the toroidal field (TF) coils of a tokamak.

The stabilizer material, high-purity aluminum or copper, will
have an increased electrical resistivity as a result of neutron
irradiation.[3] Some or all of the damage can be annealed out at
room temperature, but the warm-ups will take at least several weeks,
and their frequency must be minimized. This may be one of the
factors limiting allowable radiation at the coil.

Electrical insulation comprises a variety of materials serving
a variety of functions. Inorganic insulation could be expected
to survive the radiation anticipated in a fusion reactor's superconducting coils. However, designing a magnet with such material
could be extraordinarily difficult, and most inorganically insulated magnets to date have been on accelerator beam lines where
radiation is intense, but other system parameters (notably average
current density) can be relaxed. The great majority of superconducting magnets have been and probably will continue to be insulated with organics (varnishes, films, potting compounds) and organic-
matrix composites (such as glass-epoxy). In general, these materials are more sensitive to gamma radiation than to neutrons. The
radiation resistance at room temperature of the best organics
appears to be adequate for fusion magnets, but few data have been
available at the service temperature, 4 K.

A further possible limit on allowable radiation at the magnet--
one not related to materials damage--is nuclear heating. Because
of the low efficiency of any refrigerator operating between 4 K
and room temperature, and the limit on the fraction of reactor
power available for refrigeration, nuclear power input to the magnet must be minimized. The principal limit in this respect is on
the total power input over the whole magnet rather than the maximum local dose, as is the case for materials damage; however,
excessive local nuclear heating could also drive the superconductor
resistive. Therefore, comparisons between different limits on
radiation must be made with care; different limits may apply,
depending on the specific magnet design. A further complication
is the fact that much of the gamma-ray intensity is produced by
(n,γ) reactions that occur within the magnet itself. The relation
between the neutron dose rate (calculated from plasma and shielding
parameters) and gamma-ray dose rate, then, is a function of the
materials chosen for the magnet. The question "What limits the
allowable radiation exposure of a magnet?" cannot be answered in
general, but only for a specific design. Indeed, the answer for a
particular design is presently limited by the practical difficulty
of the flux calculation[4,5] and the uncertainty in the various
materials' radiation resistance. A calculation[5] based on the 1976
Oak Ridge Experimental Power Reactor (EPR) design, with penetrations,
projects a maximum gamma-dose rate (the most important parameter

for damage in organics) of approximately $50kW/m^3$. This is for a
first-wall loading of 1 MW/m^2; at the somewhat higher loadings
found in current reactor designs, the gamma dose rate would be
increased proportionally. Over a 20-year reactor lifetime, this
corresponds to an energy deposition of approximately 30 GGy
1 Gy = 100 rd). Obviously, more shielding is needed for this
reactor design, but it is equally apparent that designers must use
the most radiation-resistant materials available.

A number of previous investigators have studied the response
of organic materials to nuclear irradiation at room temperature,[6],[7]
but only a few have extended their efforts to cryogenic conditions.[8-12] At ambient temperature, some organic materials are
degraded at gamma doses as low as 100 Gy; the best withstand doses
in the range 10 to 100 MGy. In general, organic-inorganic composites are significantly more radiation resistant than the organic
is without the filler. The identity and form of the inorganic
filler do not appear to be especially important. For the present
problem, the ambient-temperature data are valuable chiefly for the
large number of materials that have been tested. No firm correlation has been established between ambient and cryogenic radiation
response, but it is expected that materials will fall in roughly
the same relative order. Cryogenic irradiations have been performed[11],[12] at 20 and 27 K, but not always to doses typical of
those in operating tokamaks. That work indicates that the materials that are most radiation resistant at ambient temperatures also
do well at cryogenic temperatures.

MATERIALS TESTED

The materials tested were chosen according to two criteria:
relevance to magnet design and good expected radiation tolerance.
The relevance is judged mostly by what is used in current magnets
and magnet designs. In addition, an effort was made to choose
materials so that most of the physical forms (varnish, film,
laminated sheet, etc.) commonly used in coil insulation were
represented. The purpose of this was to allow the designer as
much freedom as possible in general coil layout.

Prior information available on the irradiation behavior of
organics and composites at higher temperatures was used to select
materials that seemed likely to have the best behavior at 4 K.
Screening was necessary to make the best use of limited reactor
cold space. For example, the fluorine-substituted hydrocarbons,
like polytetrafluoroethylene, show extremely poor radiation
resistance near ambient temperature and therefore were not selected
because of space limitations. Other materials of this sort are
NEMA Grades CE and LE (textile-cloth composites), which are commonly
used in magnets but are known to have poor radiation resistance.

The materials tested in this experiment were:

1. Stycast 2850 Blue Epoxy, with 7% 24 LV hardener, Emerson and Cuming, Inc., an inorganically filled room-temperature-curing epoxy;

2. Epon 828 epoxy, 70% Epon 871 flexibilizer, 13 ppm Z curing agent, 0.5% Z6020 Silane couplant, Shell Chemical Company. (It has good strength, but requires 68°C cure.);

3. EF-527 B-stage, Synthane-Taylor Corporation, a partially cured glass cloth-epoxy composite. (It has good mechanical properties but requires curing at 175°C under pressure.);

4. National Electrical Manufacturers Association (NEMA) G-10 high-pressure glass-epoxy laminate;

5. NEMA FR-5 - like G-10, but with resin modified for temperature resistance and lower flammability;

6. Nomex paper, type 410, DuPont, aromatic polyamide sheet;

7. Kapton film, DuPont, a polyimide film frequently used in magnets;

8. Formvar insulating varnish on copper wire;

9. Aluminized Mylar, a thermal superinsulation; no tests except appearance and weight change.

IRRADIATION SCHEDULE

The experiment assembly was loaded into the Low-Temperature Irradiation Facility (LTIF),[13] a combination of a continuously operating liquid-helium refrigerator and a cryostat next to the core of the swimming-pool-type Oak Ridge Bulk Shielding Reactor. The sample chamber was then purged with dry helium gas and evacuated to about 13 Pa (0.1 torr) several times before initial cooldown.

In situ resistance measurements were made on four specimens (Table I) at a helium pressure of about 50 kPa (0.5 atm or 350 torr) and at an applied potential of 90 V. Measurements were made at room temperature and at about 100, 50, and 4 K during initial cool-down before irradiation.

With the specimen at approximately 4 K, measurements were taken with the reactor at the following power levels: 0,* 50 W,

---

*At zero power level, the residual radioactivity of the reactor core produces significant ionizing radiation.

Table I. In Situ Measurements on Irradiated Specimens

Temperature, K	Irradiation time, h	Gamma dose rate, Gy/s	Comments	Resistivity, ρ, T$\Omega \cdot$m			
				Nomex 410	2850 FT Epoxy	Epon 828 Epoxy	FR-5
305	0	0	Reactor ~2 m away	1.15	0.65	0.50	0.50
3.4	0	0		>20*	3.80	>13*	0.80
3.4	0	0.01	Reactor in place	0.13	0.45	0.14	0.60
4.8	0-1.5	40		0.0035	0.05	0.0012	0.0035
3.4	1.5	0	1 h after shut-down, reactor ~2 m	0.50	0.50	0.50	0.15
305	1.5	0	20 h later after purge	1.60	0.55	0.55	0.65
3.4	1.5	0	24 h later, cool-down	3.60	2.70	1.25	0.20
4.8	1.5-15	40	Reactor in place	0.004	0.06	0.0012	0.0035
3.4	15	0	0.5 h after shut-down, reactor away	0.30	0.32	0.15	0.09
302	15	0	24 h later, after purge	0.75	0.30	0.35	0.15
307	15	0	8 d later, reactor ~3 m away	2.10	0.60	0.90	1.15

*Current close to that of open circuit.

500 W, 5 kW, 50 kW, and 2 MW. (At full reactor power of 2 MW, the gamma ray intensity has been experimentally determined to be 40 Gy/s.) The lowest power steps (0 and 50 W) could provide information on the resistance changes that might be expected in insulation at radiation dose rates at the magnet in an operating fusion reactor ($\sim$3 mGy/s).

The irradiation at 2 MW was interrupted after 1.5 h, and the reactor was moved away from the LTIF. Resistance measurements were made at 4 K and during a step-annealing procedure at 50, 100, and 300 K. Measurements were made at the step temperature and at 4 K following each annealing step.

A total of 4 h was required to reach the last anneal at 300 K, and the experiment was held at this temperature overnight. Then the experiment assembly was again cooled to 4 K and the irradiation at 2 MW continued for 13.5 h (total time, 15 h).

At this point, the step-annealing procedure outlined above was repeated. Following the final warm-up, the experiment remained at room temperature in the sample chamber at a helium-gas pressure of about 50 kPa for about 3 weeks to allow for decay of radioactivity in preparation for subsequent testing.

TESTING PROCEDURE

The effects of irradiation on both the electrical and the mechanical properties of the specimens were measured. Electrical tests comprised breakdown and in situ resistivity and postirradiation resistivity. Breakdown tests were performed after irradiation at room temperature. The need for tests in situ is recognized and being studied. Equipment originally designed for ASTM D 877-67 (breakdown testing of transformer coils)[14] was modified to use solid specimens of 18.8 mm diam. They were either cast or punched, depending on the material.

Electrical resistance was measured in situ at intervals during the irradiation, both as a convenient diagnostic and to determine whether leakage during operation of a fusion reactor was sufficient to be of concern. Specimens 12.5 mm in diameter were designed according to specification ASTM D 257-76.[15] They were either cast, by use of copper electrodes and guard rings, or punched, in which case electrodes and guard rings were painted on by hand with electrically conductive silver-epoxy paint. Typical specimens of both types are shown in Fig. 1. The specimens were mounted in the top half of the apparatus shown in Fig. 2. Stycast 2850 epoxy, Epon 828 epoxy, Nomex 410, FR-5, and a twisted pair of Formvar-covered wires were included in this test. In addition, one pair of leads was left open to measure leakage induced by ionizing radiation.

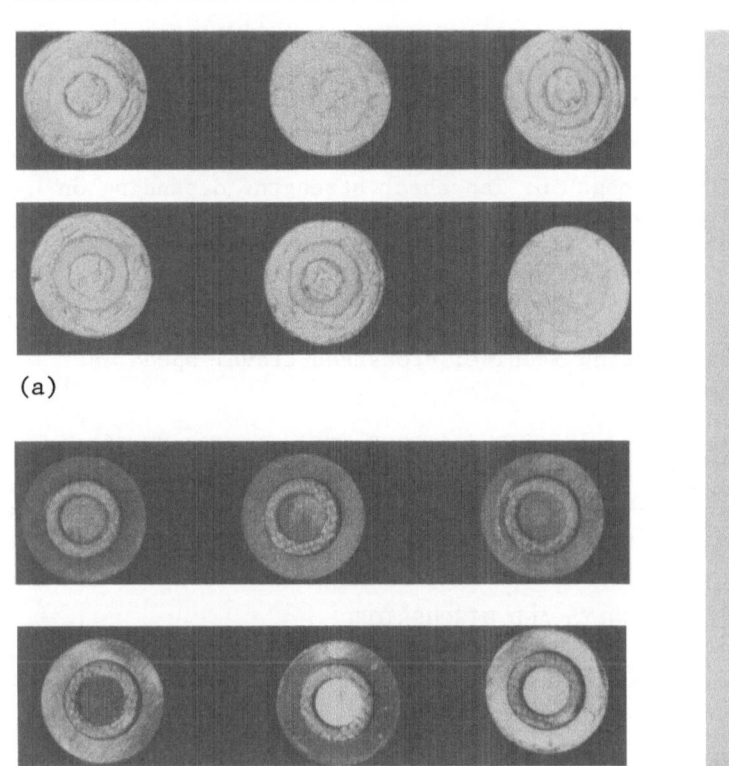

(a)

(b)

Fig. 1. Typical in situ electrical resis-
         tivity specimens. (a) Nomex Aramid
         made with silver-epoxy paint.
         (b) B-stage glass cloth made with
         copper contacts.

Fig. 2. Irradiation apparatus with the
         cadmium shield removed. In situ
         resistivity specimens are
         already in place at the top;
         the bottom will hold the other
         specimens. Total height:
         about 15 cm.

The leakage current in the open circuit was assumed to be present in the other circuits as well; therefore, it was subtracted from all currents before the resistances were calculated.

Mechanical tests, performed after the irradiation, were either three-point bending strength or lap shear strength. depending on the sample. Typical specimens for lap shear tests are shown in Fig. 3. Three-point bend (flexure) strength was measured on 2850 FT epoxy and the G-10 and FR-5 composites. Specimens about 50 x 3 x 1.5 mm were cast from the epoxy and machined from the composites. Testing was in a three-point loading fixture on a standard mechanical testing machine; crosshead travel speed was 7.5 mm/s.

Lap shear specimens were made of 2850 FT epoxy and EF-527 B-stage composite, with beryllium-copper pull-tabs. A molding jig ensured specimen uniformity. These specimens were nested to save space in the irradiation capsule.

RESULTS AND DISCUSSION

Electrical Tests

Results of the in situ resistivity measurements are shown in Table I. Because of the long leads and intense ionizing radiation, the resistivity values quoted are minima. After irradiation, the Formvar twisted pair displayed a resistance greater than 1 TΩ. The radiation flux at a reactor power of 50 MW is comparable to that at the worst location of a toroidal-field coil in an operating tokamak.

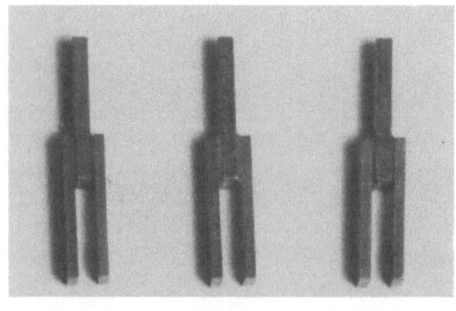 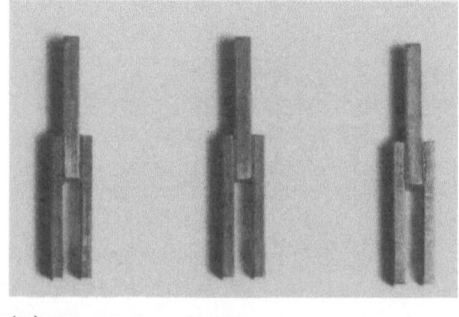

(a)   (b)

Fig. 3.   Typical lap-shear specimens. The test material is cured between the copper loading tabs. (a) 2850 FT epoxy. (b) B-stage glass cloth.

Some decreases in resistivity are apparent, but the materials are still excellent insulators relative to their intended usage.

Electrical properties measured after irradiation on the un-monitored specimens are shown in Table II. The higher postirradiation resistivity values are attributed to better measuring conditions in the laboratory than in the cryostat, which required 12 m electrical leads. The material is still entirely adequate for the purpose intended.

Table II also contains data on electrical breakdown potentials. Only for the Nomex 410 did a true through-thickness breakdown occur. For the others, a surface discharge was observed, and the potential gradient at which it occurred was calculated. These numbers, therefore, represent minimum values for the material property of interest. At the fluence reached in this exposure, no effect was observed.

Mechanical Measurements

Results of mechanical property measurements are given in Table III. Any changes that have occurred are obviously within the scatter of the measurements.

Other Effects

The apparent resistance of the in situ resistivity specimens dropped by a factor of 2 to 50 on warm-up above 200 K after irradiation of either 0.2 or 2 MGy. The original resistance was

Table II. Postirradiation Electrical Measurements

Material	Resistivity, $T\Omega \cdot m$		Electrical Breakdown, MV/m	
	Control	Irradiated	Control	Irradiated
Nomex 410	47	41	39	37
2850 FT epoxy	19	17	14	14
Epon 828 epoxy	11	11	25	24
FR-5	10	8	21	20
EF-527, B-stage glass cloth	29	24	41	43
Kapton film	50	80	80	81

Table III. Postirradiation Mechanical Measurements

Material	Shear Strength, MPa (ksi)		Flexure Strength, MPa (ksi)	
	Control	Irradiated	Control	Irradiated
2850 FT epoxy	12 (1.8)	15 (2.15)	83 (12)	83 (12)
EF-527, B-stage glass cloth	36 (5.2)	30 (4.3)		
G-10 fiberglass			390 (57)	400 (58)
FR-5			370 (54)	420 (61)

restored by purging the sample chamber with clean helium gas.
This result could be explained by electrical leakage owing to
contamination of the chamber atmosphere by a species that is
immobile at 4 K. Its identity and source are presently unknown
but will be investigated.

Weight changes were not measureable on any sample (<0.2 mg in
samples weighing 62 to 500 mg). Minor appearance changes occurred
in the epoxy and epoxy-matrix materials. No changes were observed
in the aluminized Mylar superinsulation. Radioactivity was
measured when the specimens were removed from the irradiation
capsule 18 days after irradiation. Only two specimen types exceeded 0.1 nC/kg·s. The lap shear specimens (15 g each, mostly
Cu) measured 0.4 to 0.7 nC/kg·s. Each of the 0.33 g FR-5 specimens measured 2 nC/kg·s. Neither a large gas release nor a thermal
excursion was observed during warm-up to room temperature after a
gamma dose of 0.18 MGy (after previous warm-up) and 2 MGy (total).

CONCLUSIONS

1. For the organic and composite materials (filled and unfilled epoxies, NEMA G-10 and FR-5 glass-epoxy composites, Nomex
polyamide paper, Kapton polyamide film, and Formvar coating)
tested, neither mechanical nor electrical properties are appreciably degraded by a gamma dose of 2 MGy at 5 K.

2. Electrical resistivity in these materials drops by 1 to 2
orders of magnitude under a dose rate of 40 mGy/s at 5 K.

3. Aluminized Mylar is apparently able to withstand 2 MGy at
5 K.

4. The specimen chamber atmosphere is contaminated by an
unknown substance, which makes its presence known by increased
electrical leakage upon warm-up above 200 K.

ACKNOWLEDGMENTS

The authors very much appreciate the help of J. M. Shoopman,
R. Wallace, J. Hendrix, and C. M. Fitzpatrick, who helped with
specimen and equipment fabrication; and J. M. Williams, C. E.
Klaburde, and J. K. Redman, who operated the Low-Temperature
Irradiation Facility. They also thank S. T. Sekula and W. C. T.
Stoddart, who reviewed this paper; S. Peterson, who edited it;
and J. L. Bishop, who typed it.

The research was sponsored by the Office of Fusion Energy,
U. S. Department of Energy, under contract W-7405-eng-26 with the
Union Carbide Corporation.

REFERENCES

1. J.N. Luton, F.N. Haubenreich, and P.B. Thompson, "Design of
    Superconducting Toroidal Magnet Coils and Testing Facility
    in the U.S.A.," paper presented at the 6th International
    Conference on Magnet Technology, Bratislava, Czechoslovakia,
    August, 1977.

2. J.F. Guess, R.W. Boom, R.R. Coltman, Jr., and S.T. Sekula,
    A Survey of Radiation Damage Effects in Superconducting Magnet
    Components and Systems, Report No. ORNL/TM-5187, Oak Ridge
    National Laboratory, Oak Ridge, Tennessee (1975).

3. R.R. Coltman, Jr., C.E. Klabunde, J.K. Redman, J.M. Williams,
    and R.L. Chapman, "The Effects of Irradiation on the Copper
    Normal Metal of a Composite Superconductor," paper presented
    at the Applied Superconductivity Conference, Pittsburgh,
    Pennsylvania (1978).

4. R.T. Santoro, V.C. Baker, and J.M. Barnes, Neutronics and
    Photonics Calculations for the Tokamak Experimental Power
    Reactor, Report No. ORNL/TM-5466, Oak Ridge National Labora-
    tory, Oak Ridge, Tennessee (1977).

5. R.T. Santoro, J.S. Tang, R.G. Alsmiller, Jr., and J.M. Barnes, Monte Carlo Analysis of the Effects of a Blanket-Shield Penetration on the Performance of a Tokamak Fusion Reactor, Report No. ORNL/TM-5874, Oak Ridge National Laboratory, Oak Ridge, Tennessee (1977).

6. R.W. King et al., in Effects of Radiation on Materials and Components (J.F. Kircher and R.E. Bowman, eds.), Reinhold, New York (1964), p. 84.

7. C.L. Hanks and D.J. Hamman, "Electrical Insulating Materials and Capacitors," Section 3 of Radiation Effects Design Handbook, NASA CR-1787, National Aeronautics and Space Administration, Washington, D.C. (1971).

8. S. Nishijima and T. Okada, Teion Kogaku 12, 224 (1977).

9. E.T. Smith, Investigation of Combined Effects of Radiation and Vacuum and of Radiation and Cryotemperatures on Engineering Materials -- Annual Report (9 November 1961-8 November 1962), Volume II: Radiation Cryotemperature Tests, Report FZK-161-2, General Dynamics, Fort Worth, Texas (1963).

10. E.E. Kerlin and E.T. Smith, Measured Effects of the Various Combinations of Nuclear Radiation, Vacuum and Cryotemperatures on Engineering Materials -- Annual Report (9 November 1962-30 April 1964), Volume I: Radiation Cryotemperature Tests, Report FZK-188-2, General Dynamics, Fort Worth, Texas (1964).

11. E.E. Kerlin and E.T. Smith, Measured Effects of the Various Combinations of Nuclear Radiation, Vacuum, and Cryotemperatures on Engineering Materials, Biennial Report 1 May 1964 through 1 May 1966, Report FZK-290, General Dynamics, Fort Worth, Texas (1966).

12. E. Bonjour, P. Brauns, R. Lagnier, and M. Van de Voorde, "Low Temperature Behavior of Organic Materials in a Radiation Field," in: Low-Temperature Irradiation Effects on Materials and Components for Superconducting Magnets for High-Energy Physics Applications (M. Van de Voorde, ed.), CERN-77-03, European Organization for Nuclear Research, Geneva, Switzerland (1976).

13. C.E. Klabunde and B.C. Kelley, "Irradiation Facilities in the Oak Ridge National Laboratory Bulk Shielding Reactor," paper presented at the International Symposium on Developments in Irradiation Capsule Technology, TID-4500, CONF-660511, Technical Information Center, U.S. Department of Energy, Oak Ridge National Laboratory, Oak Ridge, Tennessee (1966).

14. ASTM Standard D877-67, "Dielectric Breakdown of Insulating
     Liquids Using Disk Electrodes," in: 1975 Book of ASTM
     Standards, Part 40, American Society for Testing and Materials,
     Philadelphia, Pennsylvania (1975), p. 168.

15. ASTM Standard D257-76, "DC Resistance or Conductance of Insula-
     ting Materials," in: 1967 Book of ASTM Standards, Part 38,
     American Society for Testing and Materials, Philadelphia,
     Pennsylvania (1976), p. 86.

EFFECT OF LOW TEMPERATURE REACTOR IRRADIATION ON

ORGANIC INSULATORS IN SUPERCONDUCTING MAGNETS

S. Takamura and T. Kato

Japan Atomic Energy Research Institute
Ibaraki-ken, Japan

INTRODUCTION

When superconducting magnets are utilized in fusion reactors, the magnet components are exposed to fast neutrons and γ rays. Among the magnet components, organic materials used as insulating materials are the least radiation resistant;[1] therefore, their mechanical property changes after irradiation at low temperatures are studied in the present work.

The main requirements[2] of insulating materials in a fusion reactor magnet are:

1. High mechanical strength and high elastic modulus.

2. The capability to withstand thermal-induced differential stress.

3. Good electrical properties.

4. High radiation damage resistance.

This paper describes results of tensile and compression tests in liquid nitrogen after fission reactor irradiation at about 5 K.

EXPERIMENTAL PROCEDURES

Specimens

Tensile specimens of polypropylene, polycarbonate, Mylar (polyester), Nomex (nylon paper), and Kapton (polyimide) had dimensions of 3 mm in width and 10 mm in guage length. The specimen thicknesses were 65 μm for polypropylene, 75 μm for polycarbonate, 85 μm for Mylar, 200 μm for Nomex, and 50 μm for Kapton. Their tensile axes were perpendicular to the rolling direction of the sheets. The molecular structures of these specimens are shown in Table I. Epoxy resins were tested by compression. The specimens of epoxy resin were made of Epikote 828 (bisphenol A type) hardened by K61B (tridimethylaminophenol) or polyamide in a teflon mold of 2 mm in diameter and about 6 mm in height.

Irradiation

The irradiation was performed at about 5 K in LHTL of the Japan Atomic Energy Research Institute. The fast neutron flux is $1.15 \times 10^{12} \text{n/cm}^2\text{s}$ (>0.1 Mev) with fission neutron spectrum, which was measured by means of metal foil threshold detectors. The thermal neutron flux was of the order of $10^8 \text{n/cm}^2\text{s}$. The γ-dose rate was estimated to be roughly $1.1 \times 10^7 \text{R/h}$.

Tests

After irradiation at about 5 K, the capsule was transferred into the transporting container without warm-up. Specimens were taken from the capsule without warm-up in a cryostat and then transferred into a liquid nitrogen bath. Tensile and compression tests were carried out in this liquid nitrogen bath using an Instron type testing machine. The crosshead speed in the tensile test was 0.5 mm/min for Kapton and Nomex and 0.2 mm/min for other tensile specimens; in the compression test it was 0.5 mm/min.

EXPERIMENTAL RESULTS

Examination of the external appearance was made in liquid nitrogen after fast neutron irradiation of 1.7×10^{17} nvt and a γ dose of 4.5×10^8 R. The polypropylene changed in color from transparent to yellow and broke into small pieces. Polycarbonate and Mylar became very brittle. Nomex and Kapton were not brittle after irradiation. The tensile tests of polypropylene, polycarbonate, and Mylar were performed at a neutron doese of 9×10^{15} nvt and a γ dose of 2.4×10^7 R, since the specimens irradiated to a neutron

Table I. The Structural Formulae of Testing Materials

Specimens	Constitutional formula
Polypropylene	
Polycarbonate	
Polyethylene telephthalate (Mylar)	
Aromatic nylon paper (Nomex)	
Polyimide (Kapton)	

dose of 1.7×10^{17}nvt and a γ dose of 4.5×10^7R were too brittle to be mounted in the tensile machine. Since the stress-strain curves are a little different in each specimen, they are all shown in the following figures.

   Polypropylene, polycarbonate, and Mylar (polyethylene tereph- thalate). Figure 1 shows the stress-strain curves in liquid nitro- gen after neutron irradiation of 9×10^{15}nvt with a γ dose of 2.4 $\times 10^7$R. The elongation of preirradiated specimens was 40 to 45% for polypropylene, whereas it reduced to about 2% after irradiation. In general, crosslinking leads to an increase of both the elastic modulus and breaking stress, whereas degradation leads to a decrease of them. Therefore, the decrease in the elongation and breaking stress in the present experiments seems to indicate a degradation (probably of the main chain) owing to irradiation.

   Nomex. Nomex is a nylon paper with a benzene ring. Figure 2 shows the stress-strain curves after neutron irradiation of

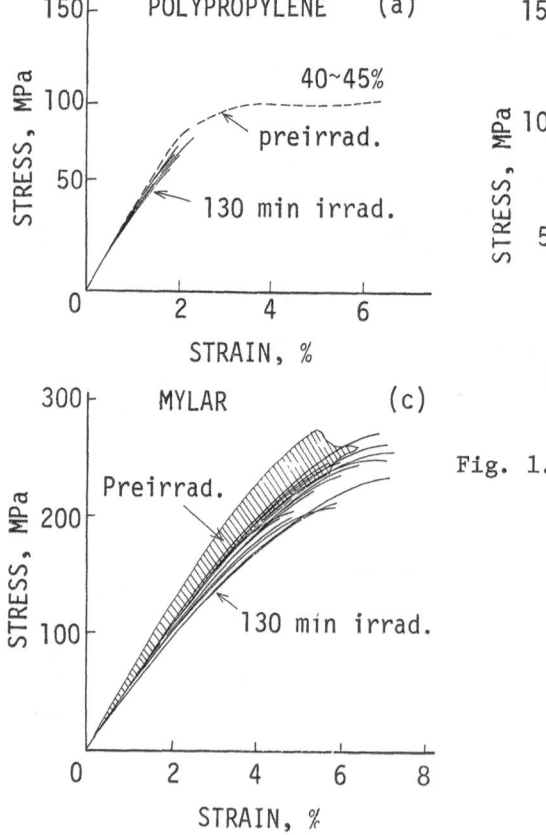

Fig. 1.   Stress-strain curves of polypropylene, polycar- bonate, and Mylar at 77 K before and after irradi- ation of 9×10^{15}nvt with a γ dose of 2.4×10^7R at 5 K. The stress-strain curves of polycarbonate and Mylar before irradi- ation are in the hatched region.

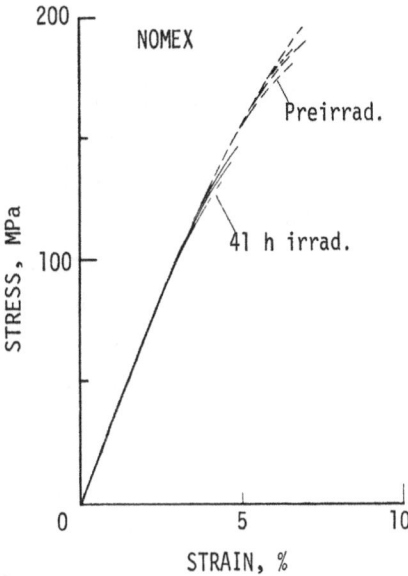

Fig. 2.   Stress-strain curves of Nomex at 77 K before and after
          neutron irradiation of 1.7×10^{17} nvt with a γ dose of
          4.5×10^{8} R at 5 K.

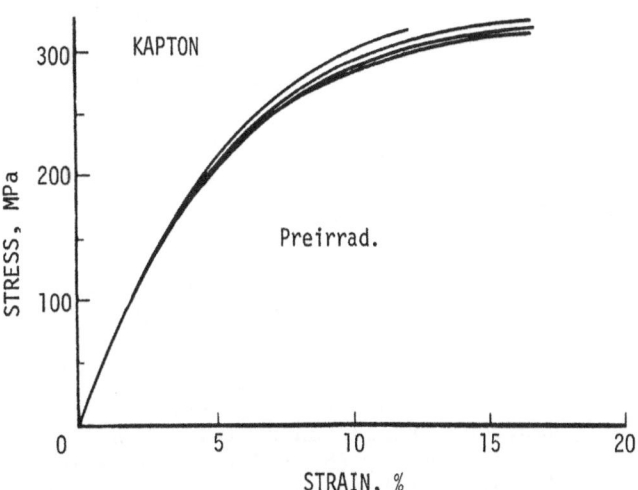

Fig. 3.   Stress-strain curves of Kapton at 77 K before irradiation.

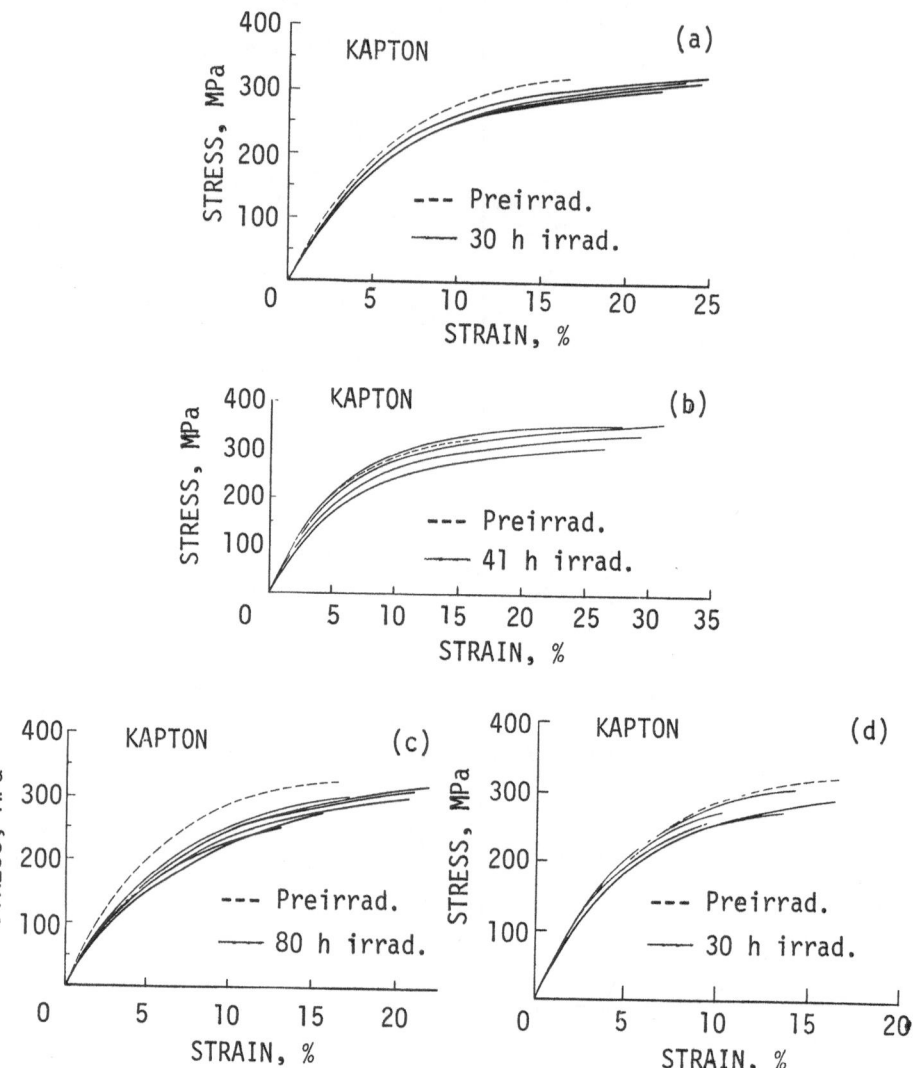

Fig. 4.   Stress–strain curves of Kapton at 77 K before and after
          neutron irradiation of (a) 1.2×10^{17}nvt with a γ dose
          of 3.3×10^{8}R, (b) 1.7×10^{17}nvt with 4.5×10^{8}R, and
          (c) 3.3×10^{17}nvt with 8.8×10^{8}R.   Stress–strain curves
          of Kapton at 77 K after irradiation at 5 K and warm-up
          to room temperatures (d).

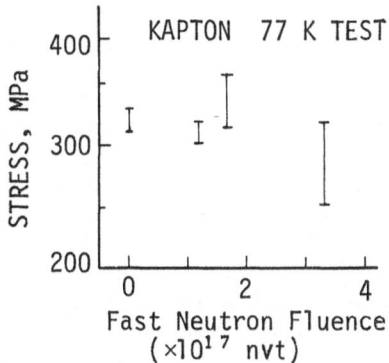

Fig. 5. Relation between the breaking stress and the fast neutron
         fluence.

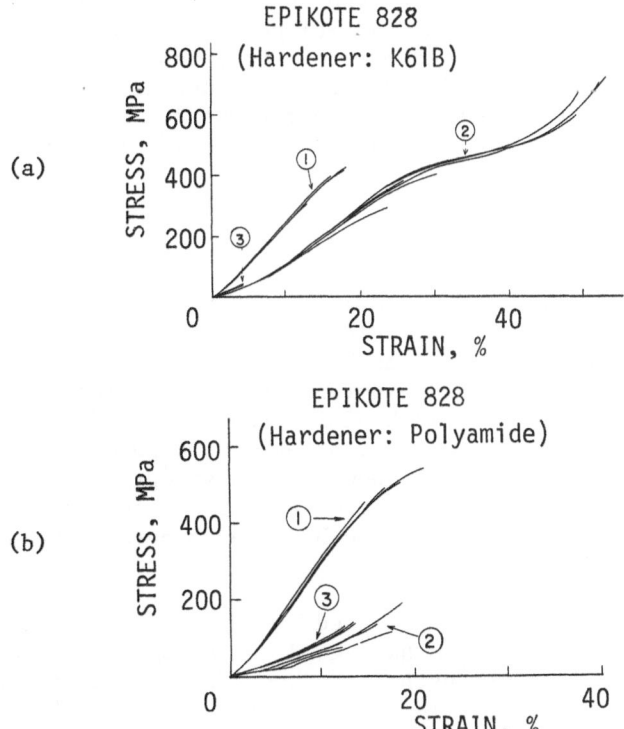

Fig. 6. Stress-strain curves of Epikote 628 hardened by (a) K61
         and (b) polyamide hardener at 77 K before and after
         irradiation. 1 - preirradiation; 2 - 30 h irradiation
         at 5 K; 3 - 8 h irradiation at 5 K and warm-up to R.T.

1.7 x 10^{17}nvt with a γ dose of 4.5 x 10^8R. The breaking stress and elongation decreased, but the elastic modulus did not change.

    Kapton. Kapton is a polyimide. Specimens used in these experiments were of commercial grade and marketed under the description of H-film. Figure 3 shows the stress-strain curves of preirradiated specimens at 77 K. The effect of irradiation on the stress-strain curves at 77 K is given in Fig. 4. Figures 4(a), (b), and (c) show the results after neutron irradiation of 1.2 x 10^{17}nvt with a γ dose of 3.3 x 10^8R, 1.7 x 10^{17}nvt with a γ dose of 4.5 x 10^8R, and 3.3 x 10^{17}nvt with a γ dose of 8.8 x 10^8R, respectively. The elongation at break increases with irradiation up to the dose of 1.7 x 10^{17}nvt but decreases at the dose of 3.3 x 10^{17}nvt, at which the breaking stress is also observed to decrease. Figure 4(d) shows the curves tested in liquid nitrogen after warm-up to room temperature after neutron irradiation of 1.2 x 10^{17}nvt with a γ dose of 3.3 x 10^8R at about 5 K.

    Epoxy resins. Figure 6 shows the stress-strain curves of Epikote 628 hardened by (a) K61B and (b) polyamide. The curves for preirradiation are indicated by 1.   The curves of the specimens after a neutron irradiation of 1.2 x 10^{17}nvt with a γ dose of 3.3 x 10^8R and of those warmed up to room temperature after a neutron irradiation of 3.3 x 10^{17}nvt with a γ dose of 8.8 x 10^8R are shown by 2 and 3, respectively.

DISCUSSION

    The experimental investigations of the mechanical properties of various organic materials after low temperature irradiation are very few, although many studies have been made after room temperature irradiation. Van de Voorde[3] studied the influence of 77 K and 20 K irradiations on mechanical properties and reported that polyimides and aromatic-based epoxies have a good radiation resistance. The present studies show that polypropylene, polycarbonate, and Mylar have low radiation resistance due to chain fracture, and these materials are too brittle to permit mechanical testing.

    Epoxy resins and composites are used in superconducting magnets, and aromatic-based epoxy resins have good radiation resistance. Van de Voorde[4] found the strength of Epikote 828 decreases sharply after low temperature irradiation of 10^9 rad. Nishijima and Okada also studied the mechanical properties of epoxy resin after neutron irradiation and γ-ray irradiation at low temperature. The present experiments show that the stress-strain curves after irradiation are strongly dependent on the kind of hardener. The strength of epoxy resin hardened by polyamide shows a remarkable decrease after low temperature irradiation.

ACKNOWLEDGMENTS

The authors would like to thank the members of the LHTL group and JRR-3, Japan Atomic Energy Research Institute, for their invaluable help. They would also like to thank Drs. S. Mori, Y. Obata, and T. Iwata of Japan Atomic Energy Research Institute and Prof. S. Okuka of Tsukuba University for their encouragement in this study.

REFERENCES

1. C.A.M. van der Klein, The Organic Insulation in Fusion Reactor Magnet Systems, RCN-240, Reactor Centrum Nederland, Petten, Netherlands (1975).

2. R.D. Hay and E.J. Rappaport, A Review of Electrical Insulation in Superconducting Magnets for Fusion Reactors, Oak Ridge National Laboratory, Oak Ridge, Tennessee (1976).

3. M.H. Van de Voorde, IEEE Trans. Nucl. Sci. 18, 784 (1970), and Ibid., 20, 693 (1973).

4. M.H. Van de Voorde, CERN ISR-MA/72-14 (1972).

5. S. Nishijima and T. Okada, paper 121 presented at 6th International Conference on Magnet Technology (MT-6), Bratislava, Czechoslovakia (1977).

CRYOGENIC FOAM INSULATIONS:

POLYURETHANE AND POLYSTYRENE

L. L. Sparks

National Bureau of Standards
Boulder, Colorado, U.S.A.

INTRODUCTION

Increased use of cryogenic temperatures in the recent past
may be credited in large part to military and aerospace endeavors.
However, the impending energy shortage, advanced surgical techniques, refrigeration of food stuffs, and scientific needs are
rapidly projecting the use of cryogenics into the mainstream of
modern living. In order for these applications of cryogenics to
be economically feasible, associated technologies have had to be
developed or extended. Thermal insulation is one of the areas
where both materials and techniques have been developed to meet
the growing needs. Indeed, the requirements for thermal insulation
are sufficiently diverse that several rather independent technologies or methods of insulation have developed. Generally speaking,
the particular application dictates the generic type of thermal
insulation or, at least, reduces the number of viable candidates.
Some of the widely used materials and their attendant thermal
conductivity ranges are depicted in Fig. 1.[1] As seen in this figure, the evacuated insulations are the most thermally efficient.
Most large-scale needs, however, preclude vacuum insulation on the
grounds of cost and durability. The three classes of unevacuated
insulation are reasonably competitive in thermal effectiveness at
low temperatures; all are used, for instance, in large liquefied
natural gas (LNG) storage tanks and LNG ship tankers. The materials
to be discussed in this paper, polystyrene (PS) and polyurethane
(PU) foam, are widely used in the present generation of LNG ships.
These foams, also known as cellular plastics or expanded plastics,
offer both a high strength-to-weight ratio and low thermal conductivity. The principal disadvantages are low maximum temperature
limits, lack of mechanical integrity upon thermal cycling[2] and

165

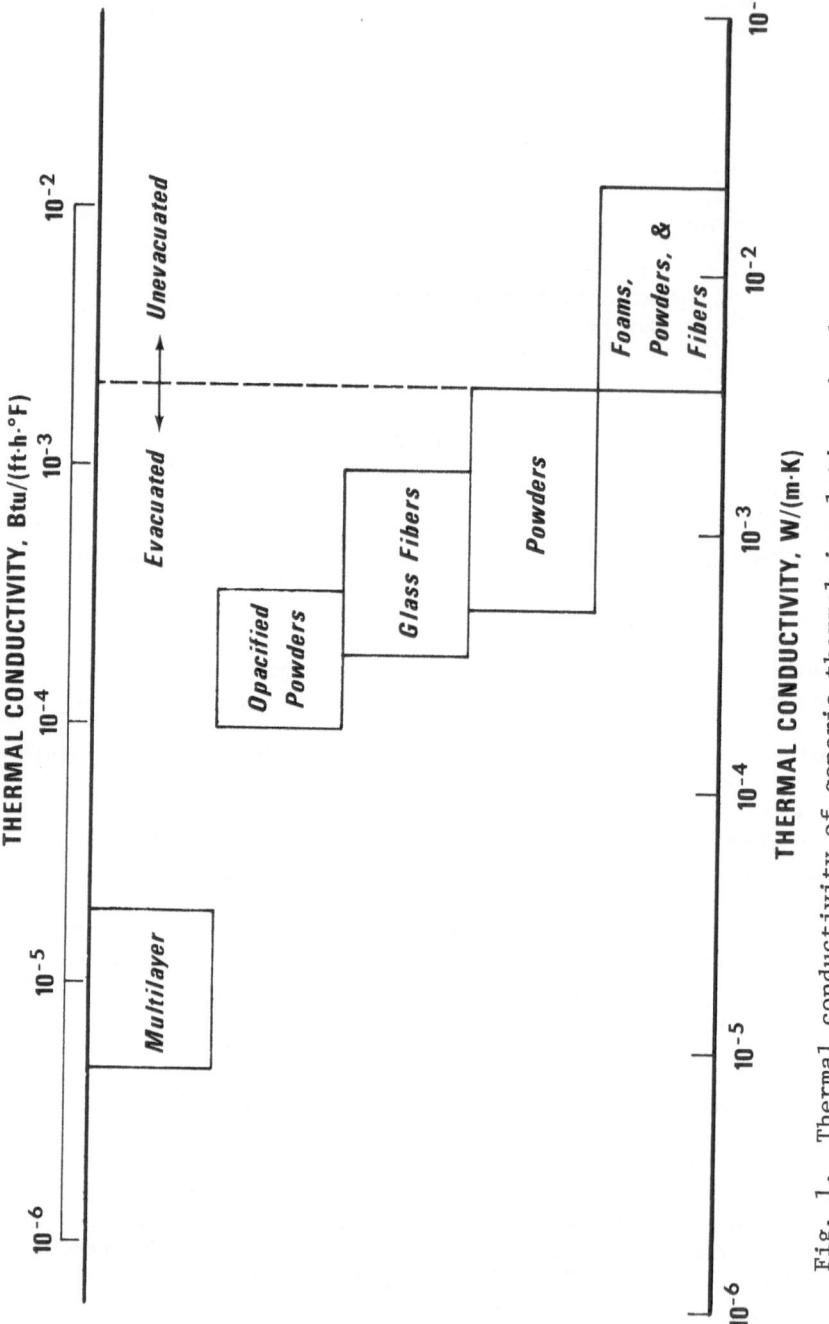

Fig. 1. Thermal conductivity of generic thermal insulations in the temperature range $77 \leq T \leq 238$ K and densities in the range $30 \leq \rho \leq 80$ kg/m^3.

degradation of insulating properties with time. Each of these
items will be addressed below. The low temperature thermal and
mechanical behavior of these two foam systems can be better understood if the basic structure, chemistry, and processing techniques
are briefly reviewed. The latter two areas tend to be highly proprietary, so that their discussion is based on well-known principles
rather than on specific, contemporary production practices.

MATERIALS

The two types of material being considered here share general
structural forms, although, as will be seen below, the chemistry
and processes necessary to generate each are completely different.
Both PU and PS are amorphous, organic polymers. Polystyrene is a
thermoplastic material, i.e., it softens at elevated temperatures
as the crosslink branches are broken, and its polymer structure is
not changed during foaming. Polyurethane, on the other hand, is
considered a thermosetting material. Thermosetting materials do
not soften at elevated temperatures but rather decompose directly,
and polymerization occurs as the material is foamed. In practice,
PU foam is found to soften to some extent. Essentially all insulating foams are of the closed-cell, rigid variety and these are the
forms that are considered throughout this paper.

A foam is a composite consisting of a skeleton of solid materials, and, in the case of closed-cell types, voids filled with
various gases. The term "closed cell" is used to describe a foam
where the voids form discrete, noninterconnected chambers. The
resulting discontinuity in the gas phase is important in establishing the thermal conductivity, as is illustrated in Fig. 2.[3] The
fraction of open cells also affects permeability of gases. The
effects of permeability will be considered in the appropriate sections to follow.

A rigid foam will be considered one that is ordinarily used
below its glassy transition temperature, T_g. The transition from
glassy to rubber behavior is a second-order transition that involves
a change of specific heat of the resin. Generally, the softening
temperature of PS and PU appears as a band rather than a discrete
temperature and depends on the composition and rate of temperature
change. For PS,[4] $T_g \simeq 370$ K, whereas PU[5] softens between 400 and
485 K.

There are two additional aspects of cell morphology that must
be considered: cell size and cell shape. In the case of low density
foams, $\rho \leq 160$ kg/m^3 (10 lb/ft^3), the cell geometry should be such
that the cellular surface area is minimized, since this is the condition for minimum surface tension. Figure 3(a)[6] illustrates the
pentagonal-dodecahedrol (a figure with twelve plane faces) cell

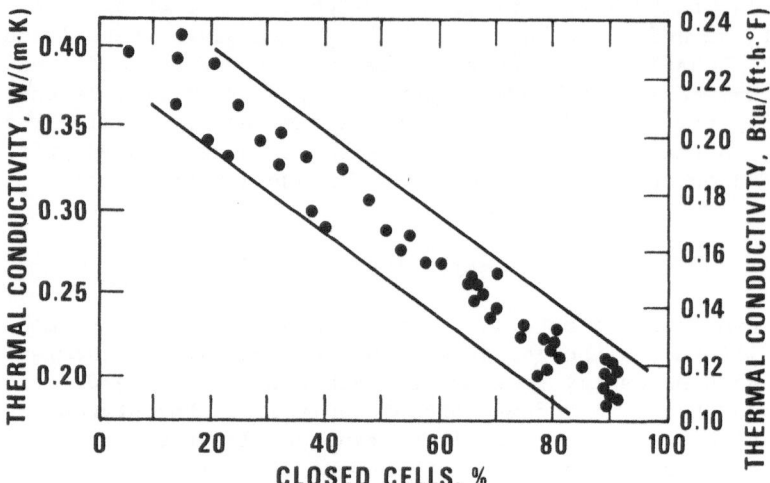

Fig. 2.   Thermal conductivity (at 277 K) as a function of closed-
          cell fraction for PU foam.

shape, which results when cellular packing is considered along with
the surface tension.   Figure 3(b)[6] schematically illustrates the
predicted cell packing geometry and the unpredicted cell distortion.
Figure 3(c)[7] is a photograph of the cell structure actually found
in a 32 kg/m³ PU foam.   The obvious differences in predicted and
actual cell geometry make it very difficult to describe the cell
structure analytically for thermal and mechanical properties analysis.   The cell structure is also affected by mechanical constraints
and method of processing.   The effect of the anisotropy seen in
Fig. 3(b) will be discussed later in relation to the thermal and
mechanical properties.

     Cell size is an important parameter in assessing the relative
strength and thermal conductivity of a given foam.   The average
size of cells determines the cell wall thickness for a given density.   Mechanical strength is dependent on this thickness.   The
radiative and convective contributions to the thermal conductivity
also increase as the cell size increases, thus causing poorer insulating quality.   Probably the most meaningful determination of
cell size is one which measures cell width along three mutually
perpendicular axes, so that the anisotropy of the cells can be
considered.   Figure 4[8] shows the relationship between cell diameter
and cell-wall thickness for various densities of PU and PS foams.

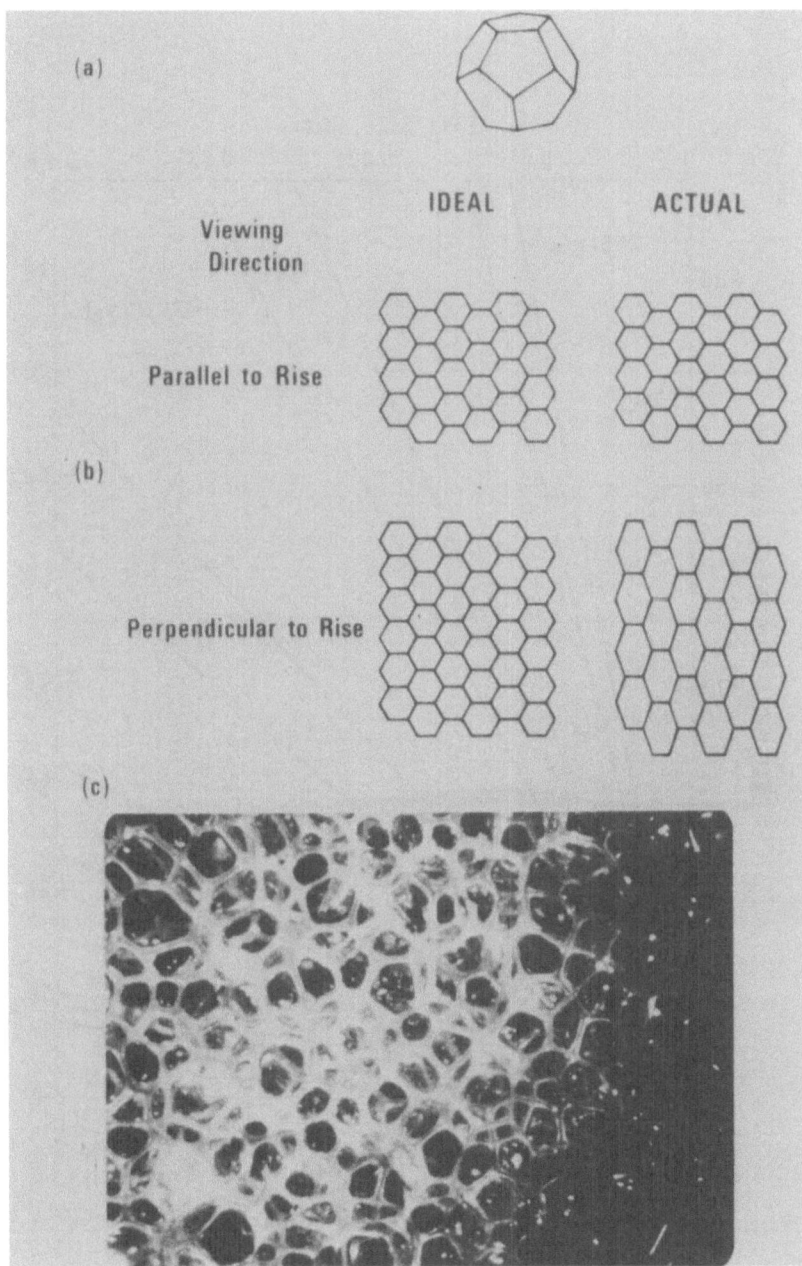

Fig. 3. (a) Predicted pentagonal-dodecahedral cell shape.
        (b) Schematic of distorted cells.
        (c) Photograph of cellular structure of PU foam.

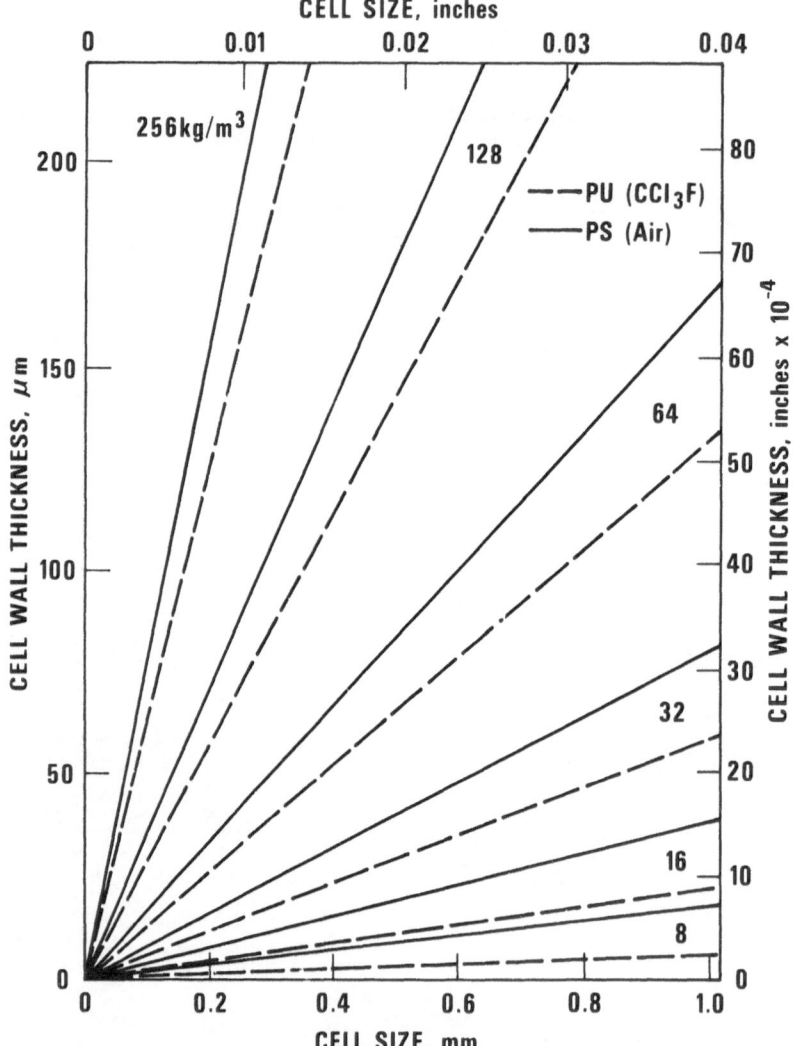

Fig. 4.   Cell wall thickness as a function of cell size for
          various densities of PS and PU foam.

CHEMISTRY AND PROCESSING

Polystyrene

The styrene monomer, CH_2, was first obtained by Bonastre in 1831.[9] The resin solidified upon aging, however, and little development occurred until 1922 when Dufraisse and Moureu developed agents that would inhibit polymerization. Large-scale production of polystyrene began in 1935 in both Germany and the United States. Two dominant processes have developed for the production of PS foam: extrusion and expansion of foamable beads. For polystyrene, the polymerization process is endothermic and, therefore, requires external heat. This requirement essentially eliminates the in situ application of insulation as is possible with the exothermic PU foams.

Expanded beads are generally formed by heating a mixture of liquid styrene monomer, an aqueous solution containing a blowing agent, and a catalyst. The hydrocarbon pentane is most often used as a blowing agent, although ketones, esters, and ethers can also be used. Controlling the mixture, heating time, and temperature allows beads with average diameters between 0.25 to 2.5 mm to be formed.[10] The solid beads thus formed are preexpanded to a predetermined density by application of heat. The heat is supplied by steam, hot water, hot air, or infrared radiation and causes the thermoplastic polymer to soften and expand under pressure from the volatile pentane gas. At this stage, the product consists of individual, hollow particles. Pentane diffuses rapidly from the beads and presents a possible fire hazard. As the particles cool, the gas pressure inside the cells drops, and a saturated liquid-vapor condition exists. This equilibrium condition creates a partial vacuum within the cell until environmental gases can permeate the cell wall and create an equilibrium pressure. The permeability of PS to various gases is shown in Table I.[9] The final step in the production of expanded bead foam involves partially filling a mold with the preexpanded beads and applying heat, generally in the form of steam. This process causes further expansion and fusion of the individual beads into a continuous mass of welded, closed cells. Expandable beads may also be extruded to form continuous lengths of low-density foam sheet.

The second major process used in PS foam formation is a single-step extrusion technique.[10] Molten polystyrene, dissolved blowing agent, and various additives are combined in a temperature-controlled, pressurized vat. The fluid is expanded through a die and slowly cooled. Temperature, pressure, component ratio, and cooling rate are carefully controlled to produce uniform cell size and orientation in low-density foams. Additives are frequently used to control cell size and reduce flammability. Hydrocarbons, methyl chloride, and other proprietary blowing agents are used in this process. Extruding produces a strained plastic state, which

Table I. Permeability of Gases through PS and PU Films.

| Gas | Permeability, 10^{-15} x $\frac{cm^3 gas \cdot mm}{s \cdot cm^2 \cdot Pa}$ | | |
| | PS | PU | |
		Polyester	Polyether
CCl_3F		2.2	
N_2	7.5 to 30.0	20.3	36.8
O_2	45.0 to 128	80.3	112 to 360
CO_2	900	30.0	1050 to 3000
Air		32.3	

Water vapor transmission:

$$PU \text{ (polyester)} = 22.5 \text{ to } 750 \times 10^{-15} \frac{g}{cm \cdot s \cdot Pa}$$

$$PS = 60 \times 10^{-15} \frac{g}{gm \cdot s \cdot Pa}$$

results in higher strength properties. If the cooling of the ex-truded product is not controlled, residual strains can cause rupture of the cell walls.

Polyurethane

The basic reaction involved in the formulation of PU is between hydroxy compounds and isocyanates. The first record[9] of this re-action was made by Wurtz and Hoffmann in 1848. Commercially signi-ficant advances did not occur until 1937, when O. Bayer of Germany developed diisocyanate. Bayer and co-workers continued the develop-ment of polyester-based urethane polymers as nylon substitutes in the 1940-45 time period. Polyester toluene-diisocyanate (TDI) foams became commercially available in the United States in 1952-53. Two further developments occurred that allowed the use of PU foam to reach the present level. The first of these was the development of polyether polyol and the subsequent "one-shot" foaming process. Rigid polyether-based foams were available in 1957. The second development was the incorporation of halocarbons as blowing agents in 1958.

The fundamental urethane unit is made up of H, C, N, and O. The chemical reaction leading to urethane is the combination of a polyisocyanate (R-NCO) and a polyhydroxyl compound (R'-OH):[10,11]

$$R - NCO + R' - OH \longrightarrow R - NH - \overset{\overset{\displaystyle O}{\|}}{C} - O - R' + heat$$

isocyanate alcohol   urethane

The reaction is exothermic, but, as shown here, no volatiles are released. When water is included in the mixture, the water and isocyanate react as shown below to produce an amine and CO_2:

$$R - NCO + H_2O \longrightarrow R - NH - \overset{\overset{\displaystyle O}{\|}}{C} - OH \longrightarrow R - NH_2 + CO_2$$

isocyanate   carbamic acid   amine

The resulting amine reacts with another isocyanate group to form a urea:

$$R - NH_2 + R' - NCO \xrightarrow{fast} R - NH - \overset{\overset{\displaystyle O}{\|}}{C} - NH - R'$$

amine   isocyanate   urea

Isocyanate-urethane and isocyanate-urea reactions then yield biurets and allophanates, which produce crosslinking and branching:

$$R - NH - \overset{\overset{\displaystyle O}{\|}}{C} - OR' + R'' - NCO \longrightarrow R'' - NH - \overset{\overset{\displaystyle O}{\|}}{C} - \underset{\underset{\displaystyle R}{|}}{N} - \overset{\overset{\displaystyle O}{\|}}{C} - O - R''$$

urethane   isocyanate   allophanate

$$R - NH - \overset{\overset{\displaystyle O}{\|}}{C} - NH - R' + R'' - NCO \longrightarrow R' - NH - \overset{\overset{\displaystyle O}{\|}}{C} - \underset{\underset{\displaystyle R}{|}}{N} - \overset{\overset{\displaystyle O}{\|}}{C} - NH - R''$$

urea   isocyanate   biuret

The thermal conductivity of foams expanded with CO_2 changes rapidly owing to diffusion of gases into, and CO_2 out of, the cells. The permeability of PU to various gases is shown in Table I. Expansion of the foam with a fluorocarbon, such as trichlorofluoromethane (CCl_3F) or dichlorodifluoromethane (CCl_2F_2), lowers the thermal conductivity of the foam. The fluorocarbon also

remains in the cells because of a very low diffusion rate through PU.
In fluorocarbon-blown foams, the initial gas pressure after the foam
has cured and cooled is 5 to 6 x 10^4 Pa (0.5 to 0.6 atm).[12] Air
rapidly permeates the unprotected foam until it establishes a partial
pressure of approximately 10^5 Pa (1 atm), which results in a net
total pressure of approximately 1.6 x 10^5 Pa (1.6 atm) in the cells.

In addition to the polyol, isocyanate, and blowing agent,
several other materials are added in commercial operations. These
additives include catalysts, surfactants, and flame retardants.[11]
As indicated above, the basic reaction is exothermic and will occur
without catalysts. However, the uncatalyzed reaction is too slow
for commercial production and frequently does not generate enough
heat to fully cure the foam. Metallic salt catalysts induce rapid
reaction and work well in spray processes. Tertiary amines are
sometimes used alone in CO_2-blown foams and in combination with tin
catalysts for fluorocarbon-blown foams. The catalysts used have a
pronounced effect on the final product, because the reaction rate
influences density and cell size.

Surfactants are used to control cell development. Both cell
size and shape are affected to the extent that, without surfactant
additives, large irregular cells may develop or the foam may collapse altogether. The most widely used type of surfactant for
rigid-closed cell PU foams is based on silicone copolymers, although
organic types, such as sulfonated castor oil and amine esters of
fatty acids, are also used.

The need for flame-resistant foams is obvious. Three methods
have been utilized to produce nonflammable foams: (1) coat the
foam with a nonflammable surface, (2) physically incorporate additives in the foam, and (3) chemically incorporate additives in the
foam. The first method calls for an additional process and materials
and can cause significant cost increases. It can, however, isolate
the foam from gases, including water vapor. Physical incorporation
has been used with a number of additives, such as phosphates and
oxides. These additives uniformly degrade the physical properties
of the foam and so, except for special situations, are not widely
used. Chemical incorporation[13] appears to be effective and shows
promise of further development. Flame retardant foams of this type
are made by chemically incorporating bromine or chlorine containing
monomers into the polyol. Efforts are being made to lower flammability in low-density foams by reducing the energy content of the
constituent materials.

In the early stages of development, the only polyol available
was polyester, and its use involved combining the polyol and isocyanate in what is known as a prepolymer. The polyester reacts
with an excess of isocyanate (no catalyst) to form an isocyanate
terminated prepolymer. A second component, consisting of water,

catalyst, and other additives, is added to effect the foaming. This relatively slow process is not generally used for rigid foam development and has been largely replaced by "one-shot" and quasi-prepolymer processes.

The "one-shot" process, made possible by the development of polyether polyol and active catalysts, consists of combining the polyol, isocyanate, water and/or blowing agent, catalyst, surfactants, and other additives at the same time. This is an extremely fast process, making it very attractive from a commercial point of view. Continuous development of this method has led to the production of foams with properties equal to or superior to those formed by the slower prepolymer techniques. A major disadvantage of the "one-shot" process is that the highly exothermic reaction tends to cause charring when used to form thick sections. A combination of this method and the prepolymer method is now used extensively to generate rigid foams. This process is known as quasi-prepolymer foaming.

This technique involves combining a portion of the polyol with all of the isocyanate to form component A; the remainder of the polyol, water and/or blowing agent, catalyst, etc., is combined to form component B. By making components A and B about the same viscosity and volume, it is relatively simple to obtain excellent foams by mixing commercially available components. Since a portion of the polymerization occurs in developing component A, heat generation upon foaming is not likely to cause charring, even in thick sections. Both "one-shot" and quasi-prepolymer procedures are used in spray and pour-in-place processes. These methods of generation are used extensively for preparing machinable slabstock and in in situ covering of storage vessels.

PROPERTIES OF FOAMS

There is a large amount of ambient-temperature foam property data in the literature and company files compared with that available at cryogenic temperatures. This is particularly true in the area of mechanical properties. Although the thrust of this paper is properties at low temperatures, it is necessary to make use of ambient temperature data to stress certain characteristic behavior. Typical values will be used in many of these illustrations so that the curves would not, in general, represent a specific material. The properties of all foamed plastics, including PS and PU, are functions of several nonindependent parameters. Density, cell size and shape, composition, method of processing, and fill gas must be considered; the interdependence of these parameters is illustrated, for example, by the cell size and shape depending, to varying degrees, on each of the items mentioned above.

It was mentioned earlier that foams are anisotropic. This is
due to unequal internal and external forces acting in the three perpendicular axes during foam rise. The results are elongated cells,
as shown in Fig. 3(b). In certain circumstances, such as molded
PS beads, the anisotropy may be minimal. The following discussion
will often include references to the parallel ($\parallel$) and perpendicular
($\perp$) foam axes. It should be understood that these specimen axes
are usually only approximations and depend on container dimensions
and resin viscosity. Only in the indicated area of the pour-in-
place bun (shown in Fig. 5)[14] is the rise direction reasonably vertical. An appreciable amount of the scatter found in foam-related
data may be attributed to incomplete characterization of the specimen axis relative to the actual rise direction.

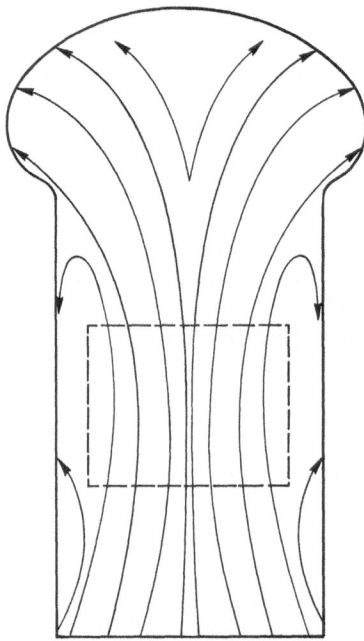

Fig. 5.   Illustration of turbulent flow occurring
          during foam rise in a narrow open-topped
          mold. The boxed area indicates material
          which could be used for determining maxi-
          mum or minimum properties, i.e., $\parallel$ or $\perp$
          to foam rise; other specimens cut from
          the bun would have some average properties
          between $\parallel$ and $\perp$, depending on specific
          directions of foam rise.

Mechanical Properties

Density is by far the most studied variable in PU and PS foam systems. The empirically developed formula,[15]

$$\text{mechanical property} = K\rho^a \qquad (1)$$

where ρ is density, is widely used to describe the mechanical properties of PU foam. Ordinarily, both a and K are considered to be functions of the specific foam system and of the temperature. Recent work[5] shows that more general relationships can be developed, at least for the compressive properties of PU foams. Specifically, in experimental work done on CO_2-blown polyester and polyether-based foams, it was found that a could be taken to be 1.75 (density in kg/m^3, compressive property in MPa, temperature in K) for the compressive property of any PU foam. The K term was considered to be a function of temperature only, so that

$$\text{compressive strength} = (8.728 \times 10^{-4} - 1.717 \times 10^{-6} \, T) \, \rho^{1.75} \qquad (2)$$

for $T \geq 298$ K (77°F) and

$$\text{compressive modulus} = (1.939 \times 10^{-2} - 3.571 \times 10^{-5} \, T) \, \rho^{1.75} \qquad (3)$$

for $T \geq 219°K$ (-65 F). The linear temperature dependence of K for compressive strength holds only to 298 K. Further work at lower temperatures might enable a similar predictive capability to be developed for the low temperature regime.

At ambient temperatures, best-value, analytical relationships have been developed for the properties shown in Table II.[16,17] The large uncertainties given in this table reflect the variations caused by differences in composition, foaming techniques, age and storage condition of the specimens, actual direction of foam rise, and experimental uncertainties. The coefficient a is reasonably constant for a given property regardless of the specific PU foam, so that a single value of the property of a foam will allow K to be determined for that particular type of foam. The uncertainties shown in Table II are reduced considerably when dealing with a specific resin, process, etc.

Density is also the dominant factor in determining the mechanical properties of PS foams. These properties are strongly dependent on the volume percent of the polystyrene present in the foam. The experimental data shown in Fig. 6 illustrate the dependence of strength on density for both PS and PU. Caution must be exercised, however, when attempting to apply generalities to specific foams. For instance, in Fig. 6, the tensile strength of the polyether PU specimen is considerably higher than the compressive strength;

Table II.   Analytical Representation of the Mechanical and Elastic
            Properties of PU Foam.

Property, MPa	K	a	90% Confidence limits		
			32 kg/m^3	64 kg/m^3	128 kg/m^3
Compressive strength	0.00173	1.416	± 0.141 MPa	± 0.751 MPa	± 2.013 MPa
Compressive modulus	0.03681	1.445	±10.859	±30.957	±81.013
Tensile strength	0.00727	1.112	± 0.407	± 0.876	± 1.896
Tensile modulus	0.16248	1.151	±23.373	±52.193	±61.191
Shear strength	0.00518	1.077	± 0.203	± 0.441	± 0.924
Shear modulus	0.05848	1.081	± 2.241	± 5.033	±10.170

The relationship, property = $K\rho^a$, yields ambient temperature
values for typical PU foams.

Traeger,[15] on the other hand, found the compressive strength of his
polyester PU foams to be consistently higher than the tensile
strength for a given density. Figure 7 further emphasizes the
individuality of particular foams, depending on axis of measurement, manufacturing process, and composition.

    As the temperature is lowered, the mechanical and elastic
properties of PS and PU foams are affected. In general, the properties reflect increased strength and stiffness as the temperature
decreases. Figures 8 through 13 contain experimentally observed
low temperature mechanical and elastic characteristics for these
foam systems.

    The ambient-temperature, density dependence suggested in
Figs. 6 and 7 is retained at lower temperatures, although, as will
be pointed out below, the generating process can mask the density
dependence. Another general attribute of PU and PS is that the
elastic and mechanical properties are higher in the direction of

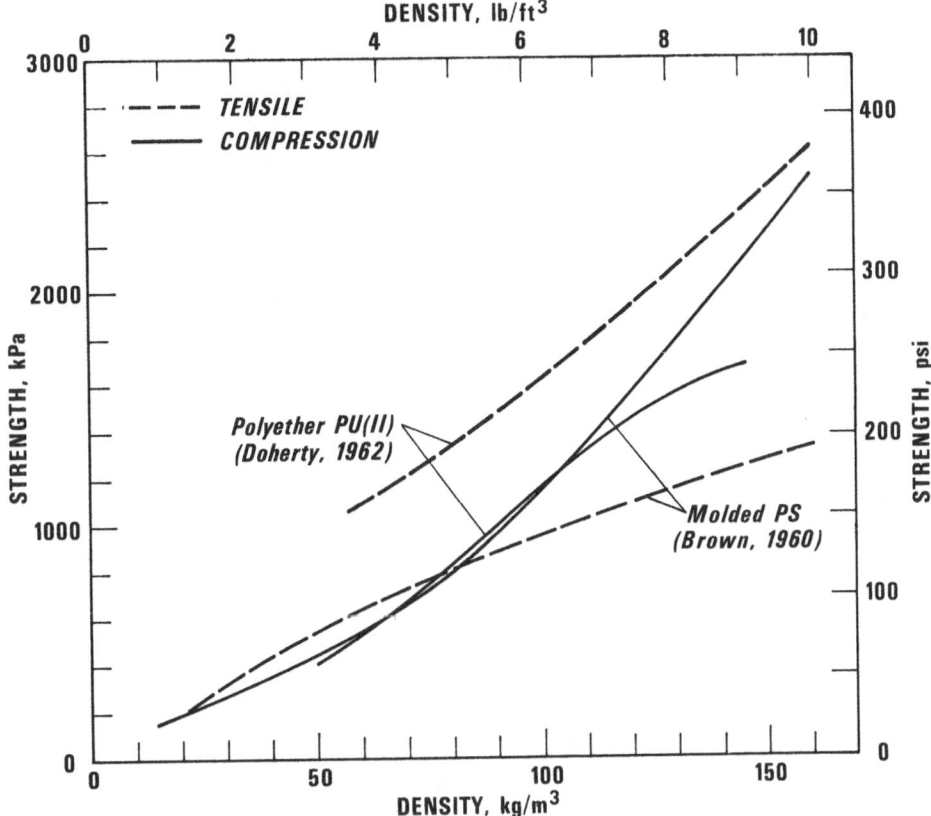

Fig. 6. Ambient temperature strength versus density for tensile
         and compressive strength of PS and PU foams.[14,18] The
         relationship between tensile and compressive strengths
         depends on actual processing techniques and composition.

foam rise (∥) than perpendicular to the foam rise (⊥). Reed et
al.[29] found little difference in the ∥ and ⊥ tensile strength
(Fig. 8) of 96 kg/m^3 (6 lb/ft^3) PU for T $\gtrless$ 150 K. Below that
temperature, however, the strength of the ⊥ specimen continued to
rise while that ∥ to the rise decreased dramatically. Smith[20]
found that the tensile strength and Young's modulus in tension
(Fig. 11) for his 28.8 kg/m^3 (1.8 lb/ft^3), extruded PS specimen
showed not only the expected parallel-perpendicular relationship,
but also that the perpendicular width and extrusion directions were
also anisotropic. These data for extruded PS have the unusual
characteristic of decreasing tensile strength with decreasing temperature. In compression (Fig. 9), both the molded and extruded
PS indicate the expected increasing strength with decreasing

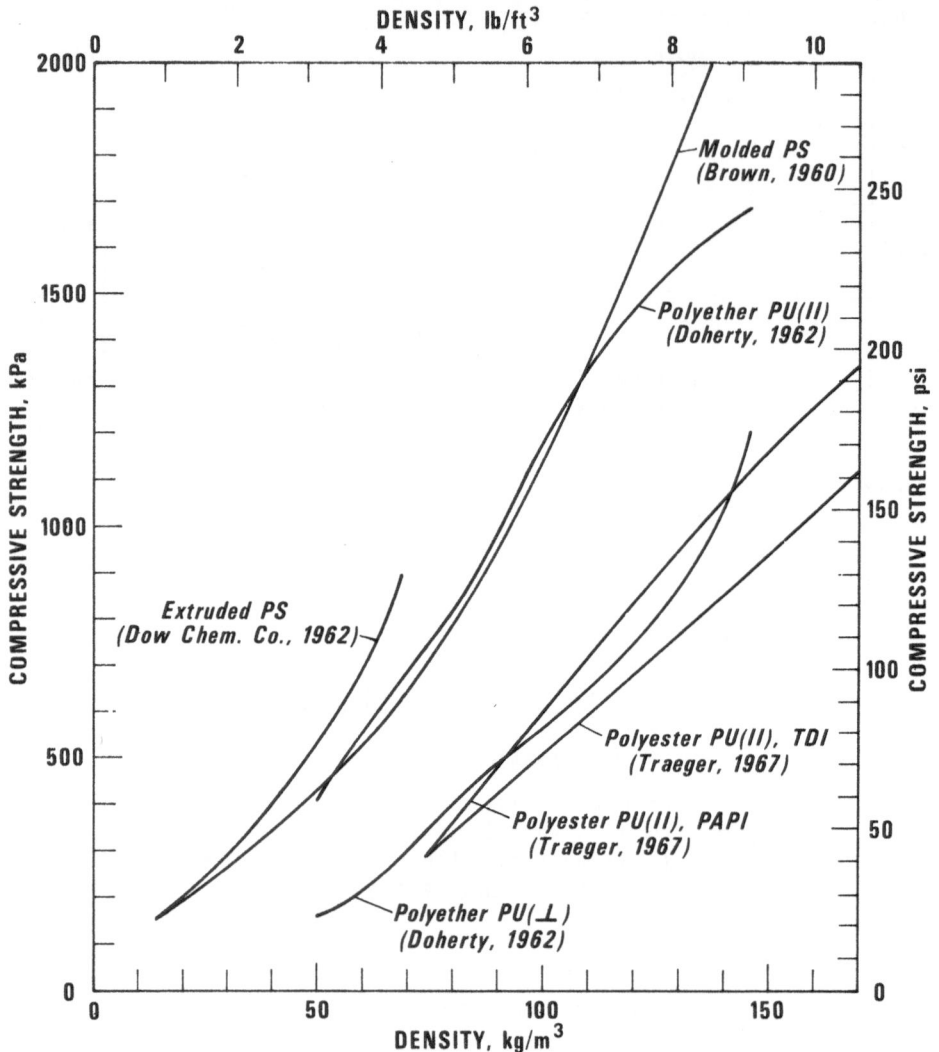

Fig. 7.   Ambient temperature compressive strength versus density
          for several PU and PS foams.[14,15,18,19] The compressive
          strength is significantly different for ∥ and ⊥ orienta-
          tion (Doherty),[14] for TDI and PAPI isocyanate (Traeger),[15]
          for polyether and polyester polyol (Doherty),[14] and for
          molded and extruded materials (Brown and Dow Chemical).[18,19]

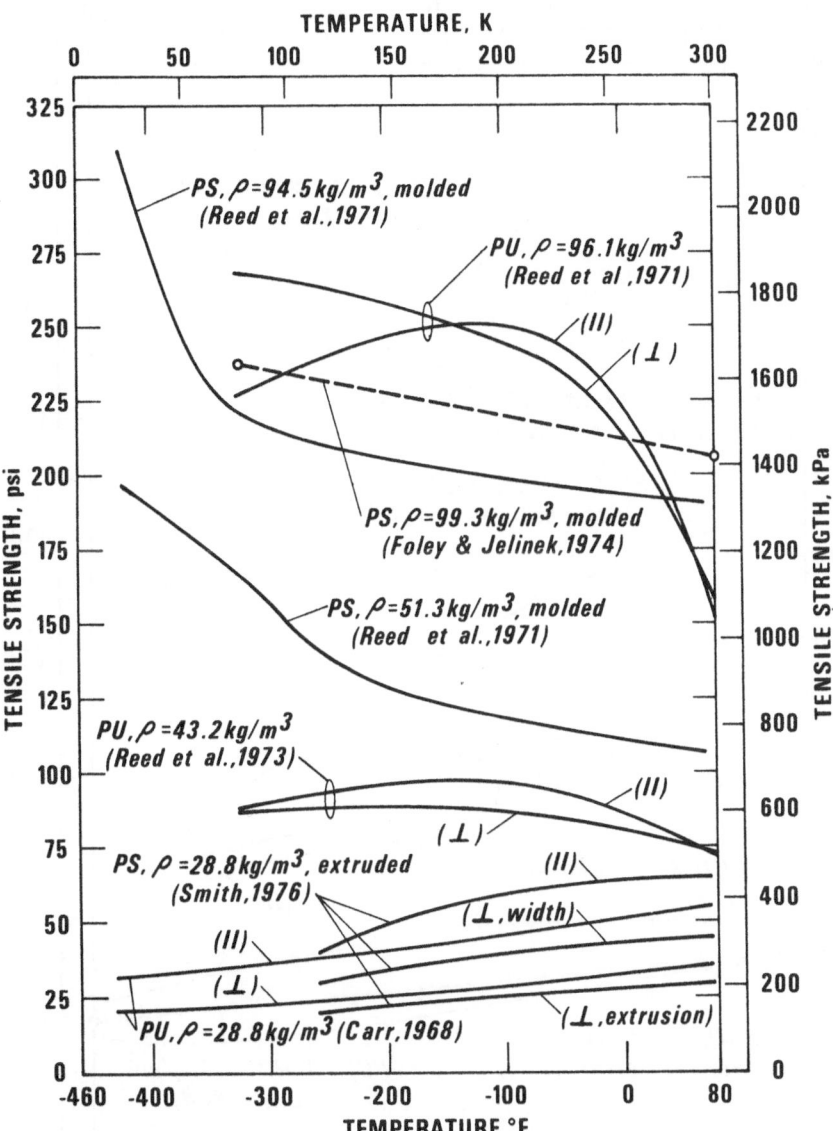

Fig. 8.   Tensile strength as a function of temperature for PU and
           PS foams with density, process, and orientation as
           parameters.[20-23]

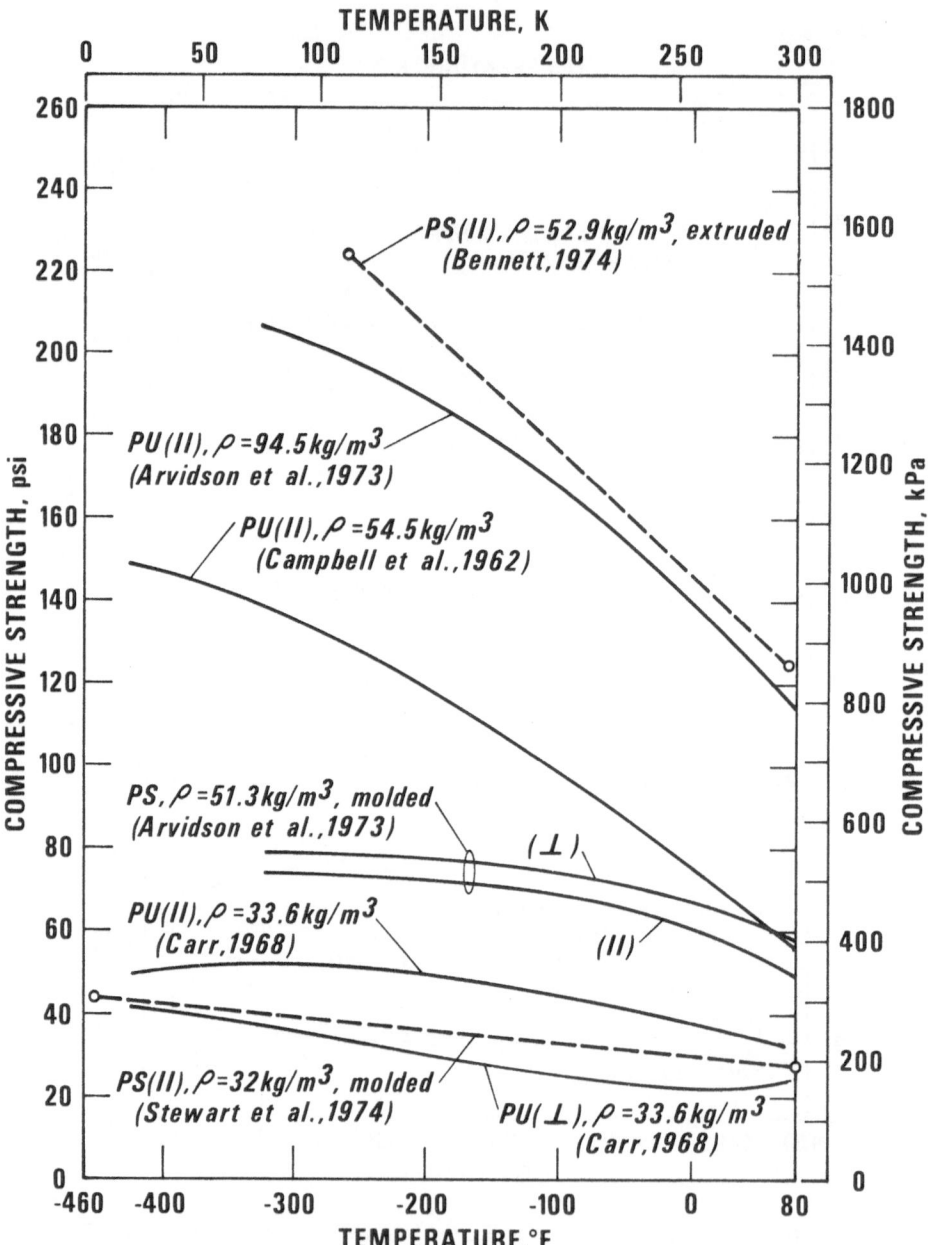

Fig. 9. Compressive strength as a function of temperature for PU
and PS foams with density, process, and orientation as
parameters.[21,24-27]

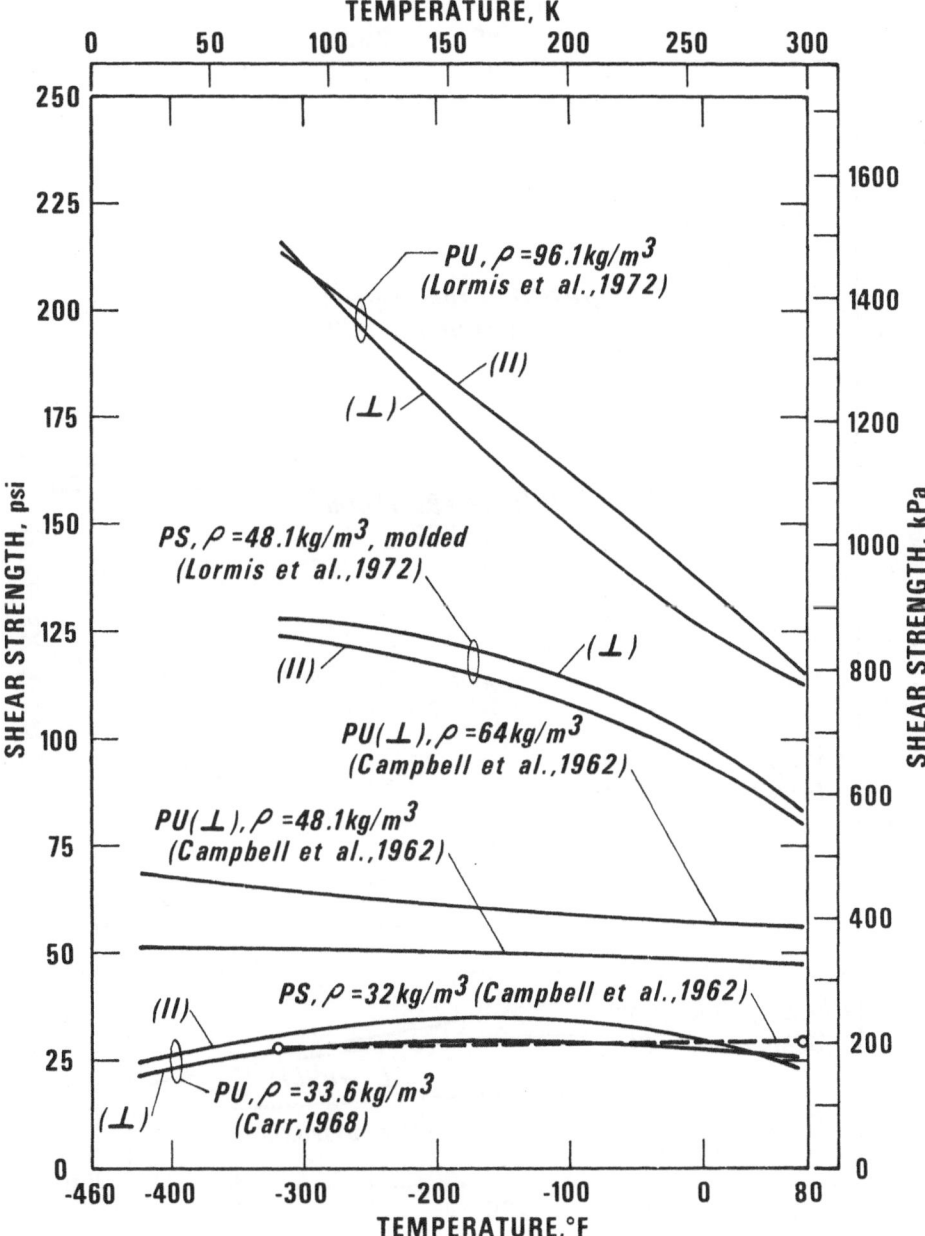

Fig. 10. Shear strength as a function of temperature for PU and PS foams with density, process, and orientation as parameters.[21,26,28]

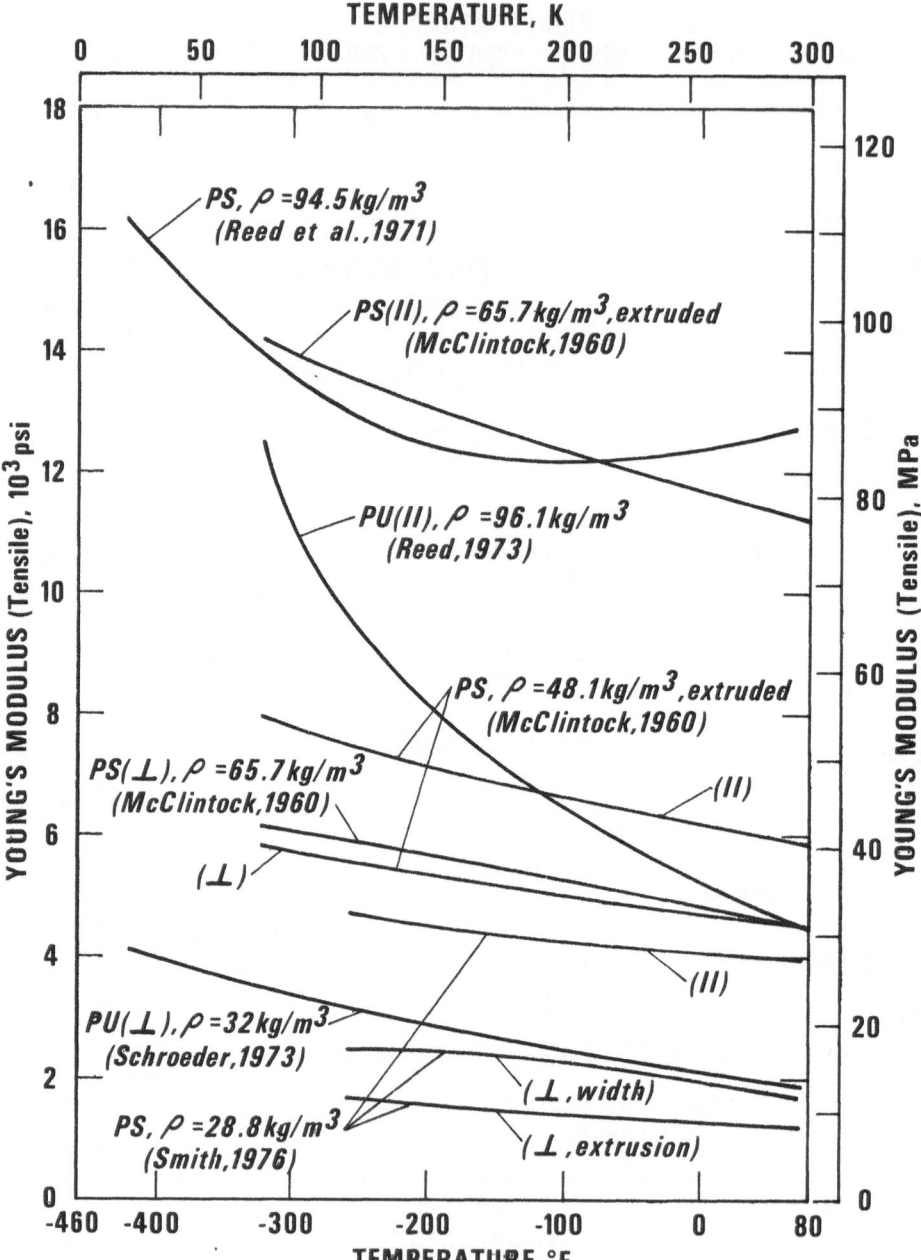

Fig. 11.   Young's modulus measured in tension as a function of
temperature for PU and PS foams with density, process,
and orientation as parameters.[20,22,29-31]

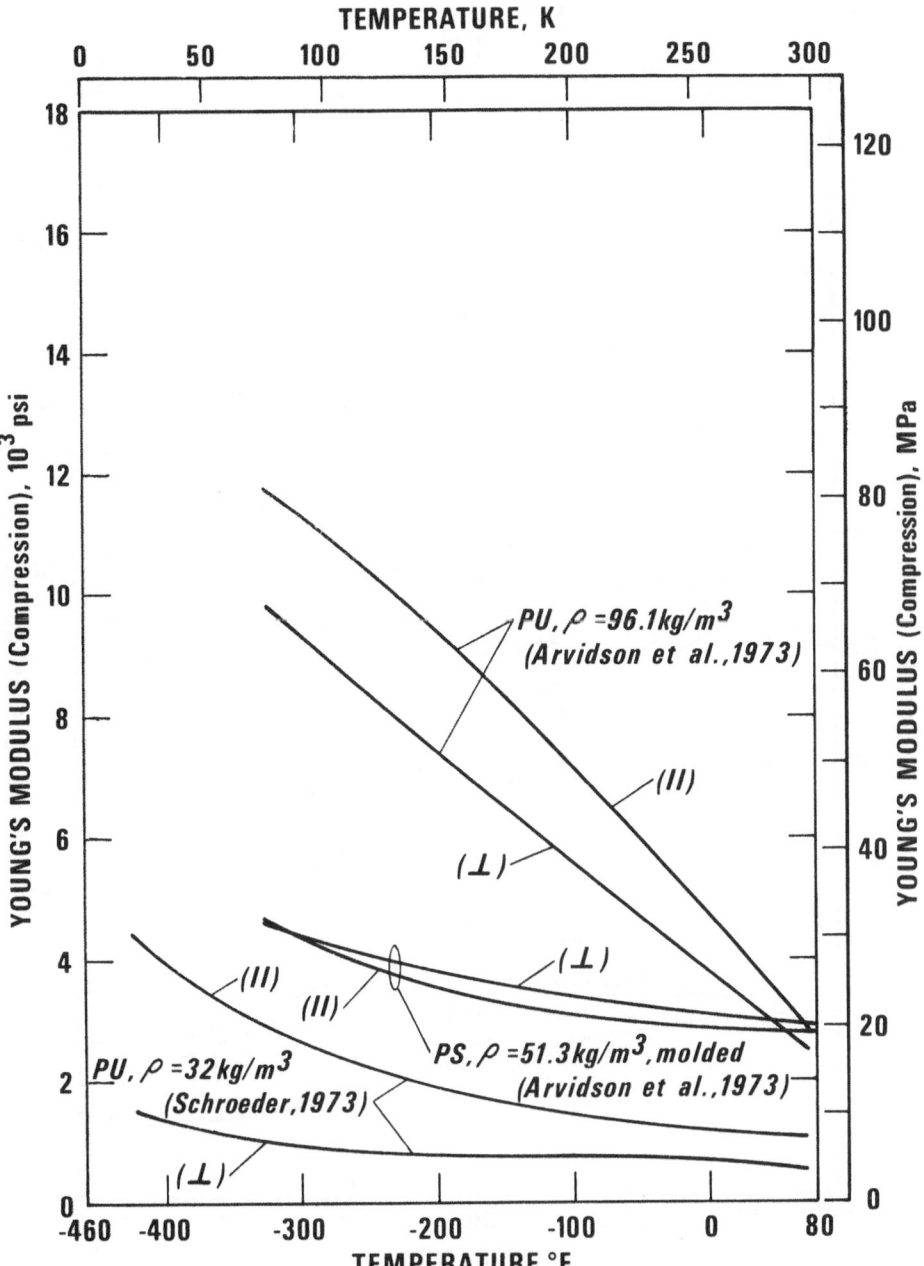

Fig. 12. Young's modulus measured in compression as a function
of temperature for PU and PS foams with density, process, and orientation as parameters.[24,31]

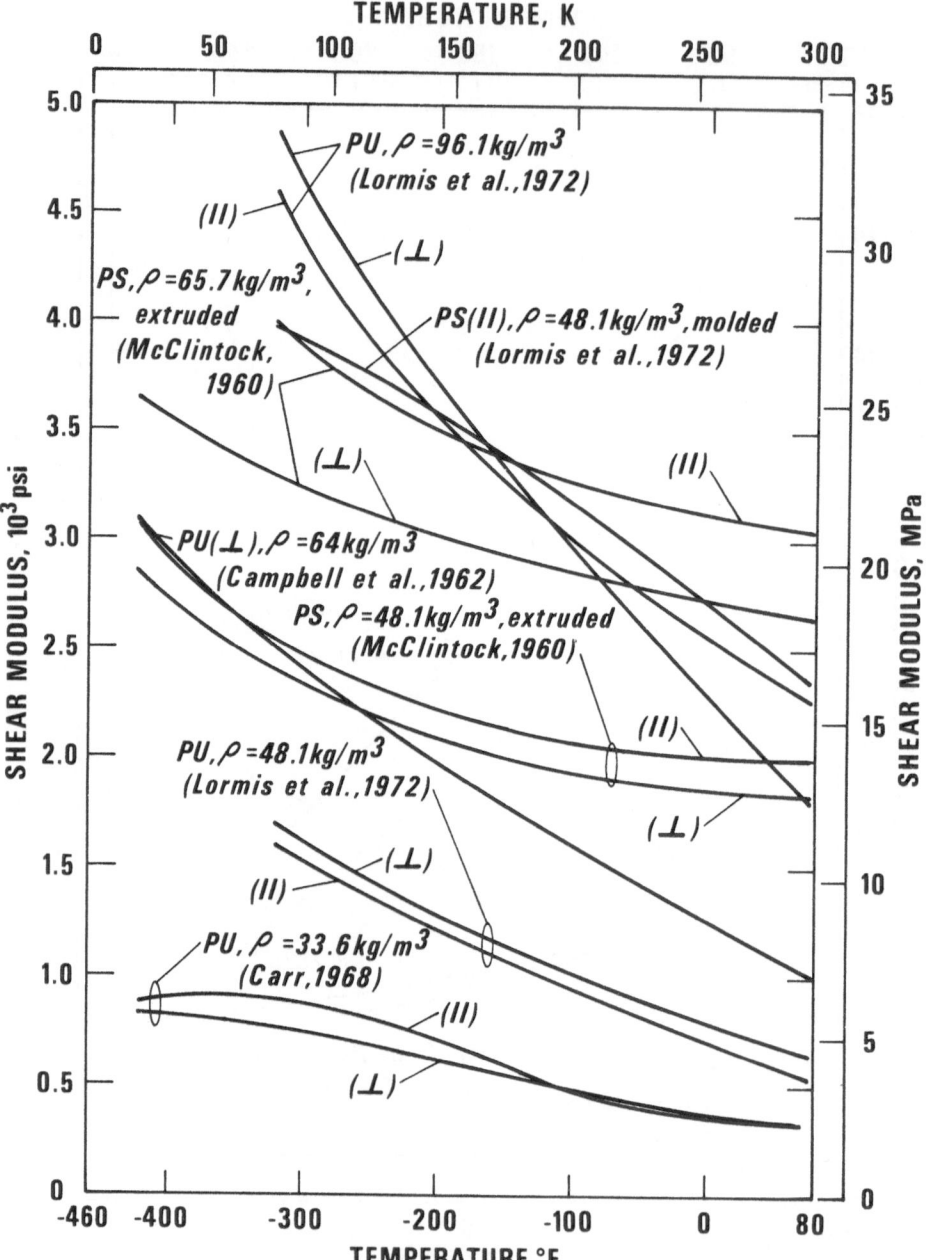

Fig. 13.   Shear modulus as a function of temperature for PU and
PS foams with density, process, and orientation as
parameters.[21,26,28,30]

temperature. Comparing the compressive data by Arvidson[24] and
Bennett[25] for nominal 53 kg/m^3 PS, one finds the extruded material
is nearly a factor of 3 stronger. McClintock[30] found the Young's
modulus in tension (Fig. 11) to be extremely sensitive to orientation for his extruded PS specimens. The ‖ orientation is seen to
be about a factor of 2 higher than the ⊥ orientation for his 65.7
kg/m^3 (4.1 lb/ft^3) specimen. Arvidson,[24] on the other hand, found
little difference in the compressive Young's modulus for his molded
PS specimens. In this case, the small difference between the ‖
and ⊥ orientations indicated the ⊥ to be stiffer, which is not
generally true. The shear modulus (Fig. 13) of the 48 kg/m^3 extruded PS tested by McClintock[30] is significantly lower than the
molded material of the same density tested by Lormis.[28] In fact,
for temperatures less than about 200 K, the shear modulus of the
48 kg/m^3 molded PS is nearly equal to that of the 65.7 kg/m^3 extruded material shown in this figure.

Although aging is generally thought of as affecting the thermal
properties, it also produces some changes in the mechanical properties. Two environmental aspects of aging seem to be particularly
important: water absorption and temperature. At ambient temperatures, moisture is absorbed by PU and PS foams until the moisture
content is in equilibrium with the surroundings. When low temperatures are encountered, however, water and ice can build up to the
extent that cell damage takes place. Repeated thermal cycling can
cause deterioration of mechanical properties if the foam environment is such that water vapor can continually permeate the cells.
It has been shown that particular compositions of PU may affect the
equilibrium moisture content. Density does affect the permeability,
but has little or no effect on the equilibrium moisture content.
The effect of moisture in the cells is to change the foam dimensions and weight and, as will be discussed later, the thermal conductivity.

Storage at relatively high temperatures tends to further cure
the foam and increases the number of polymer crosslinkages. The
effect of storage at 275 F for 2 weeks was found to increase the
ambient temperature compressive properties by 5%, and after 9 months
the increase was 10 to 15%.[15] Dimensional stability can be severely
affected by low temperature aging, particularly at lower densities.[32]
On the other hand, if a particular foam is dimensionally stable for
a few days in low temperature service, it will probably remain so.
Low temperature stability problems are caused by the pressure differential created by condensation of the cellular gases. This pressure
differential across the foam membranes can cause deformation and,
for lighter materials, extensive cellular collapse. Figure 14 shows
volume change as a function of density under three common aging conditions.

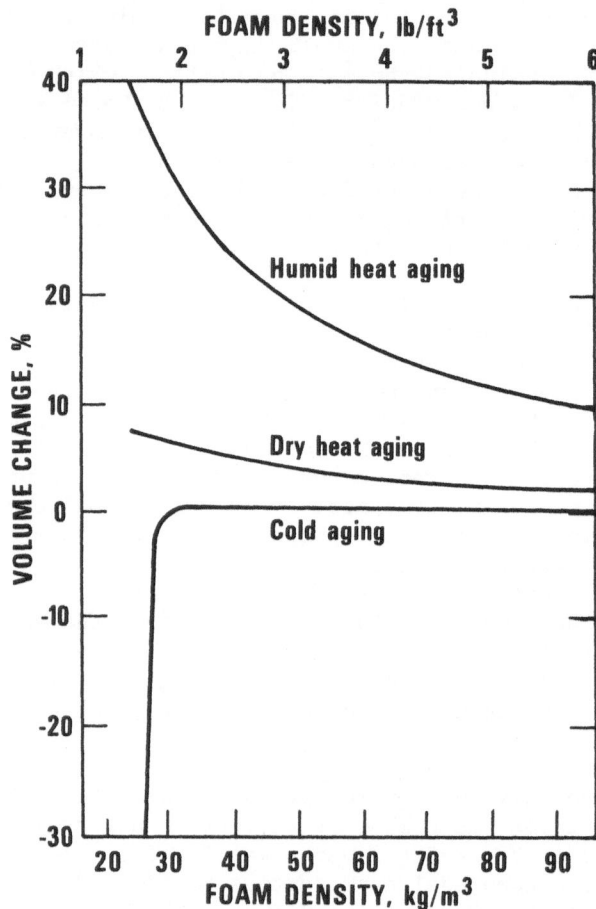

Fig. 14. Effect of environmental conditions on dimensional
         stability of PU foams of various densities.[32]

Thermal Properties

A property of primary importance in thermal insulations is, of course, the thermal conductivity (the term thermal conductivity might more appropriately be called apparent thermal conductivity, since radiation and convective components are included along with the true conductive modes). The effect of aging on thermal conductivity is of interest, because many applications of PU and PS are expected to be in service for 20 or more years. Linear thermal expansion is also of importance in insulation system design, since the opening of cracks around slabs or within the foams can cause the convective transfer of heat to become prohibitively high. It has been found that convection was the culprit in causing actual heat leak into refrigerator ships to be 77 to 126% higher than the calculated value.[33,34] Other thermal properties of interest include specific heat and thermal diffusivity. Specific heat of PS and PU foams is needed to compute the quantity of liquid needed in cooldown, and thermal diffusivity is used to determine the transient thermal response of the system. As was the case in the mechanical properties, density is found to affect each thermal characteristic.

Heat is transferred through foams via four mechanisms: (1) gas conduction, k_g, (2) solid conduction, k_s, (3) radiation, k_r, and (4) convection, k_c. These mechanisms function simultaneously and interact with one another so that experimental determination yields apparent rather than true thermal conductivity. The total heat flow, q_T, via the four parallel modes may be expressed

$$q_T = q_g + q_s + q_r + q_c \qquad (4)$$

The dominant mode of heat transfer for the foams we are considering is gas conduction until the temperature becomes low enough to effect a phase change in the gas; solid conduction and radiation components are relatively small but not negligible. Convection, on the other hand, is negligible for small-cell foams (cell diameter 3 to 4 mm),[12] but becomes critically important for cracked foams and slab installations where vertical cracks allow natural convection to exist.

The type of gases found in PS and PU foams was discussed earlier. However, due to different permeability rates for different gases, as shown in Table I, the final gas composition will be quite different than the initial gas composition. Kinetic theory for gases[35] indicated that

$$k_g = f(m) \qquad (5)$$

where m is the molecular weight of the intercellular gas. This relationship is shown in Fig. 15 for commonly used fill gases.

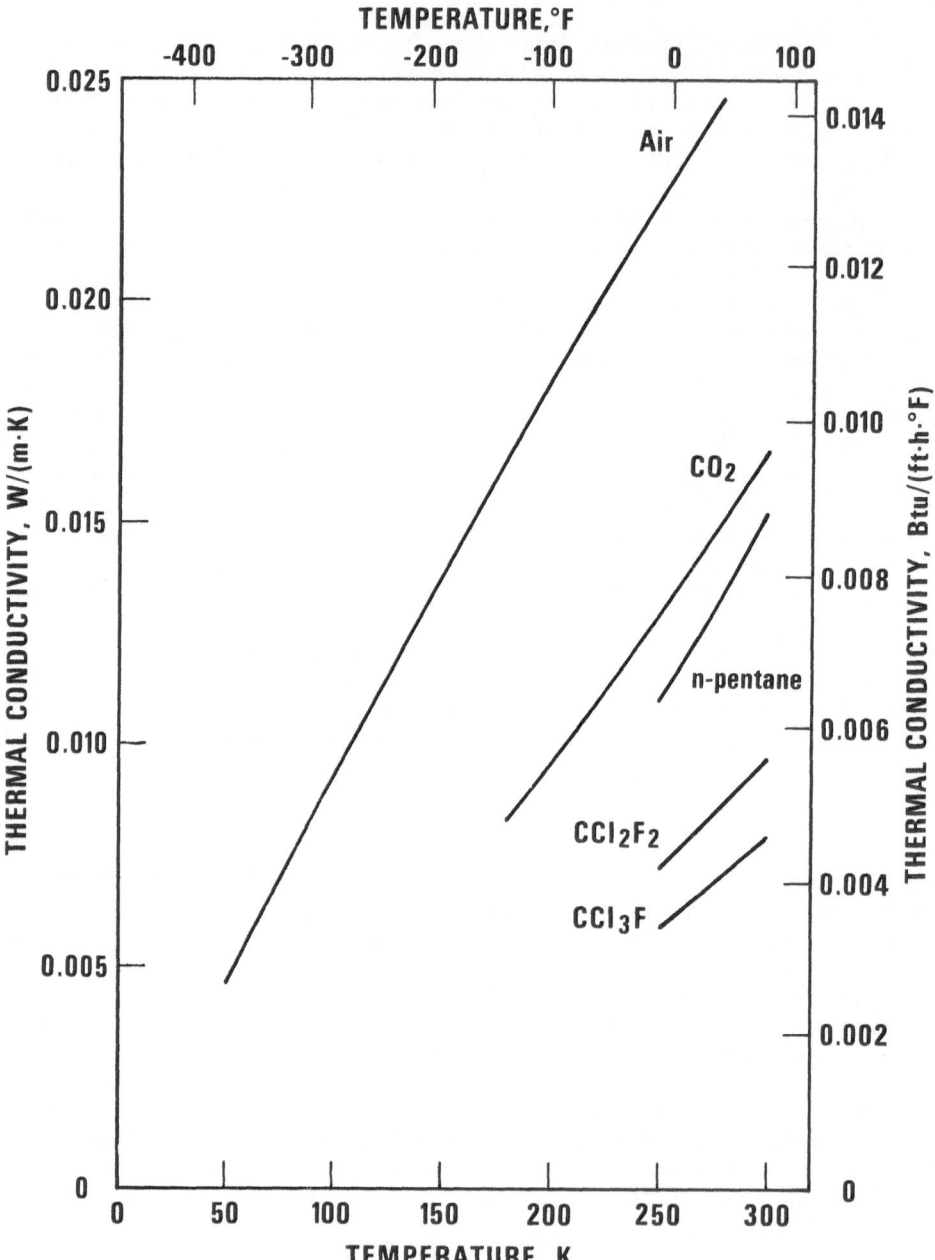

Fig. 15.   Thermal conductivity of gases commonly found in PS and
           PU cells as a function of temperature.[36]

When a mixture of gases is present, k_g can be estimated by using[37]

$$k_{mixture} = 0.5 \ (k_\alpha + k_\beta) \tag{6}$$

where

$$k_\alpha = \sum_{i=1}^{n} X_i k_i \tag{7}$$

and

$$\frac{1}{k_\beta} = \sum_{i=1}^{n} \frac{X_i}{k_i} \tag{8}$$

The thermal conductivities of the individual species are given by k_i, X_i represents the corresponding mole fraction, and n is the number of species present.

The effect of aging on thermal conductivity is almost entirely the result of changes in the fill gas. If allowed to age in air, the equilibrium-state gas content of both PS and PU will be air at 10^5 Pa (1 atm). The rates of diffusion of gases into and out of PS and PU, however, are quite different (see Table I). Polystyrene is permeable to all fill gases, and, unless the foam is sealed, the equilibrium state will be reached in a very short time. Polyurethane, on the other hand, is reasonably impermeable to the heavy hydrocarbon gases and will retain these gases for an extended time. The aging process for unprotected PU foam initially consists of air diffusing into the cells until the partial pressure of air in the cells equals the external air pressure. A net positive gas pressure is established until, eventually, the fluorocarbon diffuses out of the cells, leaving only air. Figure 16 illustrates the effect of aging on the thermal conductivity of PU.

Two additional factors must be considered in the gas conduction component for foams in cryogenic service. They are the effect of gas composition and condensation temperatures on k_g. Consider, for example, an aged, CCl_3F blown PU foam with one face at 19 K (-425°F) and the other at 300 K (80°F). This configuration will present three distinct gas-conduction regimes. Beginning at the 300 K face, gas conduction will be due to the mixture of air and CCl_3F. The conductivities of several possible mixtures are shown in Fig. 17. If the mixture is assumed to be 50 mole % air and 50 mole % CCl_3F, the conductivity will decrease along this constant composition line until the temperature is reached where CCl_3F begins

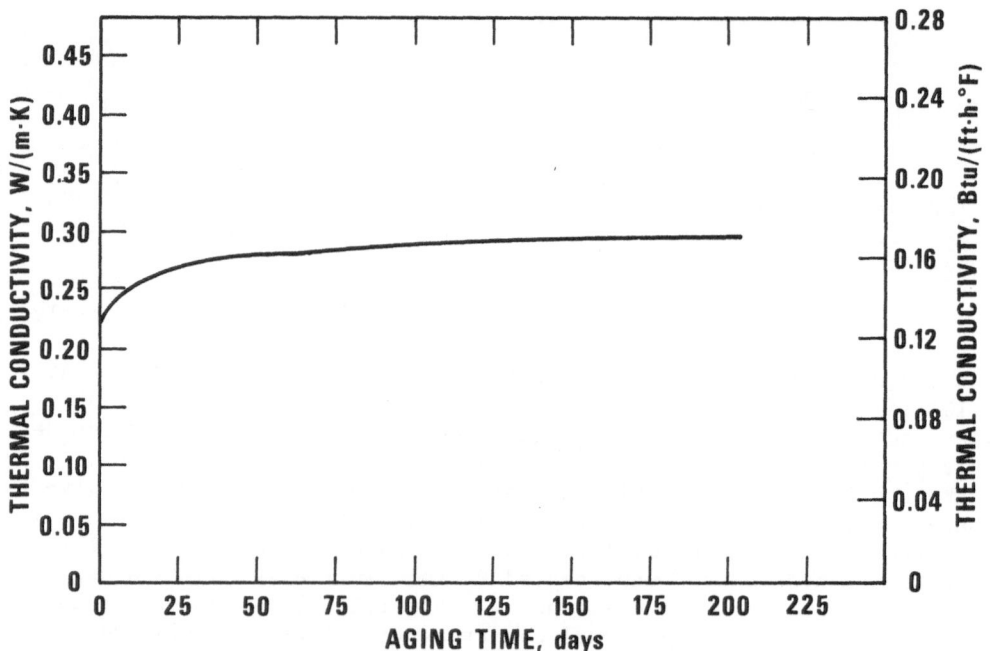

Fig. 16.   Effect of aging on PU foam.[12]   The PU was aged at 333 K (140°).

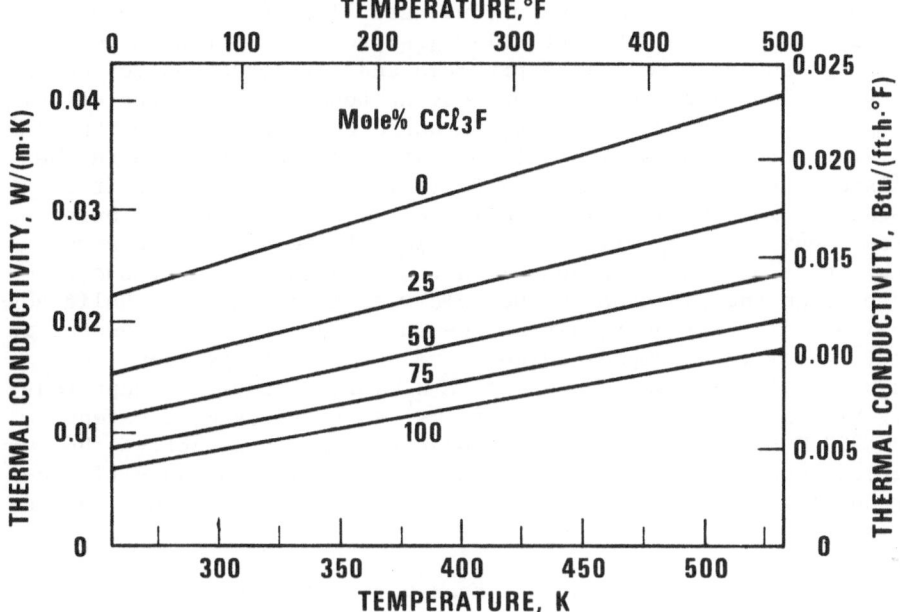

Fig. 17. Thermal conductivity as a function of temperature for mixtures of CCl₃F and air.[38]

to condense. The actual temperature at which this occurs depends
on the partial pressure of the CCl_3F in the cells, but, for purposes
of illustration, it is taken to be 280 K. This corresponds to the
280 K inflection in the thermal conductivity of Schroeder's 32 kg/m^3
PU shown in Fig. 18.[31] As the temperature is lowered, the mole
fraction of CCl_3F gas will decrease until essentially only gaseous
air is present. This decreasing CCl_3F gas content causes the observed conductivity to have a negative slope in the range
$230 \leq T \leq 280$ K. This behavior would be expected from the trends
shown in Fig. 17. The second regime of gas conduction is encountered as the distance to the cold wall is decreased. The gas conduction here is entirely due to air and decreases smoothly until
the components of air begin to condense. The third regime begins
at this temperature. The cellular-gas pressure continues to decrease as the cold wall is approached and air condenses (of course,
pressure also decreases with decreasing temperature in the cells).
The gas conductivity decreases, as would be expected from the conductivity of air shown in Fig. 15. At 50 K, the pressure in the
cells will be approximately 133 Pa (1 mm Hg) and the mean free path
of the gas molecules will approximate the cell diameter. Under
these conditions, k_g will be a function of pressure. Pressure
within the cells and the concomitant gas conduction continue to
decrease as the distance to the cold wall is decreased. Solid conduction and radiation remain as the only modes of heat transfer in
the very low (cryopumped) temperature section of the foam. Gas
conduction in PS foams decreases in a similar manner, except that
only air is present in the aged foams, so that regime 2 extends
from 300 K to air condensation temperature. Consequently, the
thermal conductivity does not contain inflections as does fluorocarbon-blown PU. A representative curve for total thermal conductivity of PS is included in Fig. 18.

The second mode of thermal transport, solid conduction, takes
place through the cell walls of the foam. Both PS and PU resins
are made up of disordered (noncrystalline) materials. This structure results in a low-phonon or lattice conduction at low temperatures and the conductivity is nearly a linear, decreasing function
of temperature. Figure 19 illustrates this behavior for solid PS.
The heat conducted by the solid phase[40] can be estimated by

$$k_s = V_s k_{ss} \qquad (9)$$

where V_s is the relative volume of resin in the foam and k_{ss} is the
thermal conductivity of the nonexpanded polymer. Using this and
the PS data at 110 K from Fig. 18, $k_s = 0.03 \times 0.125 = 0.0038$ W/m·K,
since the resin only occupies approximately 3% of the volume of a
32 kg/m^3 (2 lb/ft^3) foam. This contribution must be further reduced
by a steric factor,[38] i.e., the portion of the solid that is

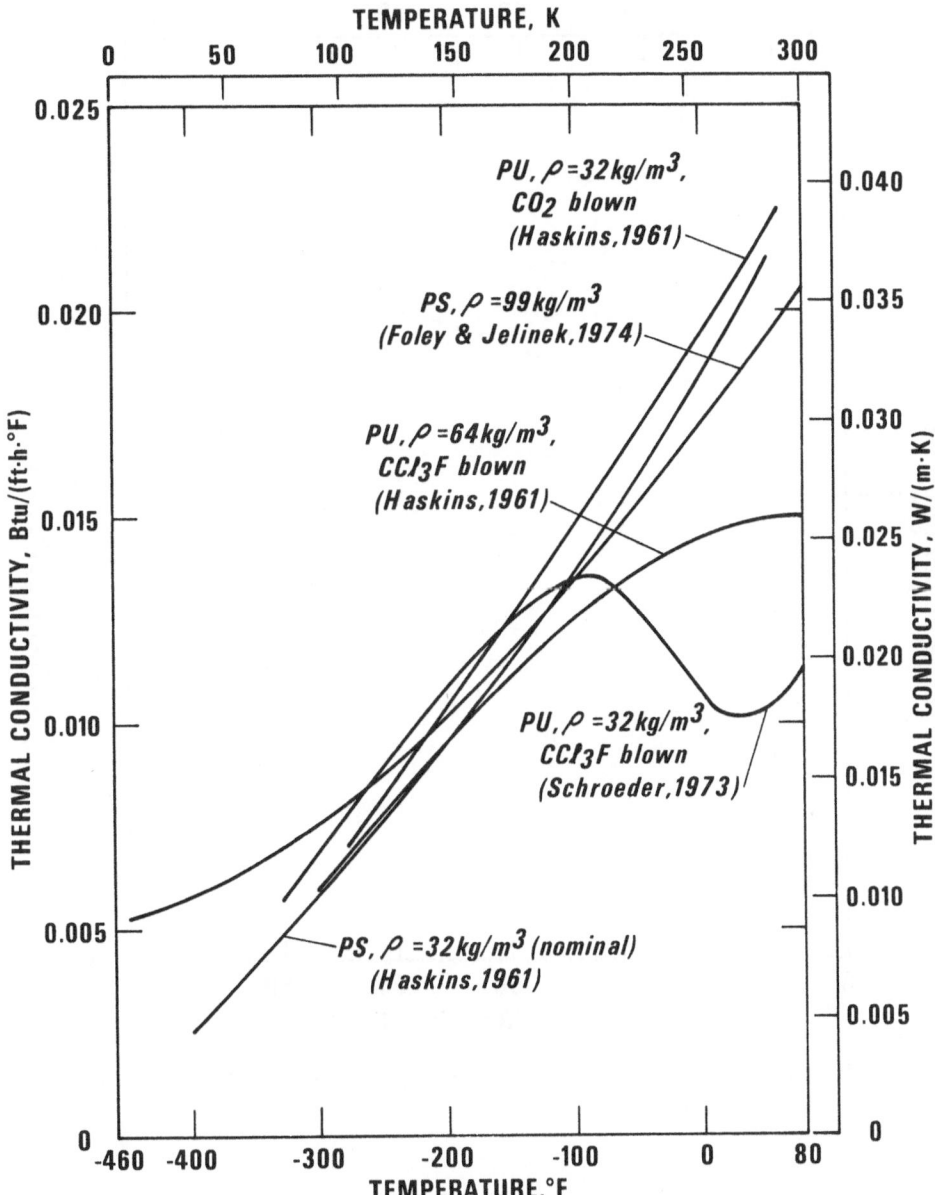

Fig. 18. Thermal conductivity as a function of temperature for PU and PS foams with fill gas and density as parameters.[23,31,39]

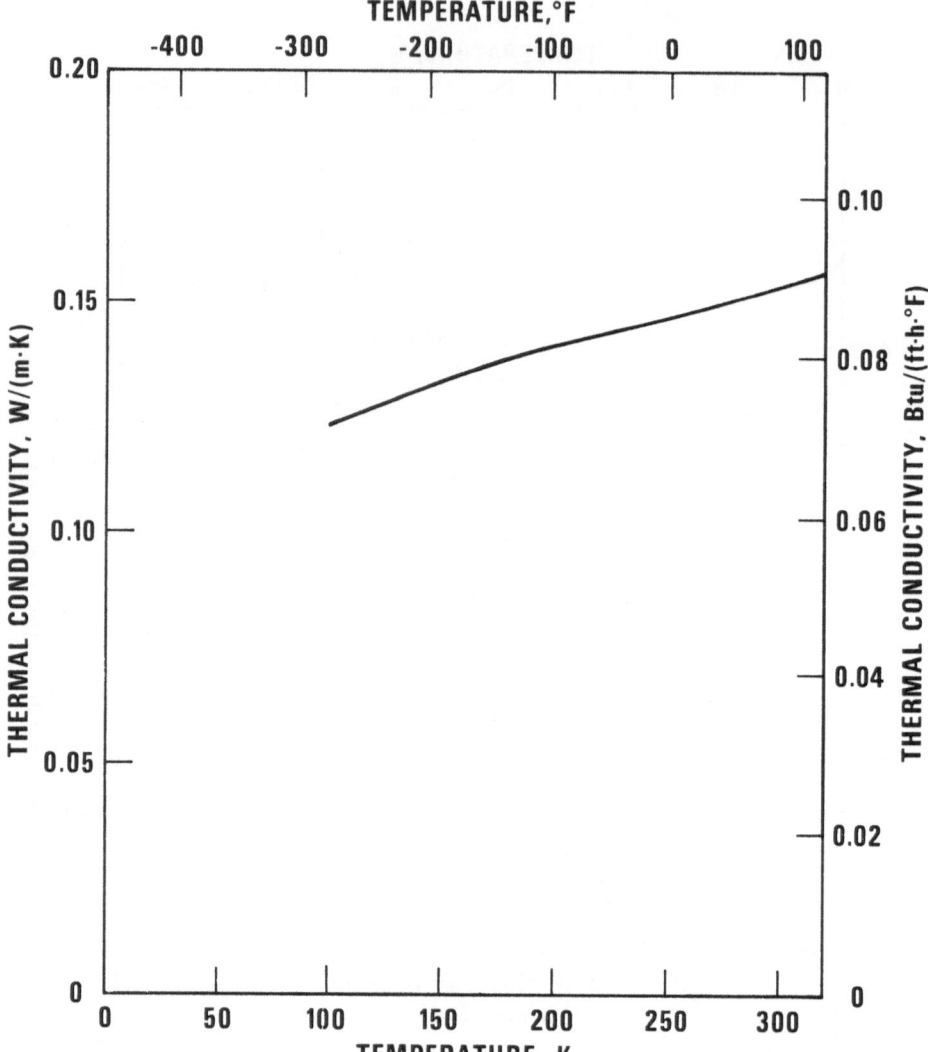

Fig. 19.   Thermal conductivity of nonexpanded PS as a function
of temperature.[4]

perpendicular to the direction of heat flow cannot contribute directly to the conduction.   The relative contribution of k_s to the
total conductivity is small, except at very low temperatures where
k_g is essentially zero as a result of cryopumping.

Radiant heat flow through foam insulators is due to transmission and absorption and reradiation. This complex energy transfer is generally assessed by looking at the difference between q_g + q_s + q_c and the total observed heat transfer. Resins used for making foams are partially transparent in the 0.05 to 0.75 μm range of wavelengths.[12] Sections associated with thermal insulations are normally optically thick (thickness >> photon mean free path), however, so that there ordinarily will not be direct transmission. The effective k_r in this case is proportional to T^3. The radiant component is a function of both density and cell size. Higher density foams present more energy-absorbing material, and smaller cell size presents more cell walls to act as reflectors. The fact remains that, at low temperatures, k_r represents a small portion of the overall conductivity.

The fourth mode of heat transfer possible in PS and PU foams is gaseous convection. It has been shown experimentally[12] that this component will be near zero for cell diameters less than 3 to 4 mm. This is much larger than the cells ordinarily found in undamaged insulating foams. Serious convective problems can develop, however, because of cracks between slabs and cracks in the foam itself.

Thermal contraction observed in PS and PU foams is quite large when compared with that of metals, e.g., $\Delta L/L_{293} \simeq$ -12.5 x 10^{-3} at 110 K (-262°F) for PS and -3.5 x 10^{-3} for commonly used aluminum alloys. The linear thermal expansion (LTE) of the expanded plastics does not differ significantly[41] from that of the solid plastics for densities above about 32 kg/m^3 (2 lb/ft^3). Figure 20 shows the experimentally determined LTE for PS and PU at low temperatures. The mechanical strength of low density foams, $\rho \lesssim$ 32 kg/m , is such that deformation may occur as a result of changes in internal gas pressure at low temperatures.

Specific heats of expanded plastics may be determined using the Kopp-Neumann additivity rule.[42] The total C_p is given by the sum of specific heats of each constituent times the respective weight fraction. This method was used to calculate C_p for the two densities of PS shown in Fig. 21. A second method of computing C_p for expanded plastics makes use of the relationship

$$C_p = k/\alpha\rho \tag{10}$$

where k is thermal conductivity, α is thermal diffusivity, and ρ is density. Dynamic methods were used to experimentally determine both k and α for the PU foam specimen shown in this figure.

Thermal diffusivity depends on both the fill gas and the density, as seen in Fig. 22. The values by Ho et al.,[4] were calculated whereas those by Kudryacheva and Kozhevnikov[44] and by Sacchi et al.,[45] were established experimentally.

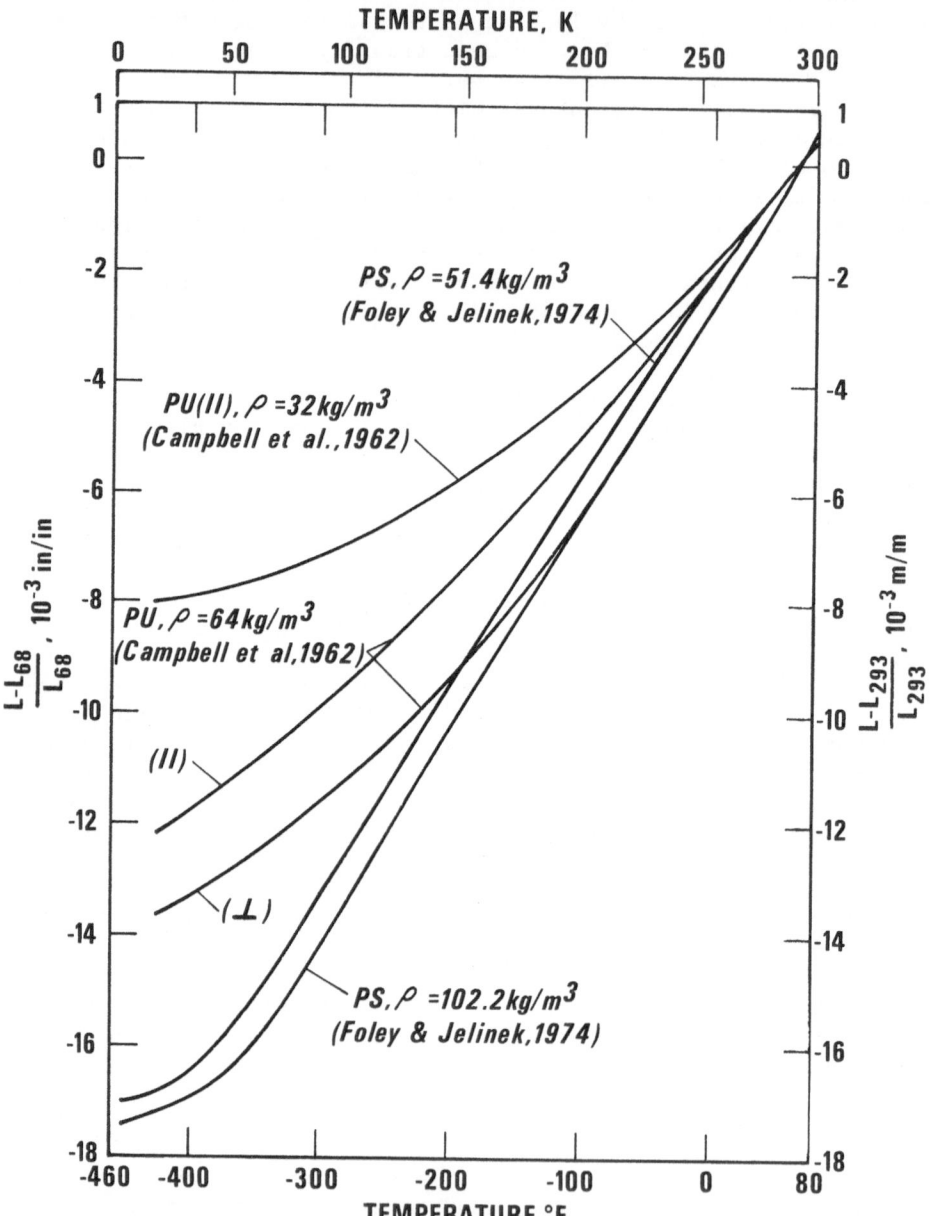

Fig. 20. Linear thermal expansion as a function of temperature for PU and PS foams with density and composition as parameters.[23],[26] $\Delta L/L = 0$ when T - 293 K (68°F).

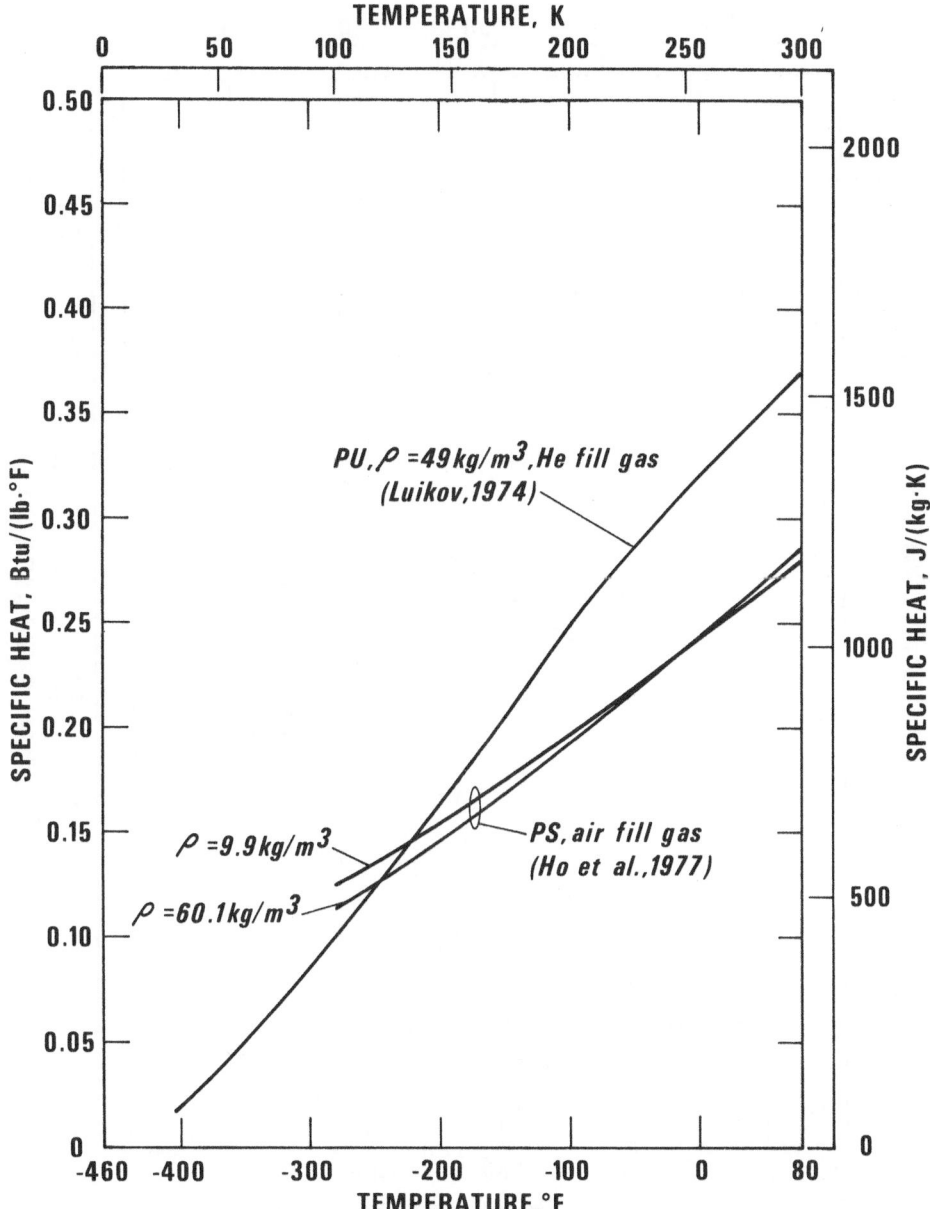

Fig. 21.   Specific heat as a function of temperature for PU and
           PS foams with density and fill gas as parameters.[4,43]

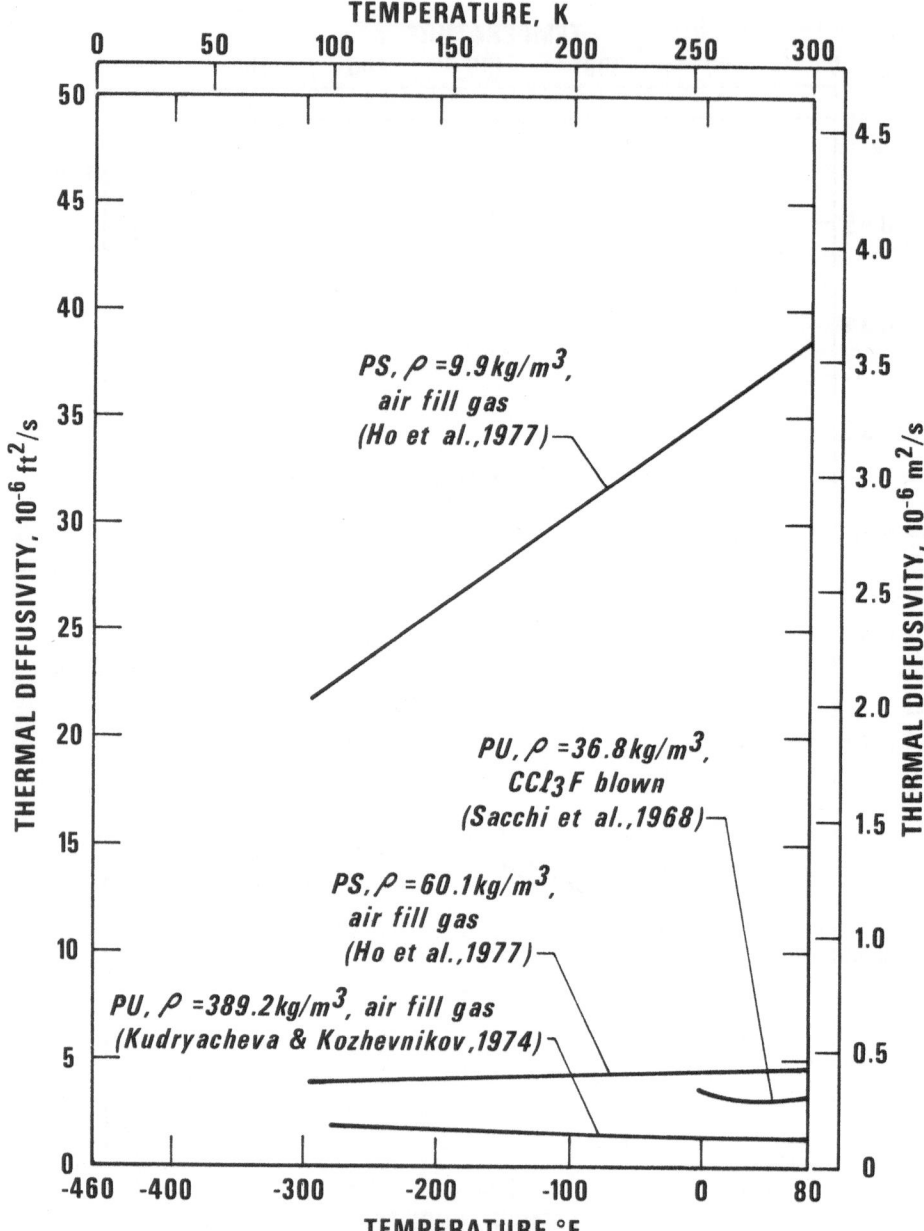

Fig. 22. Thermal diffusivity as a function of temperature for PU and PS foams with fill gas and density as parameters.[4,44,45]

SUMMARY

The mechanical and thermal properties of the expanded plastics PU and PS make them excellent choices for many low temperature thermal insulation systems. Rapid methods of application have been developed which, when coupled with low material cost, make thermal insulations of this type economically attractive.

The mechanical and elastic properties are, in general, inverse functions of the temperature, i.e., the strengths and moduli increase with decreasing temperature. These properties are also proportional to density. These general trends are illustrated in Figs. 8 through 13. The mechanical and elastic behavior for a specific foam may defy the general trends, since these properties reflect many complex dependencies. A consequence of the foaming action is an elongated cell structure in the direction of foam rise. This, in turn, gives rise to anisotropic behavior. The degree of anisotropy depends on the particular process and composition. Again, a trend is observed in that the longitudinal (parallel to the direction of rise) strengths and moduli are greater than those in the transverse direction (perpendicular to the direction of rise). In many cases, a specimen cannot be characterized as purely parallel or perpendicular, and the properties will lie somewhere between those for longitudinal and transverse directions.

The thermal conductivity of PS and PU depend to a large extent on the gas occupying the cells. For PS this will be air, unless the foam is somehow sealed soon after manufacture, because PS is very permeable to both commonly used blowing agents and air. PU, on the other hand, will retain the usual heavy fill gases for long periods of time but will also readily absorb air if left unprotected. The net result is that PU has a lower thermal conductivity than PS until condensation removes the gas conduction component. At lower temperatures, the conductivities are nearly equal. As with the mechanical and elastic properties, the thermal conductivity is a function of density. The tendency in this case is for higher conductivity with higher density except at densities less than about 32 kg/m^3. At very low densities, increased cell size allows k_r and k_c to become significant.

Aging and conditions of aging are significant factors in the mechanical and thermal performance of PS and PU. It is particularly important for foams in cryogenic service to be shielded from water vapor. If this is not done, moisture will permeate the cells, form ice, and result in damage to the cells in addition to increasing the thermal conductivity. Insulation systems for use on LNG tankers enclose the foam in membranes of plastic or metal. Accelerated tests on these systems indicate a functional lifetime in excess of 20 years.

REFERENCES

1. P.E. Glaser, I.A. Black, R.S. Lindstrom, F.E. Ruccia, and A.E. Wechsler, NASA Spec. Pub. SP-5027 (1967).

2. E.L. Sharpe and R.G. Helenbrook, in: Nonmetallic Materials and Composites at Low Temperatures, Plenum Press, New York (1979), p. 207.

3. R.E. Knox, ASHRAE J. 4, 43 (1962).

4. C.Y. Ho, P.D. Desai, K.Y. Wu, T.N. Havell, and T.Y. Lee, Report No. GCR-77-83, National Bureau of Standards, Washington, D.C. (1977).

5. S.L. DeGisi and T.E. Neet, J. Appl. Polym. Sci. 20, 2011 (1976).

6. R.H. Harding, ASTM Tech. Pub. 414, American Society for Testing and Materials, Philadelphia (1967), p. 3.

7. Scott Paper Company, Foam Division, Chester, Pennsylvania.

8. F.O. Guenther, SPE Trans. 2, 243 (1962).

9. J.R. Scott and W.J. Roff, Handbook of Common Polymers, Batterfield & Co., Ltd., London (1971).

10. C.J. Benning, Plastic Foams: The Physics and Chemistry of Product Performance and Process Technology 1, John Wiley & Sons, New York (1966).

11. A.H. Landrock, Report No. 37, Plastics Technical Evaluation Center, Picatinny Arsenal, Dover, New Jersey (1969).

12. R.E. Skochdopole, Chem. Eng. Prog. 57 (1961).

13. H. Ulrich, Polyurethane, Modern Plastics Encyclopedia 54, 86 (1977).

14. D.J. Doherty, R. Hurd, and G.R. Lester, Chem. Ind. (Lond.), 1340 (1962).

15. R.K. Traeger, J. Cell. Plast. 3, 405 (1967).

16. R.H. Harding, F. Hostettler, T.J. Mahoney, Handbook of Foamed Plastics, Lake Publishing Co., Libertyville, Illinois (1965), p. 151.

17. Properties of Rigid Urethane Foams, Bulletin E-08934, DuPont Company, Wilmington, Delaware.

18. W.B. Brown, Plastics Progress 1959, 149 (1960).

19. "Styrofoam" expanded polystyrene, Dow Chemical Co., Midland,
     Michigan (1962).

20. H.S. Smith, in: Proceedings Fourth International Cellular
     Plastics Conference, Technomic Publishing Co., Westport,
     Connecticut (1976), p. 271.

21. W.A. Carr, Saturn S-II Materials and Processes Development
     during the First Half of 1968, Report No. SID-63-600-10, Rock-
     well International Space Division, Downey, California (1968).

22. R.P. Reed, J.M. Arvidson, and R.L. Durcholz, in: Advances
     in Cryogenics Engineering, Vol. 18, Plenum Press, New York
     (1973) p. 184.

23. R.F. Foley and F.J. Jelinek, in: Proceedings Fifth Interna-
     tional Cryogenic Engineering Conference, IPC Science &
     Technology, Sussex, England (1974), p. 439.

24. J.M. Arvidson, R.L. Durcholz, and R.P. Reed, in: Advances in
     Cryogenic Engineering, Vol. 18, Plenum Press, New York (1973),
     p. 194.

25. R.B. Bennett, in: Advances in Cryogenic Engineering, Vol. 19,
     Plenum Press, New York (1974), p. 393.

26. M.D. Campbell, J.F. Haskins, J. Hertz, H. Jones, and J.L.
     Percy, Thirty Day Evaluation of Foams and Honeycomb for
     Centaur Intermediate Bulkhead, Report No. MRG-312, General
     Dynamics/Astronautics (1962).

27. W.F. Stewart, D.T. Eash, and W.A. May, in: Advances in
     Cryogenic Engineering, Vol. 19, Plenum Press, New York (1974),
     p. 385.

28. F.E. Lormis, Mechanical and Physical Properties of Organic
     Foams, Report BDX-613-562-REV, Bendix Corp., Kansis City Div.,
     Kansas City, Missouri (1972).

29. R.P. Reed, R.L. Durcholz, and J.M. Arvidson, in: Advances in
     Cryogenic Engineering, Vol. 16, Plenum Press, New York (1971),
     p. 37.

30. R.M. McClintock, in: Advances in Cryogenic Engineering, Vol. 4,
     Plenum Press, New York (1960), p. 132.

31. C.J. Schroeder, Insulation Commonality Assessment (Phase II), Report No. SD72-SA-0157-2, Rockwell International Space Division, Downey, California (1973).

32. J.C. Hilado, J. Cell. Plast. 3, 161 (1967).

33. G. Lorentzen, Proc. Int. Inst. of Refrig. (Nantes), 127 (1957).

34. G. Lorentzen and E. Brendeng, in: Proceedings Tenth Int. Congress of Refrigeration, Pergamon Press, New York (1960), p. 294.

35. D.K. Edwards, V.E. Denny, and A.F. Mills, in: Transfer Processes, Hemisphere Publishing Corp., Washington, D.C. (1976), p. 206.

36. Y.S. Touloukian, P.E. Liley, and S.C. Saxena, in: Thermal Conductivity: Nonmetallic Liquids and Gases, Thermophysical Properties of Matter, IFI/Plenum, Yew York (1970).

37. R.C. Reid and T.K. Sherwood, in: The Properties of Gases and Liquids, McGraw Hill Book Co., Inc., New York (1958), p. 240.

38. M.B. Hammond, Jr., "An Analytical Model for Determining the Thermal Conductivity of Closed-Cell Foam Insulation," presented at AIAA 3rd Thermophysics Conference, Los Angeles, California (June 1968).

39. J.F. Haskins, Thermal Conductivity of Plastic Foams from -423 Degrees F to 75 Degrees F., Report No. MRG-242, General Dynamics Corp., San Diego, California (1961).

40. R.L. Gorring and S.W. Churchill, Chem. Eng. Prog. 57, 53 (1962).

41. L. Vahl, in: Proceedings Tenth Int. Congr. of Refrig., Pergamon Press, New York (1960, p. 267.

42. E.S.R. Gopal, Specific Heats at Low Temperatures, Plenum Press, New York (1966), p. 21.

43. A.V. Luikov, A.G. Shoshkov, L.L. Vasiliev, S.A. Tanaeva, P.Yu. Bolshakov, and L.S. Domorod, in: Heat Transmission Measurements in Thermal Insulations, STP 544, American Society for Testing and Materials, Philadelphia, Pennsylvania (1974), p. 290.

44. G.M. Kudryacheva and I.G. Kozhevnikov, Int. Polym. Sci. Technol. 1, T/105 (1974).

45. A. Sacchi, F. Zerro, and C. Codegone, in: <u>Proceedings 7th</u>
     <u>Thermal Conductivity Conference</u>, NBS Spec. Publ. SP-302,
     National Bureau of Standards, Washington, D.C. (1968), p. 151.

DURABILITY OF FOAM INSULATION

FOR LH$_2$ FUEL TANKS OF FUTURE SUBSONIC TRANSPORTS

E. L. Sharpe

NASA, Langley Research Center
Hampton, Virginia, U.S.A.

and

R. G. Helenbrook

Bell Aerospace
Buffalo, New York, U.S.A.

INTRODUCTION

The potential short-supply of petroleum-based fuels has led to activities by NASA to establish technical characteristics of air transportation systems that would use hydrogen-fueled aircraft. These activities cover sources and production of liquid hydrogen,[1] aircraft configurations,[2] and air terminal modifications, as affected by the introduction of liquid-hydrogen-fueled aircraft.[3,4]

These studies show that, among many technical areas that require attention before a hydrogen-fueled aircraft can be a reality, one of the most important is thermal protection for the fuel tanks. Both the subsonic studies and studies of a hypersonic vehicle application[5] indicate that low-density foams are attractive insulations for the LH$_2$ fuel tanks, if reliability and adequate life can be established. Although this type of insulation has been used extensively for large throw-away booster stages for space application, the lack of reuse requirements has not demanded extensive investigation of the cyclic life of foam insulations.

The object of this study was to experimentally investigate the suitability of commercially available organic foams as cryogenic insulation for liquid-hydrogen tanks under extensive thermal cycling typical of subsonic airline type operations. A goal of 2400 flight cycles was established for insulation survival (equal to approximately 9 years of airline service).

Fourteen insulations were examined in this study. The insulations included three types of polyurethane, three varieties of
polymethacrylimide, one polybenzimidazole, two polyisocyanurates,
two polymetric isocyanate foams (one with chopped fiberglass reinforcement, one without), a modified isocyanate foam, and two
insulation systems (one of polymeric isocyanate foam, one of
toluenedi isocyanate foam, both with two vapor barriers and fiberglass reinforcement).

Certain commercial organic foam insulations and bonding
agents are identified in this paper in order to specify adequately
which materials were used in the research effort. In no case does
such identification imply recommendation or endorsement of the
materials by NASA.

INSULATION DESIGN CONSIDERATIONS

Aircraft Application

The study of hydrogen fueled long-range subsonic transport
aircraft (see Fig. 1) documented in Ref. 2 indicates that tanks
should be kept at liquid-hydrogen temperatures, except during
overhaul periods, to reduce the possibility of fatigue failures
due to thermal stresses in both the insulation and the aluminum
tank. Because of this mode of operation, the fuel boil-off during
ground hold becomes an important factor in the determination of
optimum insulation thickness. The economic impact of insulation
thickness and the individual factors that contribute to the
selection of insulation thickness are shown in Fig. 2. Figure 2
was taken from Ref. 2; the properties of an efficient insulation
were used in the calculations. Amortized airplane costs, cost
of block fuel, and cost of the hydrogen boiled off both during
flight and on the ground are shown in terms of dollars per flight
hour as a function of insulation thickness. The costs were based
on the aircraft of Fig. 1 with: a 15-year life, an average usage
of 3285 hours per year, a liquid hydrogen cost of $3 per 1 054 GJ
(10^6 BTU), an insulation with a density of 35 kg/m^3 (2.2 lbm/ft^3),
and a conductivity of 0.0173 W/mK (0.010 BTU/ft-h°F). The minimum
point of the top curve of Fig. 2, which represents the cumulative
effect of all factors, occurs at an insulation thickness of about
16.5 cm (6.5 in). These results were obtained on the basis of
no recovery of boiled-off hydrogen on the ground, i.e., as if the
vent gases were simply allowed to escape. In comparison, the minimum point of the second curve from the top, which does not include
ground boil-off, occurs at about 14.0 cm (5.5 in) of insulation
thickness. Since the practical way to handle ground boil-off would
be to collect the cold boil-off gas and reliquefy it, both Ref. 2

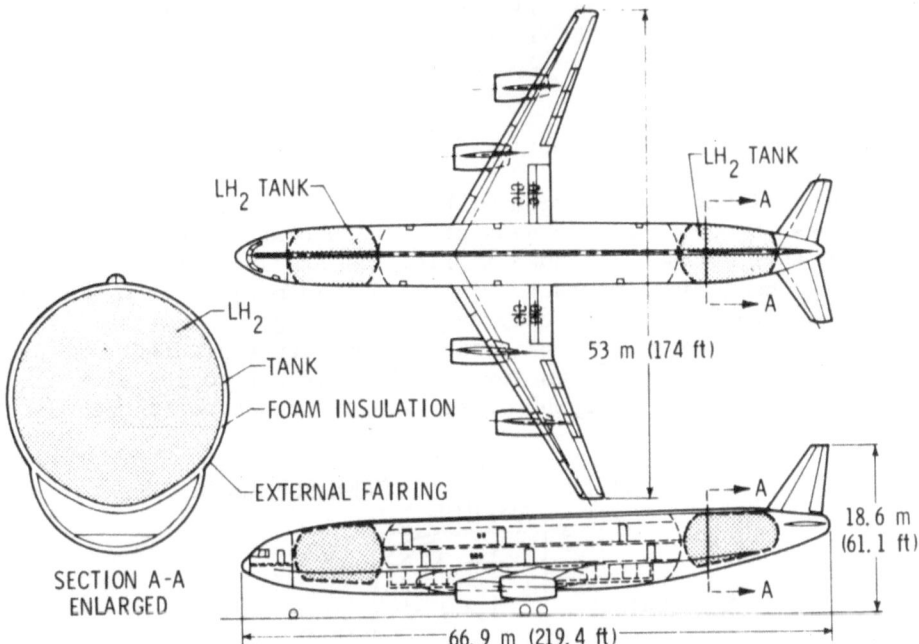

Fig. 1. Liquid hydrogen fueled subsonic transport.[2]
[400 passenger, 10 180 km (5500 nmi), M = 0.85.]

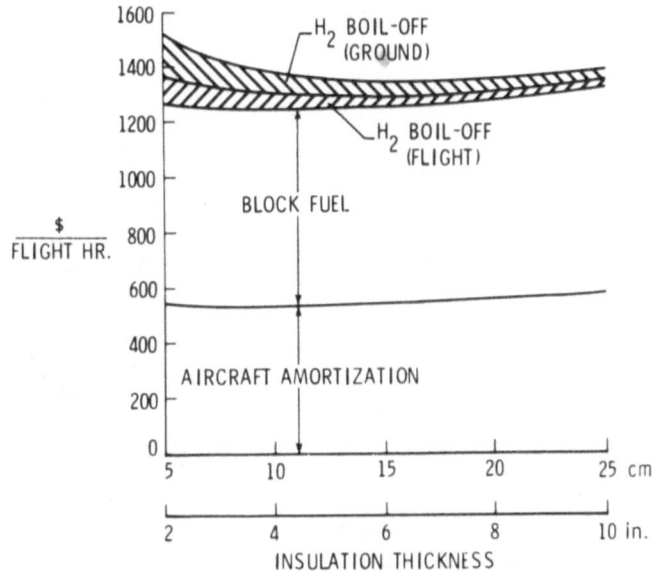

Fig. 2. Economic impact of insulation thickness.[2]

and this study assumed the optimum insulation thickness was 15.2 cm
(6 in), midway between the optimum values for the two upper curves
of Fig. 2.

Thermal Stress

The qualitative variation with time of temperature, strain,
and stress through the thickness of foam insulation bonded to an
aluminum tank surface is shown in Fig. 3. At time zero, t_0, there
is a step change in temperature in the aluminum plate from ambient
temperature to LF_2 temperature, as shown in Fig. 3a. At some short
time later, t_1, there is a very steep temperature gradient existing
in a very thin layer of foam immediately adjacent to the aluminum.
As time progresses, the temperature gradient tends to flatten out.

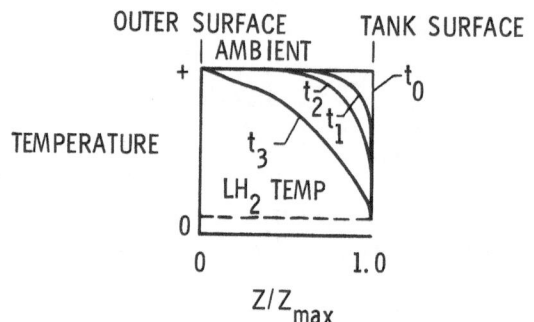

(a) Temperature Distribution

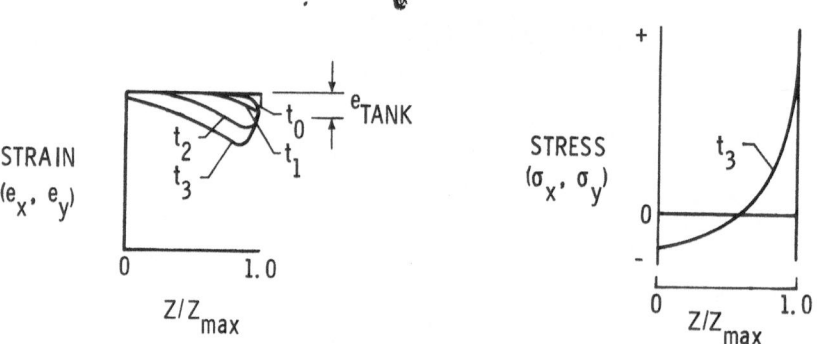

(b) Strain distribution

(c) In-plane stress
distribution

Fig. 3. Conditions in foam bonded to aluminum plate
which is cooled to LH_2 temperature. Z is the
distance into the insulation from its outer
surface, and Z_{max} is the insulation thickness.

Initially (at t_o), the aluminum plate is cooled and contracts, compressing the warmer foam, as shown in Fig. 3b. As time progresses (t_1, t_2, and t_3), the insulation cools and the foam tends to contract in accordance with the local temperature and its coefficient of thermal expansion. However, near the tank surface, the insulation is constrained from contracting by the much stiffer aluminum plate, which, because of a lower coefficient of expansion, experiences less contraction than the foam. Ultimately, at t_3 (as shown in Fig. 3c), the combination of the thermal contraction mismatch between the insulation and the aluminum tank and temperature distribution through the insulation lead to an in-plane stress pattern with large tensile stresses in the insulation near the tank and smaller compressive stresses at the free surface.

The magnitude of the tensile stresses is determined primarily by the thermal contraction mismatch between the aluminum and the foam; however, the length, width, and thickness of the insulation influences the level of the compressive stresses. For 15.2 cm (6 in) thick insulation, as shown in Fig. 4, edge effects significantly reduce compressive stresses for smaller pieces of insulation. [The stresses in Figs. 4 and 5 are based on polymethacrylimide insulation bonded on an aluminum tank at 20 K (-424°F) with a maximum insulation exterior temperature of 317 K (110°F)]. For example, a 0.3 m (1 ft) square insulation sample experiences a maximum compressive stress that is approximately 1/3 the stress in a 1.83 m (6 ft) square piece of insulation. The latter approaches the stress level for an infinite slab of insulation.

Stress Simulation

For economy in testing, small specimens are desirable. However, the stress relief produced by edge effects on small specimens precludes representative structural testing of full-depth insulation. Stress distributions representative of those in full-scale insulations may be obtained in small insulation specimens by reducing the specimen thickness while maintaining the same overall temperature gradient across the insulation thickness. Thermal stress distributions for a 5.1 cm (2 in) thick 0.3 by 0.6 m (1 by 2 ft) specimen are compared with stresses for the 15.2 cm (6 in) thick infinitely long and wide specimen in Fig. 5. The in-plane thermal stresses of the smaller specimen closely approximate the larger specimen. The maximum tensile stresses (at the cold tank wall) are equal for both cases, whereas the maximum compressive stress in the 0.6 m (2 ft) direction (σ_x), is 95% of that of the 15.2 cm (6 in) thick plate. The compressive stress in the 0.3 m (1 ft) direction (σ_y) is not as large as σ_x; this agrees with the results of Fig. 4, which indicates that the stresses decrease as the size decreases. Out-of-plane stresses

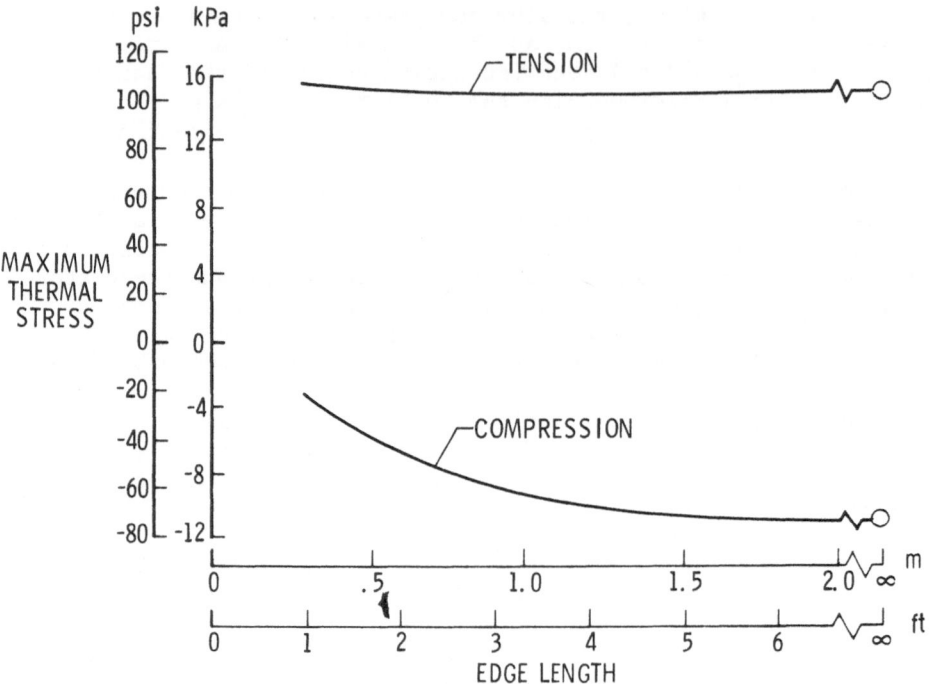

Fig. 4. Maximum in-plane tensile and compressive thermal
         stresses for square 15 cm thick polymethacrylimide
         foam bonded to an aluminum plate (foam surface
         temperature 317 K; aluminum temperature 20 K).

for the smaller specimen, although much larger than those for the
infinite plate, are much smaller than the in-plane stresses.
Comparison of the thermal stresses with allowable stress for the
two materials for which properties at cryogenic temperatures were
available (polyurethane and polymethacrylimide), indicates that
the in-plane stresses are the most critical; therefore, the stress
simulation offered by the thinner specimens was deemed acceptable.

     Based on the stress analysis, specimens 0.3 m by 0.6 m by
5.1 cm (1 ft by 2 ft by 2 in) were selected for the study. With
this specimen size, six different insulations could be tested at
the same time on a 0.6 by 1.8 m (2 by 6 ft) simulated tank surface.

APPARATUS AND INSTRUMENTATION

     The apparatus used in this study consists of an environmental
control system and a cryogenic insulation test specimen assembly.
The environmental control system, illustrated in the schematic

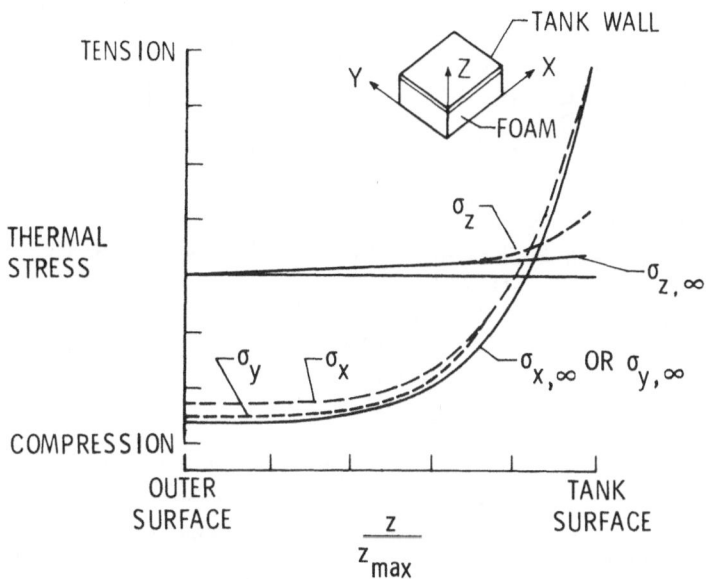

Fig. 5.   Maximum normal stress (σ) distributions for a 15.2 cm
          thick infinite slab and a 5.1 cm thick 0.3 m by 0.6 m
          sample of polymethacrylimide foam bonded to an alumi-
          num plate (foam surface temp. = 317 K; aluminum temp.
          = 20 K).

drawing of Fig. 6, is comprised of a large centrifugal blower, a
diverter valve, hot and cold heat exchangers, a test chamber, and
appropriate ducting to provide a closed circuit for flowing heated
(or cooled) air over the test specimens. Air flowing from the
blower at a rate of 0.73 m^3/s (1550 ft^3/min) is directed by the
diverter valve through the hot or cold heat exchanger, depending
upon which portion of the flight cycle is being simulated. In
the hot heat exchanger, air passes over coils which contain water
heated by an 18 kW electric heater; in the cold heat exchanger,
the air passes over coils which contain refrigerant cooled by a
17.6 kW (4.5 kg) refrigeration system.

     After passing through either one of the heat exchangers, air
travels through a duct and is manifolded into the test chamber.
The manifold consists of a tee section in the duct that allows

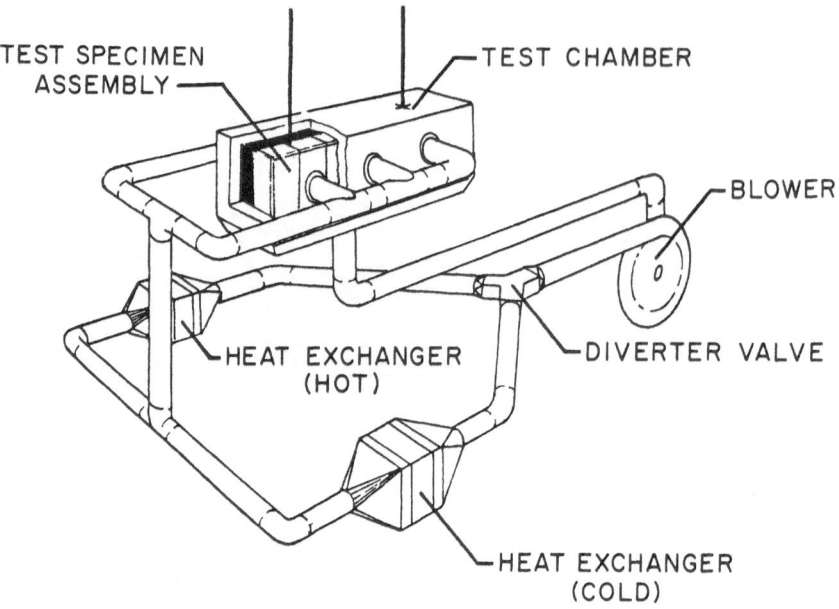

Fig. 6.   Cryogenic insulation test apparatus – environmental
          control system.

flow to go to two sides of the test changer. Both sides of the
test chamber have three ports which are connected to the incoming
air supply ducts [all ducts and ports are 20.3 cm (8 in) in
diameter]. Upon entering the test chamber, air strikes a per-
forated aluminum plate, which diffuses the flow over the insulation. After passing over the insulation specimens, the air
leaves the test chamber through three ports in the bottom of the
test chamber. Ducts from the three ports merge into one duct that
returns the air to the blower. The ducts, test chamber, and heat
exchangers in the environmental control system are insulated; the
humidity was not controlled.

    Figures 7 and 8 are drawings of the cryogenic insulation test
specimen assembly. The compartmented aluminum tank, which can
accommodate 6 insulation specimens (see Fig. 7), measures 1.8 by
0.6 m (2 by 6 ft) and is 3.9 cm (1.53 in) thick. The tank was
fabricated from 0.6 cm (2 ft) sections of an extruded web core
structural plank. The webs form a series of cells which measure
61 by 3.9 by 4.5 cm (24 by 1.53 by 1.78 in). The cells are connected at the bottom of the tank to form separate test and guard
compartments (see Fig. 8). The insulation specimens span a test
compartment and one-half a guard compartment at both ends. This

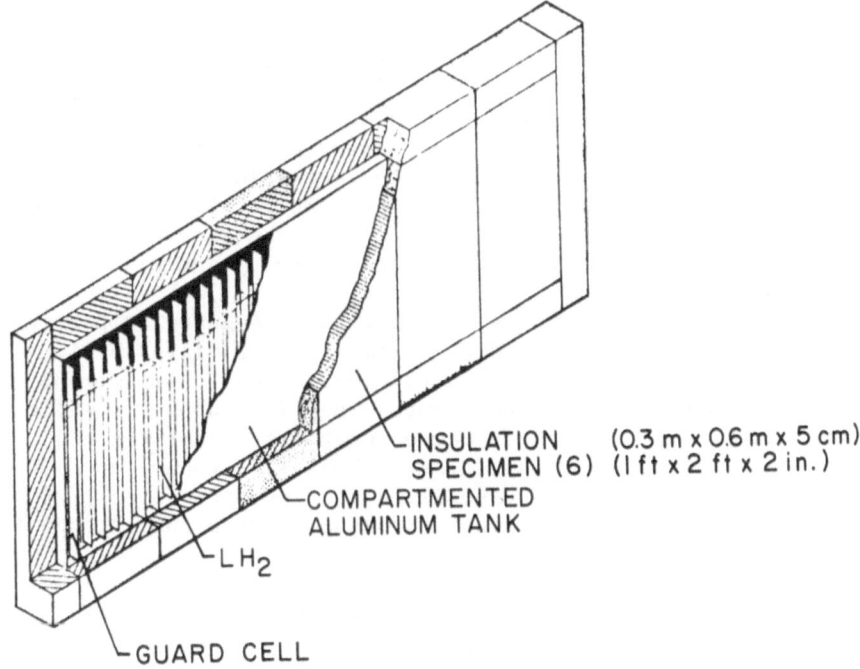

Fig. 7. Cryogenic insulation test specimen assembly.

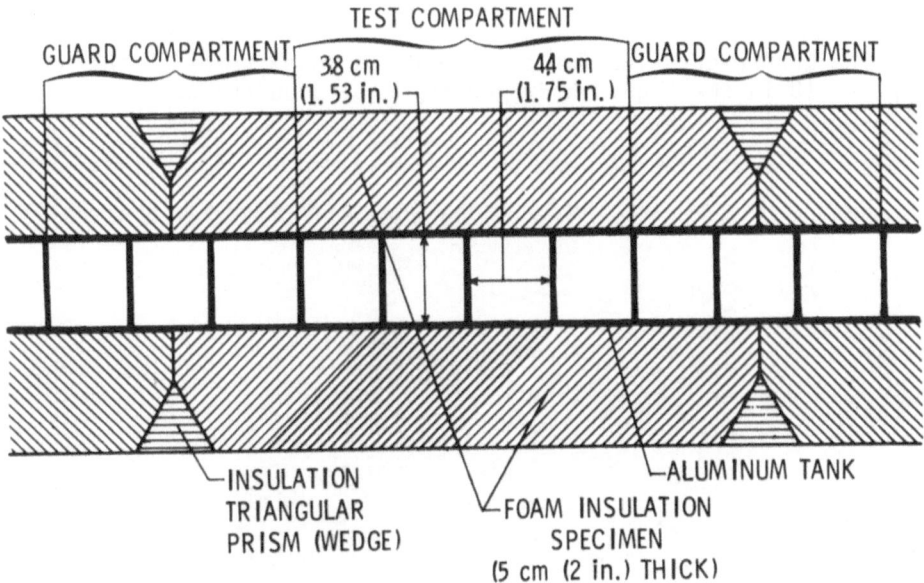

Fig. 8. Edge view of section of compartmented aluminum
         tank and insulation specimen.

arrangement makes it possible to measure the thermal performance of individual insulations by monitoring the rate of liquid hydrogen boil-off in each test compartment.

Insulation specimens were bonded to the aluminum tank with a polyurethane adhesive (Crest 7410). Triangular prism insulation sections were fitted and bonded at the insulation specimen joints (Fig. 8). This technique provided good contact at specimen joints during bonding. A photograph of six insulation samples mounted on the tank is shown in Fig. 9. (Note the narrow joint pieces.)

The six test compartments, along with five guard compartments between them and two end guard compartments, are filled by individual tubes, which enter the central cell of each compartment and go to the bottom of that compartment. The fill tubes are all connected to a common tube, which passes over the top of the cells through an open area over the cells. These fill tubes can also be used to empty the tank by pressurizing the top space over the liquid hydrogen and allowing the hydrogen to drain out through the tubes and out of the test area to a vent stack.

The open space over the top of the tank cells (see Fig. 7) acts as a manifold for the hydrogen boil-off gas and is connected to a stack, which is 12.2 m (40 ft) above the test area. This stack has a pneumatic valve, which can be controlled to maintain steady pressure in the area of the cryogen. [During testing a pressure of 55 kPa (8 psi) above ambient pressure was maintained in the cryogenic tank.]

The instrumentation for the apparatus consists of 25 thermocouples--18 in the liquid hydrogen regions, 5 on the outer surface

Fig. 9.   Insulation specimens mounted on tank.

of the insulation samples, and 2 in the ducts leading to and from
the test chamber. Each test compartment fill tube has 3 thermocouples suspended from it--one 2.54 cm (1 in) from the top of the
compartment, one at the middle, and one 2.54 cm (1 in) from the
bottom of the compartment. The combination of 3 thermocouples
per test compartment and the thin tank allows sensitive liquid
level measuring, since a small volume of hydrogen boil-off results
in a relatively large depth change.

TEST CONDITIONS AND PROCEDURE

Temperature histories for the external surface of the cryogenic insulation for a Mach number (M) of 0.85, long-range,
hydrogen-fueled transport during typical mission cycles are presented in Fig. 10. The histories, which were taken from Ref. 2
and are representative of the upper and lower limits for 95% of
the flights such a vehicle will experience, are based on the
assumption that the aircraft is refueled immediately after each
flight. The maximum thermal stresses, which were previously
presented in Fig. 4, are encountered shortly after the maximum
external surface temperature is reached. Obviously, an exact
simulation of the time-dependent temperature distribution through
the depth of the insulation for the complete life of a typical
commercial aircraft (approximately 15 years) would be costly and
time consuming. However, if the aircraft is refueled immediately
after each flight and the tank is maintained at cryogenic temperatures, the primary effect of a typical flight cycle is to impose
a perturbation on the compressive thermal stresses near the
external surface of the insulation (see Fig. 3), which can be
simulated by a relatively short (10 min or less) thermal cycle,
such as the typical test-temperature history presented in Fig. 11.
A less frequent but more severe thermal stress variation will be
encountered when the aircraft is removed from service for periodic
maintenance or overhaul and then returned to service, during which
time the tanks and insulation will be cycled from cryogenic temperatures, to ambient temperature, and back to cryogenic temperatures. Based on current airline practice, the overhaul periods
are sufficiently infrequent that they can be simulated in the
present test by simply suspending cryogenic testing, allowing the
tank to reach ambient temperature, and then resuming cryogenic
testing.

During a test period, the tank was filled with liquid hydrogen
and the temperature history of the exterior of the insulation was
cycled repeatedly as the hydrogen was allowed to boil off. Five
thermocouples strategically distributed over the insulation surfaces indicated the temperatures of the outer surface, and the
exact cycle time was controlled by the thermocouple that last
reached the desired temperature. (The initial tests were observed

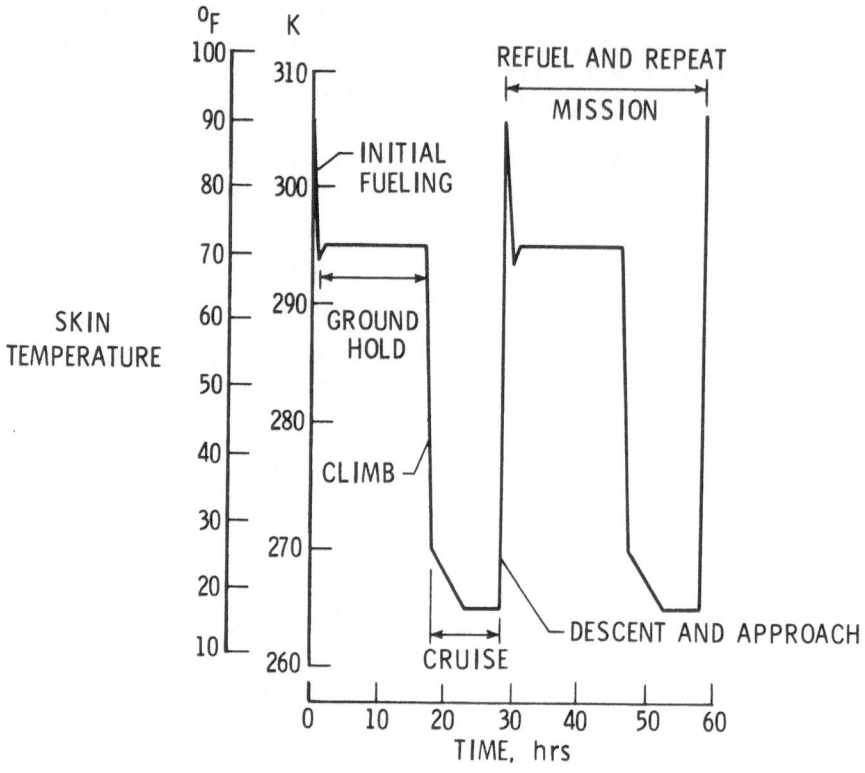

Fig. 10.   Cryogenic insulation surface temperature for typical
            LH$_2$ fueled subsonic transport mission cycles.[2]
            [10 180 km (5500 nm range), 15.2 cm (6") thick
            insulation.]

closely to prevent over heating the specimens, and the temperature
distributions over the outer surface of the specimens were found
to be uniform.) The tank was refilled when the lowest liquid level
thermocouples indicated that all tank compartments emptied to
2.54 cm (1 in) or less. Thus, in contrast to an aircraft application for which the tank would be filled and emptied once per flight,
the external temperature and hydrogen level cycled independently
during the tests. There was a wide variation between specimen
hydrogen boil-off rates, as indicated in the results section. Even
with the guard sections of the tank, which were intended to cut
down heat flow between neighboring compartments, some specimens
exhibited such poor performance that the performances of neighboring insulations were strongly affected.

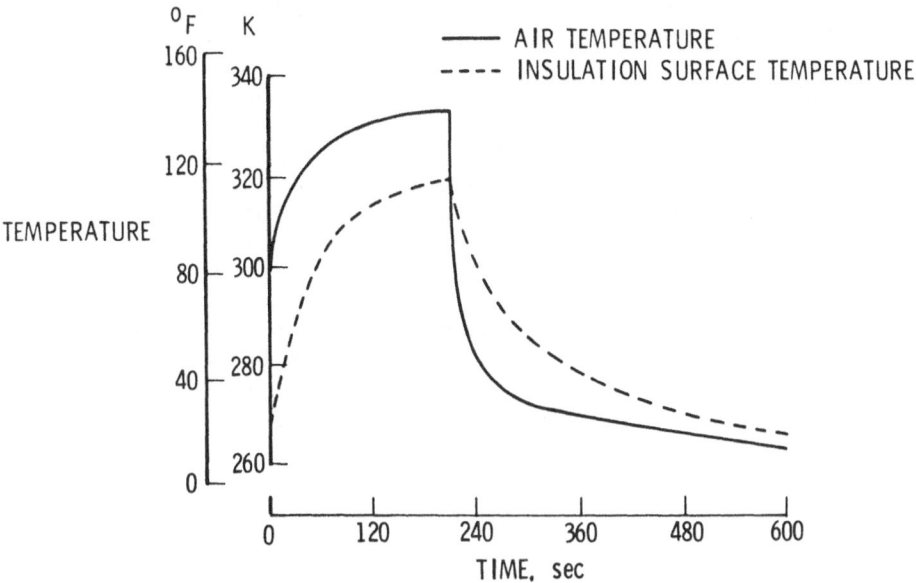

Fig. 11.   Air and insulation surface temperature histories
for a typical test cycle with tank wall at LH$_2$
temperature.

Tests were conducted on a three-shift basis, so that once a
test series began, it ran for 24 hours a day, 5 days a week, or
until deteriorating performance indicated that the tests should
stop and specimens should be examined. The shutdown periods
represented the time an airplane would be overhauled and the tank
would be allowed to warm up. While the tank was warm, decisions
were made pertaining to sample replacement or continuation of
cyclic thermal loading on the individual test specimens. The
criteria for sample replacement was poor thermal performance and/or
extensive structural damage to the insulation.

DATA REDUCTION

Even though the guard compartments were designed to prevent
interaction between adjacent test compartments, the thermal interactions between compartments have proven to be quite large. Therefore, the data have been judiciously faired to provide the curves
of normalized boil-off times as a function of number of thermal
cycles that are presented in the Results and Discussion Section.
This approach is believed to be appropriate, since the precise
measurement of conductivity is not the purpose of the paper and
the uninterpreted data would not be readily assimilated by the

reader. However, it should be emphasized that the curves presented
are qualitative rather than quantitative.

The behavior of all six compartments was considered in constructing the curves. The extent of the interpretive fairing applied to the data is illustrated by Fig. 12, which presents the raw
data for three adjacent compartments covered by insulations with
different thermal performance. Each data point on the curves of
Fig. 12 represents every fourth tank boil-off cycle. The triangles
and vertical lines on the figure represent the times when the tank
was allowed to warm up for inspection and weekend shutdown. It is
readily apparent from a comparison of the boil-off histories of the
three compartments, that boil-off times for the better insulation
(compartment 3) tends to mirror the performance of the poorer performing adjacent insulations (compartments 2 and 4). It is only

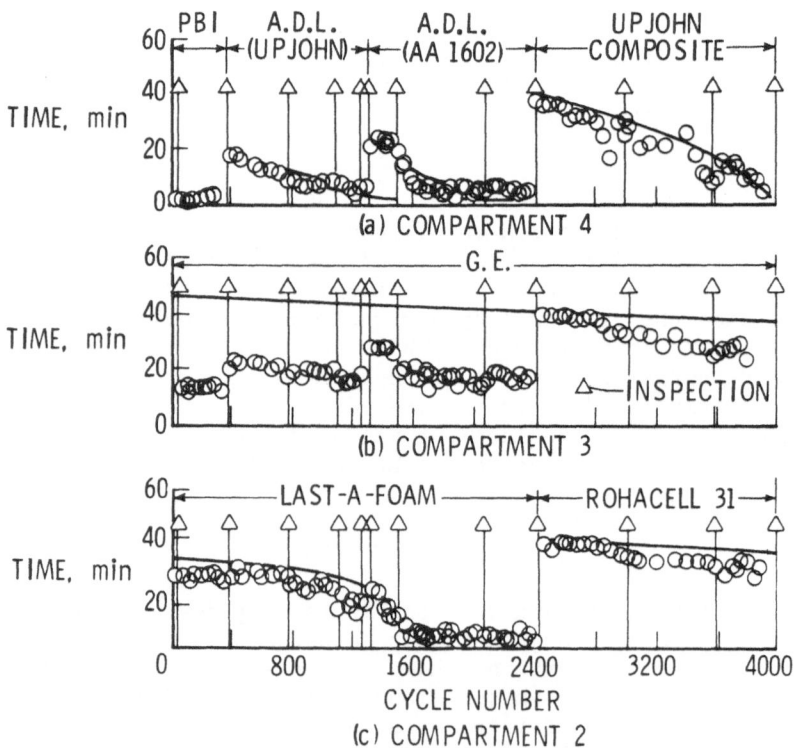

Fig. 12. Boil-off times for three adjacent compartments.

when insulations of comparable thermal performance are installed
on compartments 2 and 4 (at approximately 2400 cycles) that the
true performance of the insulation on compartment 3 begins to
emerge. It is inconceivable that the initial performance of the
insulation in compartment 3 was poorer than the performance at
2400 cycles. In fact, the thermal performance of closed-cell
Freon blown polyurethane foams decreases with time owing to the
diffusion of air into the insulation.[6] Hence the curve for the
GE foam is drawn as shown in Fig. 12b.

Since the heat going into better-insulated compartments from
poorly-insulated compartments decreased the boil-off time for the
better-insulated compartments and increased the boil-off time in
the poorly-insulated compartments, it is believed that the true
values of insulation performances are spread over an even greater
range, making the performance of the good insulations even better
and the performance of the poorer insulations even poorer.

RESULTS AND DISCUSSION

The results of the experimental investigation are summarized
in Table I. The table, which covers the initial 12 test periods,
indicates the location of the insulation specimens, the specimen
condition at the beginning and end of each test period, and the
number of warm-up cycles, thermal cycles, and hydrogen fill cycles.
Each test period entails a warm-up cycle; hence the test-period
designation indicates the number of warm-up cycles. Figure 13
presents boil-off times for the various insulations normalized by
the initial value for the best performing insulation.

Basic Foams

Polyurethane. Two polyurethane foams (Stepan Bx 250A, and
General Electric Polyurethane) exhibited the best overall performance. The thermal performance of these insulations was initially
excellent and degraded very slowly (see Fig. 13a). Both of these
insulations survived the entire test series (over 4200 thermal
cycles or the equivalent of approximately 15 years of airline
service), with no evidence of structural failure.

The third polyurethane specimen, Last-A-Foam, exhibited good
thermal performance for approximately 800 cycles (approximately
3 years of airline service) before experiencing a large degradation in thermal performance. The failure of the Last-A-Foam was
first detected by a significant increase in the hydrogen boil-off
rate. Visual examination of the warm insulation at that time
revealed only a few very fine tributary type cracks. When the
insulation was examined immediately after the next test period,

Table I. Test Summary

Test period	Number of days	Thermal cycles	Total cycles	LH$_2$ fills	Total fills	Compartments — Insulation specimen/initial condition-final condition (see codes)					
						1	2	3	4	5	6
1	1	11	11	4	4	1/N-G	2/N-G	3/N-G	4/N-G,F	5/N-G	6/N-G
2	3	360	371	81	85	1/G-G	2/G-G	3/G-G	4/G,F-G,F	5/G-M	6E/G-S 6W/G-G
3	4	403	774	83	168	1/G-G	2/G-G	3/G-G	7/N-F	8/N-J	6E/S-S 6W/G-G
4	3	335	1109	71	239	1/G-G	2/G-G	3/G-G	7/F-F	8/J-J,M	6E/S-M 6W/G-G
5	2	198	1307	47	286	1/G-G	2/G-G	3/G-G	7/F-F	8/J,M-J,M	6E/M-M 6W/G-G
6	0.5	5	1312	2	288	1/G-U	2/G-U	3/G-U	9/N-U	10/N-U	6E/N-U 6W/G-U
7	2	224	1536	47	335	1/U-G	2/U-S	3/U-G	9/U-F	10/U-J	6E/U-S 6W/U-G
8	5	497	2034	136	471	1/G-G	2/S-M	3/G-G	9/F-F	10/J-J,S	6E/S-S 6W/G-G
9	3.5	376	2409	104	575	1/G-G	2/M-M	3/G-G	9/F-F,J	10/J,S-J,S	6E/S-M 6W/G-S
10	5	616	3025	112	687	1/G-G	11/N-S	3/G-G	12/N-G 13/N-S,F	14/N-S	6E/N-G 6W/S-M
11	5	598	3623	123	810	1/G-G	11/S-S	3/G-G	12/G-S 13/S,F-M,F	14/S-S	6E/G-M 6W/M-M

Table I. Test Summary, continued

Test period	Number of days	Thermal cycles	Total cycles	LH$_2$ fills	Total fills	Compartments Insulation specimen/initial condition-final condition (see codes)					
						1	2	3	4	5	6
12	3	345	3968	72	882	1/G-G	11/S-S	3/G-G	12/S-S 13/M,F-M,F	14/S-S	6E/M-M 6W/M-M
13	4	431	4399	106	988	1/G-G	11/S-S	3/G-G	15/N-S*	14/S-S*	15/N-S*

Insulation Specimen Code – Name/Density, kg/m^3 (lbm/ft^3)

1 – Stepan Foam BX250A/37 (2.3)
2 – Last-A-Foam/63 (3.9)
3 – General Electric/68 (4.2)
4 – PBI/29 (1.8)
5 – Rohacell 41S/35 (2.2)
6 – Rohacell 51/50 (3.1)
 (6E-East Side; 6W-West Side)
7 – ADL System (Upjohn) 34 (2.1)

8 – Texthane 333/43 (2.7)
9 – ADL System (Stafoam)/43 (2.7)
10 – CPR-483-1/37 (2.3)
11 – Rohacell 31/30 (1.9)
12 – Upjohn without fibers/34 (2.1) East Side
13 – Upjohn with Fibers/35 (2.2) West Side
14 – Marvacell/96 (6)
15 – Marvacell/60 (3.7)

Condition Code

N – New
G – No visable cracks
S – Slight cracks
S* – Many small cracks

M – Major cracks
J – Cracks at "V" joint
F – Surface frost
U – Not inspected

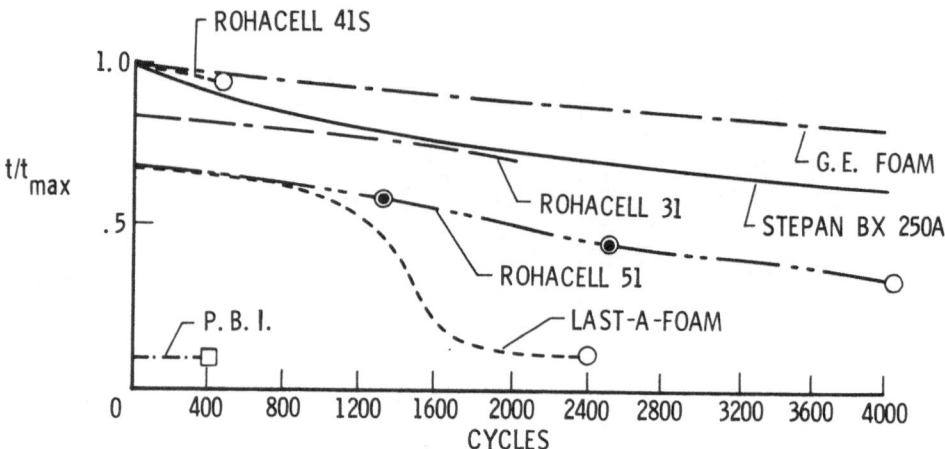

(a) Polyurethane, polymethacrylimide, and polybenzimidazole.

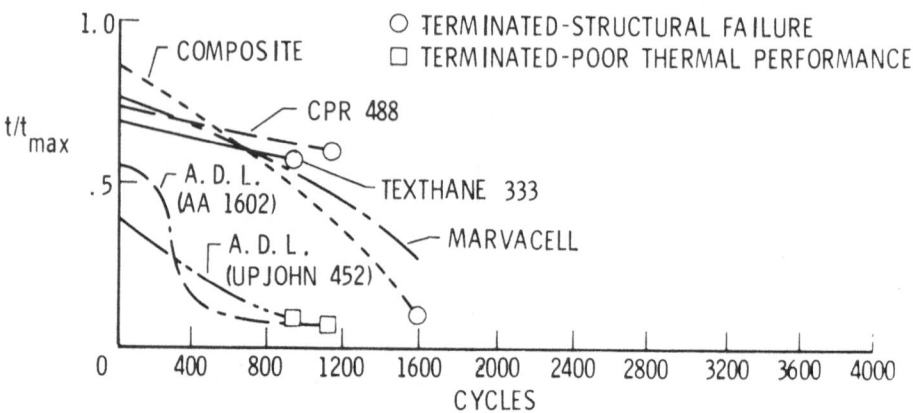

(b) Polyisocyanurate, isocyanate, and foam systems.

Fig. 13.   Hydrogen boil-off times for foam insulations as
a function of simulated flight cycles.

while the insulation was still cold, there was significant frost
buildup around these cracks as well as a stream of white vapor
flowing from these cracks. This suggested that the cracks propagated all the way to the tank surface and that air was cryopumping
to the tank surface. This was confirmed during sample removal, at
which time the sample separated along these cracks. Similar failure modes occurred for the polyurethane materials in Ref. 5.

Polymethacrylimide. Based on previous experience with cryogenic foams for a hypersonic application[5] and calculations that
indicated the highest margin of any of the foams between the ultimate stress of the foam and the anticipated thermal stress, the
polymethacrylimide foam insulations (Rohacell 31, 51, and 41S)
were leading candidates for the subsonic transport application at
the onset of the test program. However, the thermal cycle performance, as shown in Fig. 13a, was poorer than that shown for the
best polyurethane foams.

Rohacell 31, a 30 kg/m² (1.87 lbm/ft³) density foam, displayed
the best combined thermal and physical performances of polymethacrylimide materials. After the first few warm-up periods, short
hairline surface cracks were observed. However, as of the writing
of this paper, the Rohacell 31 specimen has sustained over 1600
thermal cycles with little degradation of the thermal performance.

The Rohacell 51 specimen cracked on one side (east side -
Table I) after (or during) the first 371 cycles. However, because
the thermal properties had not degraded significantly, the specimen was retained until 1296 cycles, at which time the cracked
side was removed and another Rohacell 51 specimen installed (the
other side was still unblemished and was retained). The next warm-
up cycle revealed that the new insulation specimen had also cracked,
apparently because of voids in the bond under the foam, but it was
not removed until it had undergone a total of 1104 cycles. A third
piece was bonded to the troublesome side and the cycling resumed.
After 1200 cycles on the new piece and a total of 3600 cycles on
the side with the original insulation (west side - Table I), the
insulation was cracked badly on both sides and its useful life was
over. Although the Rohacell 51 failed structurally, the thermal
performance of the insulation degraded slowly.

Rohacell 41S, which contains a flame retardant additive, was
badly damaged after 371 thermal cycles and therefore the specimen
was removed. The initial thermal performance appeared to be
excellent, but the structural failures were extensive.

The Rohacell foam insulations all failed in a similar manner.
The first indication was a curved hairline surface crack, which
had a very shallow inclination angle with respect to the surface
of the insulation. (Figs. 14a and 14b.) As the insulation was

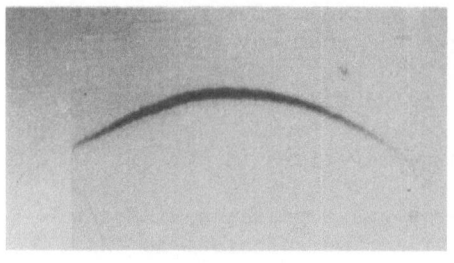

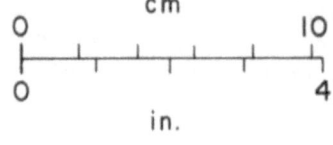

(a) Rohacell 51 (1108 cycles)

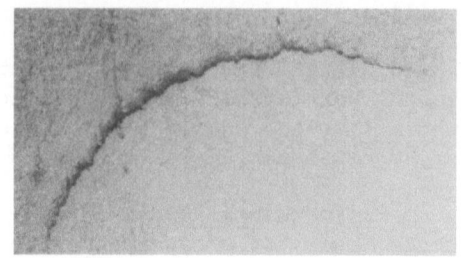

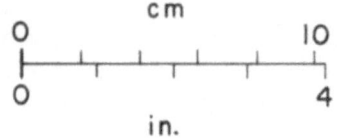

(b) Rohacell 41S (371 cycles)

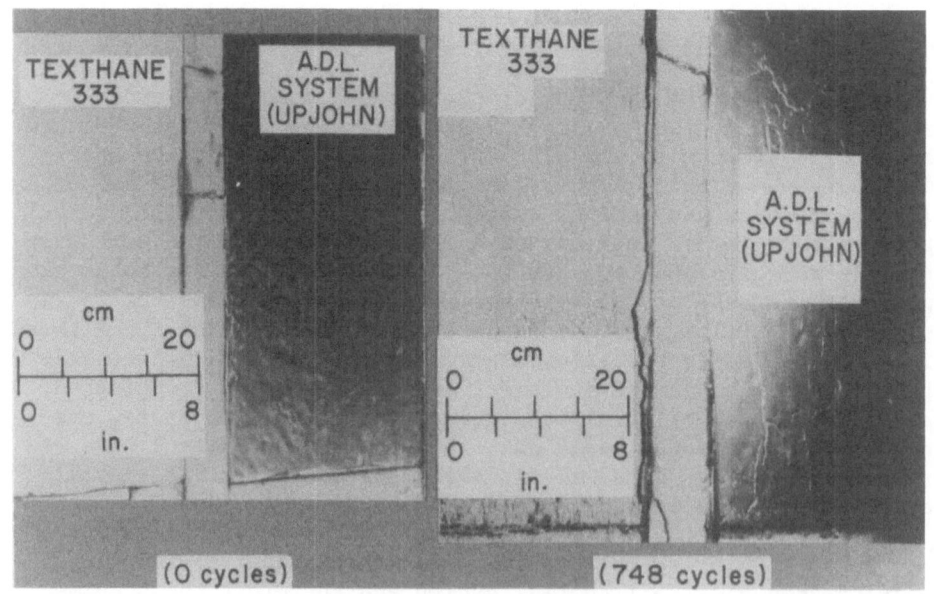

(c) Texthane and ADL (Upjohn)
    system (0 cycles).

(d) Texthane and ADL (Upjohn)
    system (748 cycles).

Fig. 14. Typical structural failures in foam insulations.

exposed to more thermal cycles, the crack grew in length and depth
and began to lift on the concave side of the crack until, after
repeated cyclic exposure, both ends of the crack met and a circular
crack was formed. The lack of an initial through crack to the tank
surface is consistent with the gradual deterioration of the thermal
properties of the polymethacrylimide foams.

Polybenzimidazole. The polybenzimidazole (PBI) material was
developed by NASA Ames Research Center to be used as a flame retardant material for helicopters. Previous experience[5] indicated
that the material is permeable. In an effort to seal the material,
two approaches were taken: a polyethylene sheet was bonded to the
outer surface of the insulation on one side of the tank and a cryogenic polyurethane adhesive, Crest 7410, was buttered on the outer
surface of the insulation on the opposite side of the tank. This
adhesive, which is the same as that used to bond the specimens to
the tank, formed a tough, tenacious skin. However, the thermal
performance of the polybenzimidazole material was still very poor,
by far the worst tested (see Fig. 13a). When the specimen was
examined after 371 cycles, it was found to be saturated with ice
crystals even under the polyethylene sheet. The performance of
this specimen was so poor (boil-off times less than 0.1 those of
the better insulations) that it affected the performance of all of
the specimens, especially those adjacent to it. Therefore, the
specimen was removed, even though no structural damage was
observed.

Modified Polyisocyanate. Marvacell is a modified isocyanate
foam which is nonflammable. A special high-density Marvacell is a
leading candidate for the insulation of the National Transonic
Facility, a cryogenic wind tunnel that is being constructed at the
Langley Research Center. Normally Marvacell is available in several densities and thicknesses; however, at the time of the investigation only 3.8 (1.5 in) thick, 96.3 kg/m^3 (6 lbm/ft^3) density
specimens were available. Consequently, the ρk product (mass
density x thermal conductivity) of this foam was high relative to
the other foams; furthermore, normalized boil-off times for this
insulation presented in Fig. 13b should be increased by a factor
of approximately 1.3 [ratio of specimen thickness to the standard
5.1 cm (2 in) specimen thickness] to provide an assessment of the
thermal performance of this insulation relative to the other insulations. Although the initial boil-off times were very high
(considering the thickness adjustment), the thermal performance
decreased markedly with thermal cycles and the insulation ultimately failed structurally after approximately 2000 thermal
cycles. Lighter densities of this material deserve further study
when they become available.

Polyisocyanurate. The two polyisocyanurate materials examined
in this study are currently the prime and backup insulations for

the single-use, throw-away LH_2 fuel tank for the boost stage of
the space shuttle. These two insulations, Texthane 333 and CPR 488,
exhibit good thermal performance (Fig. 13b). Both foam insulations
deteriorated structurally and had to be removed after a relatively
short time by aircraft standards. However, both materials survived
over 900 thermal cycles while maintaining fairly good thermal performance.

These foams were either poured or sprayed in layers. Their
failure was characterized by relatively wide and ragged cracks
along the 0.6 m (2 ft) edges of the specimen (see Fig. 14d) and
other smaller cracks that propagated under the surface of the
specimen into the interior. As the specimens were exposed to
repeated cycling, the width and depth of the cracks increased, but
no pieces of insulation separated from the main panel. Upon removal of the specimens from the apparatus, a slight handling load
caused the insulations to delaminate at the interfaces between
the poured layers. In addition, the insulation that was nearest
the tank wall was relatively spongy with a very low abrasive
resistance, suggesting a complete disintegration of the foam cells
or possibly a chemical change.

Insulation Systems

Two foam insulation systems prepared by the A. D. Little
Company of Cambridge, Massachusetts were also tested. These systems had two vapor barriers, one on the outer surface ($Z/Z_{max} = 0$)
and one at $Z/Z_{max} = 0.62$ (where Z is the distance into the insulation from the outer surface and Z_{max} is the insulation thickness).
Each vapor barrier was a laminate composed of one layer of mylar
[0.013 mm (0.0005 in) thick], two layers of aluminum [0.025 mm and
0.013 mm (0.001 in and 0.0005 in) thick], another layer of mylar
[0.013 mm (0.0005 in)], and a layer of dacron woven fabric
[33.9 g/m^2 (1.0 oz/yd^2)]. The two layers of mylar offer tensile
strength, the two layers of aluminum resist gas diffusion, and the
dacron cloth resists tearing. Both systems used foams which had
chopped fiberglass added for reinforcement. One system used Stafoam AA1602 (a toluenedi isocyanate), whereas the other used Upjohn
452 (a polymetric isocyanate). As can be seen from Fig. 13b, both
systems had fair thermal performance initially, but the performance
deteriorated rapidly with thermal cycles.

Visual examination of both insulation systems after a week of
cyclic testing showed that the specimens were completely covered
with frost within 7.6 (3 in) of the edge of the samples. The cold
surface was in complete agreement with the high boil-off rate recorded for these insulations. 'Initially the exterior vapor barrier
of the ADL Upjohn system appeared relatively smooth (Fig. 14c).

After the first set of cyclic tests, the vapor barrier was drawn
tight against the outer surface of the foam insulation (Fig. 14d)
and had a cratered appearance. This behavior suggests that the
insulation was permeable and some cryopumping was occurring. After
the specimen was removed from the test apparatus, no cracks were
detected; however, the insulation system was found to be permeable.

The ADL Stafoam system, which did not draw the vapor barrier
taut as did the Upjohn system, was found to be uncracked and non-
permeable after testing. Nevertheless, the thermal performance
was unsatisfactory.

In an effort to determine the effect of chopped fiberglass
reinforcement and vapor barriers on foam thermal performance and
strength, two specimens of Upjohn 452, a polymetric isocyanate,
were bonded to a single test compartment (one on each side). One
of the specimens had fiberglass reinforcement, but neither had a
vapor barrier. The thermal performance of this "composite specimen" is shown in Fig. 13b. Even though the foam in the composite
specimen cracked, the thermal performance was better than the ADL
Upjohn system. The fiberglass reinforced side cracked much more
than the unreinforced side. Furthermore, the reinforced side had
a great deal of frost on it, and it was steaming from cold air
after each test period while the unreinforced side had no frost.
Therefore, it was concluded that the fiberglass reinforcement degraded both the thermal and structural performance of the foam.
In contrast, vapor barriers, while not improving the thermal performance, apparently improved the structural integrity, since the
insulation specimens without a barrier cracked while the insulation systems which had barriers did not crack.

CONCLUDING REMARKS

Fourteen commercially available organic foam insulations were
examined to determine their suitability for insulating liquid hydrogen tanks of subsonic hydrogen fueled aircraft. Materials investigated were polyurethane, polymethacrylimide, polyisocyanurate, polymetric isocyanate, polybenzamidazole, toluenedi isocyanate, and
isocyanate foam. The test specimens included foams with chopped
fiberglass reinforcements, flame retardants, and vapor barriers.

Foam thickness was scaled to simulate stress conditions encountered by insulation on a large diameter tank. Insulation
specimens were bonded to a thin, flat aluminum tank, which had
separate compartments for six specimens. The tests were conducted
by filling the compartments with liquid hydrogen and exposing the
outer surface of the insulations to a cyclic thermal environment
representative of repeated subsonic aircraft flight. The boil-off
rate in each compartment indicated insulation thermal performance.

The thermal performance of all insulations deteriorated with increased flight cycles although, in some cases, the deterioration was negligible. Two unreinforced polyurethane foams survived over 4200 thermal cycles (representative of approximately 15 years of airline service) without evidence of structural deterioration. The polyurethane foam insulations also exhibited excellent thermal performance.

The addition of chopped fiberglass reinforcement or flame re-tarding materials during foam formulation proved harmful to thermal performance and/or the useful life of the foams. Vapor barriers had little influence on the thermal performance; however, they enhanced structural integrity.

Each generic foam type had a characteristic failure mode. Polyurethane foams exhibited fine hairline through cracks that grew in length and width; polymethacrylimide foams exhibited arc-shaped surface cracks that grew in depth and length; the polyisocyanurate foams became soft and mushy on the side bonded to the tank. Insula-tions poured or sprayed in layers failed at the interlayer boundaries.

REFERENCES

1. R.D. Witcofski, Alternate Aircraft Fuels - Prospects and Operational Implications, NASA TMS 74036 (May 1977).

2. G.D. Brewer, R.E. Morris, R.H. Lange, and J.W. Moore, Study of the Application of Hydrogen Fuel to Long-Range Subsonic Trans-port Aircraft, NASA CR-132559 (January 1975).

3. G.D. Brewer, LH2 Airport Requirements Study, NASA CR-2700 (March 1976).

4. Anon, An Exploratory Study to Determine the Integrated Tech-nological Air Transportation System Ground Requirements of Liquid-Hydrogen-Fueled Subsonic, Long-Haul Civil Air Transports, NASA CR-2699 (May 1976).

5. R.G. Helenbrook and J.Z. Colt, Development and Validation of Purged Thermal Protection Systems for Liquid Hydrogen Fuel Tanks of Hypersonic Vehicles, NASA CR-2829 (June 1977).

6. R.R. Dixon, L.E. Edelman, and D.E. McLain, "Effect of Aging in Thermal Conductivity of Cellular Materials," Journal of Cellu-lar Plastics (Jan.-Feb. 1970), p. 44.

STRENGTH SPECIFICATIONS AND METHODS OF TEST FOR STRUCTURAL

GRADES OF CRYOGENIC BALSA WOOD INSULATION

G. E. Padawer

Cabot Corporation
Billerica, Massachusetts, U.S.A.

INTRODUCTION

Balsa wood is a useful material for cryogenic applications owing to its relatively high structural strength and good thermal insulating properties that increase with decreasing temperatures. Balsa wood insulation has been used in overland transporters of cryogenic liquids,[1] in cargo holds of LNG tanker ships,[2] and in land-based LNG storage tanks.[3] The construction and operation of such cryogenic liquid containers in many instances are subject to international, national, or local governmental regulations or industry codes, intended to insure sound industrial practice and public safety. Such codes or regulations typically include specifications for minimum structural properties of the materials of construction and frequently require mechanical testing of coupon specimens to demonstrate code compliance.

Some difficulties arise when the material requirements, or the material test methods, must be specified for balsa wood because its physical properties are uniquely complex. Structural components typically are fabricated from a multiplicity of laminated strips or planks. These are selected and combined in such a manner that the mechanical properties averaged over the full size component will meet or exceed certain specified values. The quality assurance and quality control (QA/QC) procedures for making sure that this is the case are not simple, however, It will be shown later that conventional test coupons complying with recommended ASTM Standard Methods of Test[4] are likely to exhibit mechanical properties that are far from being representative of the whole laminate. Conventional methods of material testing,

231

inspection, and qualification thus are not generally applicable
for structural components made of balsa wood.

In this paper, the thermal and mechanical characteristics of
balsa wood and balsa wood laminates are reviewed, and it is shown
that "composite" mechanics that have been developed for the class
of synthetic fiber-reinforced plastic (SFRP) materials may be useful for describing the density and direction--dependent mechanical
properties of balsa wood in bulk or laminated form. It may be
asked whether such advanced analytical methods, perhaps combined
with specially developed methods of test, could be used effectively
towards developing more applicable QA/QC procedures that will
clearly qualify balsa wood as a structural material in applications
where strictest code compliance is a necessity. This question has
prompted the following review and discussion.

THE THERMAL AND MECHANICAL PROPERTIES OF
SELECTED CRYOGENIC INSULATING MATERIALS

Representative values of the compressive strength and thermal
conductivity (k-factor) of five selected insulating materials have
been listed in Table I. The k-factors of the foam type of thermal
insulators in general tend to be lower, by a factor of two to four,
than k for balsa wood. A higher k-factor can be compensated for
by increasing the thickness of the insulating layer, of course. A
much greater difference is apparent when the mechanical strengths
listed in Table I are compared. The compressive strength (parallel
to the grain) of balsa wood at room temperature is on the order of
ten times higher than that exhibited by the foam-type products.
In the low temperature regime, ultimate strengths at 90 K (-300°F)
have been reported as 1.25 times the strength at 300 K (80°F), for
compression both parallel and transverse to the grain.[10] In contrast, the compression strengths of the foam type of insulators
exhibit slight increases, or slight decreases, at low temperature.
It may be deduced from this comparison that in applications where
mechanical strength as well as thermal insulation at cryogenic
temperatures are required, and where physical properties alone are
the governing criteria for material selection, balsa wood might be
the preferred choice.

EXPERIMENTAL MEASUREMENTS OF THE
MECHANICAL PROPERTIES OF BALSA WOOD

The mechanical properties of balsa wood are more complex than
those of more common structural materials. In some respects, balsa
wood resembles the class of synthetic fiber-reinforced plastic
(SFRP) composites in that its strength and stiffness are highly

Table I. Representative Mechanical and Thermal Properties of Selected Insulating Materials.

Material type*	Typical ambient temperature k-factor,† W/mK ($\frac{Btu-in}{h-ft^2-°F}$)	Typical low temperature k-factor,‡ W/mK ($\frac{Btu-in}{h-ft^2-°F}$)	Typical density, kg/m³ (pcf)	Typical compression strength, MPa (psi)	Reference§
Open-cell poly-phenylene foam	0.087 (0.60)	0.100 (0.72)	40 (2.5)	1.00 (150)	5
Closed-cell poly-urethane foam	0.020 (0.14)	0.015 (0.10)	50 (3.0)	0.175 (25)	6
Closed-cell poly-styrene foam	0.022 (0.15)	#	58 (3.6)	0.860 (125)	7
Closed-cell glass foam	0.055 (0.38)	0.030 (0.20)	136 (8.5)	0.800 (120)	8
Balsa wood parallel to grain	0.090 (0.62)	0.050 (0.35)	96 (6.0)	5.20 (750)	9, 10
	0.120 (0.83)	0.074 (0.51)	176 (11.0)	13.0 (1910)	
Balsa wood perpendicular to grain	0.036 (0.25)	0.017 (0.12)	96 (6.0)	0.55 (80)	9, 10
	0.050 (0.35)	0.023 (0.16)	176 (11.0)	1.0 (150)	

* Listed by increasing density.

† Measured at 240 to 275 K (0 to 40°) mean temperature.

‡ Measured at 115 to 145 K (−250 to −200°F) mean temperature.

§ Numbers in brackets refer to the References following the text.

\# No data available.

anisotropic and depend almost linearly on its density analogous
to the SFRP materials whose mechanical properties are strong functions of both the direction and the volume fraction of the reinforcing fibers. Furthermore, balsa wood is a naturally grown
product and exhibits an unusually wide range of variability in its
properties. A random sampling from a given grade or lot may exhibit
strengths ranging over a factor of 6 or 7. This is not necessarily
evidence of poor material, however. It is, rather, the expected
mode of behavior for this particular structural material.

It is convenient, when discussing the elastic properties of
wood, to describe it as having orthorhombic symmetry, where the
longitudinal (L, parallel to the grain), the radial (R, normal to
the growth rings), and the tangential (T, tangent to the growth
rings) directions are the three principal axes of symmetry. This
is an idealization, of course. Wood is neither perfectly elastic
(hence its excellent acoustical damping characteristics) nor homogeneous (see, for example, H. Hoerig's discussion[11] of the quasi-
periodic inhomogeneity resulting from the seasonal variation of
the tree growth rate). It should be noted further that the L, R, and
T directions are ideally orthogonal only with respect to cylindrical coordinates, so that "orthorhombic symmetry" could be ascribed
at best only to quarter-sawn planks. For planks cut from the log
in any other way, the R and T directions rotate continuously with
respect to the plank face.

Experimental measurements of the elastic and cohesive properties of wood nevertheless have been found to compare reasonably
well with calculations based on the orthorhombic model. C. F.
Jenkin in 1920,[12] H. Carrington in 1921,[13] H. Hoerig in 1931,[14]
R. F. S. Hearmon and W. W. Barkas in 1941,[15] and R. F. S. Hearmon
in 1943,[16] among many others, reported such investigations on
varieties of spruce and beech. Of greater interest here are the
data for balsa wood reported by C. A. Wiepking and D. V. Doyle
in 1944,[17] D. V. Doyle et al. in 1945,[18] and by P. D. Soden and
R. D. McLeish in 1976,[19] and these will be reviewed briefly.

Balsa wood for structural uses was found to range in density
from 60 to 240 kg/m^3 (3.7 to 15 pcf). [The variation among
individual planks comprising a structural component is usually
held to narrower limits, for example, 120 to 168 kg/m^3 (7.5 to
10.5 pcf) for a footing block of 144 kg/m^3 (9 pcf) average density.][20] It was noted that "...large variations of density are
found not only between pieces from different locations in the tree
but also within the individual piece...."[17] Experimental measurements established that both the cohesive and the elastic properties
of balsa wood were approximately in linear proportion to the average density of the specimen, for all directions of loading. Wiepking and Doyle[17] formulated empirical expressions of the form

$$\sigma_{ij} = a_{ij}D + b_{ij} \tag{1}$$

$$C_{ij} = c_{ij}D + d_{ij} \tag{2}$$

to describe the mechanical properties of balsa wood, whereas Soden and McLeish,[19] incorporating the results of Doyle et al.[18] with their own experimental results, preferred the exponential form

$$\sigma_{ij} = e_{ij}D^{1.5} \tag{3}$$

$$C_{ij} = f_{ij}D^{1.5} \tag{4}$$

where the σ_{ij} are the cohesive strengths, C_{ij} are the engineering constants of elasticity E or G, D is the specimen density, and the a_{ij}, b_{ij} ....f_{ij} are material constants. Table II summarizes the numerical results of both investigating teams.

Discussion of Empirical Results

The exponential and linear forms of expressing the density dependence of balsa wood properties are nearly equivalent in the range of interest, i.e., $60 < D < 240$ kg/m^3. The exponential form realistically predicts zero values for all the properties in the limit D = 0, but the linear form is somewhat simpler. Both formulations represent best fit for the available data points which exhibited ±20 to 30 percent scatter.

Some of the discrepancies between the Wiepking-Doyle data and the Soden-McLeish data may have been systematic. The Wiepking-Doyle (linear) results[17] were based on a sample population that uniformly had a 12 percent moisture content. The Soden-McLeish (exponential) results,[19] however, were derived in part from the Doyle et al. work[18] based on samples with 9.5 percent moisture content, and in part on their own experiments for which no moisture contents were reported. It is difficult to tell whether these variations in percent moisture could have biased the results significantly. No moisture-related effects on the properties of balsa wood have been evaluated explicitly in any of the literature references surveyed here. To the extent that moisture contents were reported at all, they ranged generally from 8 to 12 percent. This range appears to be a normal condition for kiln-dried balsa wood exposed to ambient atmosphere in temperate climates. (Note: The ASTM Standard Method of Testing Small Clear Specimens of

Table II. Summary of Density-Dependent Mechanical Properties of Balsa Wood (Empirical Formulation).

Property[*]	Exponential form (after Ref. 19), MPa[†]	Linear form (after Ref. 17), MPa[†]
Compression modulus parallel to grain, E_ℓ	$(1.70)\ D^{1.5}$	$(38.4)\ D - 1516$
Shear modulus parallel to grain, $G_{\ell t}$	[‡]	$(1.70)\ D - 68.6$[§]
Compression strength parallel to grain, σ_ℓ	$(6.63 \times 10^{-3})\ D^{1.5}$	$(1.05 \times 10^{-1})\ D - 4.24$
Compression strength perpendicular to grain, σ_t	$(3.97 \times 10^{-4})\ D^{1.5}$	$(4.75 \times 10^{-3})\ D$
Shear strength parallel to grain, $\tau_{\ell t}$	$(1.55 \times 10^{-3})\ D^{1.5}$	$(1.39 \times 10^{-2})\ D - 0.165$

[*] The longitudinal "ℓ" direction is identical with the orthotropic L axis. The transverse "t" direction (or "ℓt" plane) corresponds to the average of the orthotropic R and T directions (or LR and LT planes).

[†] The density D is in kg/m^3.

[‡] No data available.

[§] The shear modulus values were taken from Ref. 18.

Timber,[4] in Section 22.5, refers to "an equilibrium moisture content of approximately 12 percent, for most species." For balsa wood, the "equilibrium" value could be somewhat lower.)

A more significant bias was probably introduced by divergent methods of testing. Soden and McLeish obtained their shear strength values from slender tensile specimens loaded along an axis inclined 7.5 or 15 degrees to the grain direction, whereas Wiepking and Doyle employed the shear tool method specified in Section 89 of

ASTM D-143.[4]   Lower nominal shear strength values should be expected from the latter method owing to the nonuniform stress distribution across the sheared surface, and this is borne out in fact when
the shear strength values predicted by the equations in Table II
are compared.

The formulations in Table II, in any case, predict the average
behavior of a sample population that exhibited a large variation.
To obtain an estimate for the lower bound, the nominal values
should be reduced by 30 percent, at least. Structural designers
have accounted for this deficiency by employing conservatively
large factors of safety (F.S.), for example, F.S. $\simeq$ 14 for the
bearing strength and F.S. $\simeq$ 7 for the shear strength, in the actual
design of balsa wood footing blocks for a large LNG tank (for
details see Ref. 20).

To summarize the density dependence of the more important
mechanical properties of balsa wood, a graphic representation of
the Soden-McLeish (exponential) equations is shown in Fig. 1. The
width of the bands is shown as approximately ±30 percent of nominal,
for illustrative purposes. In practice, proof testing of full-
scale components has been relied upon to eliminate errors on the
unconservative side.[21]

THEORETICAL MECHANICS

A Brief Historical Review

The use of theoretical mechanics for describing the mechanical behavior of wood has a long history. As pointed out earlier,
C. F. Jenkin[12] calculated the effect on the apparent modulus of
the angle between load and fiber directions, based on a simple
two-dimensional analysis, in 1920. He found fair agreement with
experimental results. In 1928, W. Voigt[22] published the three-
dimensional transformation formulae for axis rotation in orthorhombic crystals, and A. T. Price[23] described wood as an homogeneous, elastic, orthotropic continuum whose elastic properties
could be specified by nine independent constants. Three years
later (in 1931), H. Hoerig[11] adopted Voigt's transformation matrix
for orthorhombic axis rotation to calculate the variations in the
elastic constants resulting from arbitrary rotations of the load
direction with respect to the principal L, R, and T axes of the
wood grain structure. It was Hoerig who in 1931 first gave the
formula for calculating the off-axis extensional modulus E_θ.

$$\frac{1}{E_\theta} = \frac{1}{E_\ell} \cos^4\theta + \frac{1}{E_t} \sin^4\theta + (\frac{1}{G_{t\ell}} - \frac{2\mu_{\ell t}}{E_\ell}) \sin^2\theta \cos^2\theta \qquad (5)$$

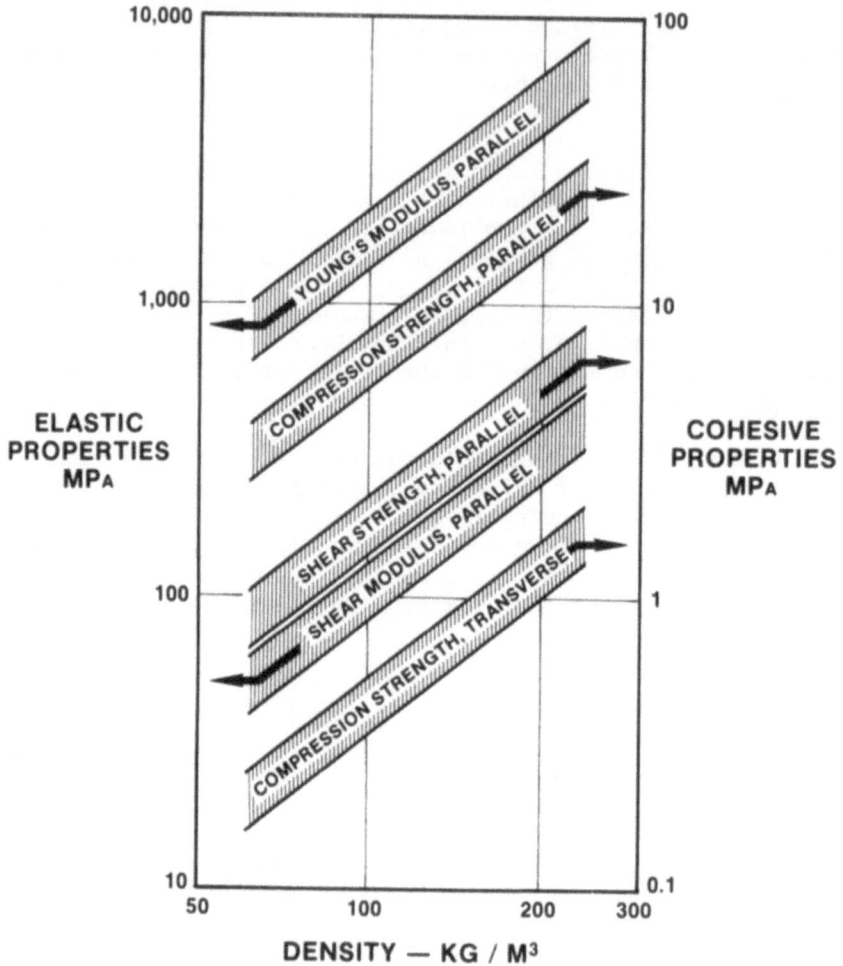

Fig. 1. The density dependence of the mechanical properties of
         balsa wood (schematic).

which is a familiar expression in contemporary composite mechanics.
Major contributions to the mechanics of anisotropic continua were
made by R. F. S. Hearmon and co-workers in the 1940's, for example,
by emphasizing the significance of coupling between shear and extension in cases where the load direction is not parallel to any
of the three principal orthotropic axes, and by investigating the
elastic properties of plywood (an early form of the "laminated
composite structure").

In the 1960's, there was a resurgence of interest in "composite mechanics," possibly stimulated by the needs of a rapidly growing aerospace technology, and also perhaps by the development of new high-strength, low-mass, fibrous reinforcement and compatible polymeric matrix materials. Important advances were achieved in the area of adopting the classical tensor calculus for computation by EDP methods whereby complex multilayer and multidirectional composite plate structures could be designed and analyzed. New developments have included methods for analyzing unsymmetrical ("unbalanced") composite configurations, and for computing thermal stresses, interlaminar shear phenomena, and free edge effects. The names of the contributors to modern composite mechanics are too numerous to mention. References to their work may be found in at least one of four relatively recent monographs: L. R. Calcote;[24] S. W. Tsai, J. C. Halpin, and N. J. Pagano;[25] J. E. Ashton, J. C. Halpin, and P. H. Petit;[26] and J. E. Ashton and J. M. Whitney.[27]

The purpose of this brief, selective, and certainly incomplete review has been to point out that applying the powers of "composite mechanics" for the analysis of wood, specifically balsa wood, is not a new notion. Modern "composite mechanics," in fact, have been derived largely from earlier work that was done to gain a better understanding of wood used as a high-performance structural material.

Practical Applications

It has been mentioned earlier that Soden and McLeish[19] employed the classical tensor calculus to compute the effects of axis rotation and thereby to deduce the shear strength of balsa wood from the results of off-axis tensile tests.

Similar computational methods may also be employed to check the veracity of experimental work. This is useful when large scatter in the data makes them doubtful. For example, orthotropic symmetry considerations ideally require that

$$E_i \, \mu_{ji}/E_j \, \mu_{ij} \equiv 1 \tag{6}$$

for i and j = L, R, T. (See, for example, equation 1-54 in Ref. 27.) The experimental measurements of elastic moduli and Poisson's ratios of balsa wood, as reported by Doyle et al.,[18] may be checked in this way. For a speciment population whose density ranged from 100 to 120 kg/m^3, they found

$$\frac{E_L \, \mu_{RL}}{E_R \, \mu_{LR}} = 1.52, \qquad \frac{E_R \, \mu_{TR}}{E_T \, \mu_{RT}} = 1.13, \qquad \frac{E_T \, \mu_{LT}}{E_L \, \mu_{TL}} = 0.978 \tag{7}$$

The second and third relationships show reasonable agreement with
the ideal orthotropic symmetry; the first shows considerable divergence. Divergence may be related to the difficulty of measuring
μ_{RL} and μ_{LR} in specimens having the usual shape of a right prism,
owing to the natural curvature of the growth rings.

Balsa wood laminates are generally designed to minimize differential dimensional changes caused by variable temperature or
humidity conditions. In practical terms, this means that on
assembly, the R and T orientations of individual planks are rotated
with respect to their four nearest neighbors to the extent that
this is practicable. Considering the laminate as a whole, the R
and T directions are thus nearly randomized. The elastic response
of a laminated component therefore approaches "plane-isotropic"
symmetry, analogous to uniaxial fiber-reinforced composites. The
elastic properties may then be described simply in the two dimensions ℓ and t, where ℓ (longitudinal) corresponds exactly to L,
but t (transverse) is the average of R (radial) and T (tangential).
It may be noted that Soden and McLeish[19] found considerable overlap
of the observed R and T data, and combined the results to obtain
"across the grain" parameters σ_t, E_t, and $G_{\ell t}$ (see also Table II).

In our laboratory, we have measured the apparent shear stiffness $G_{\ell t}$ of balsa wood specimens with D = 112 kg/m^3 by means of
the experimental method described on page 37 of Ref. 24. Using
three prismatic specimens of balsa wood cut so that the grain
directions were inclined 0°, 45°, and 90° to the specimen axis and
bonding strain gages to their prism faces, we measured E_ℓ, E_t, $\mu_{\ell t}$,
and E_{45}. The apparent shear modulus $G_{\ell t}$ was then computed by
equation (5) with θ = 45°. The result was

$$G_{\ell t} \text{ (apparent)} = 137 \text{ MPa (19.9 ksi)} \tag{8}$$

This number was in good agreement (probably fortuitously) with the
data of Doyle et al.[18] who measured (for D = 112 kg/m^3) G_{LR} = 154 MPa,
G_{LT} = 120 MPa, average = 137 MPa.

These examples illustrate that "composite" mechanics can be
an appropriate tool for describing, testing, or qualifying balsa
wood and balsa wood laminates.

STRENGTH SPECIFICATIONS AND METHODS OF TEST

The Limitations of Standard Test Coupon

The ASTM standard[4] coupon size for the compression tests is
50 x 50 x 200 m (2 x 2 x 8 in) and for the shear tests it is
50 x 50 x 62.5 mm (2 x 2 x 2.5 in). These dimensions coincide more

or less with the size of the individual balsa planks that make up a typical laminate. As a result, individual test coupons, taken from random locations of a structural laminate, are likely to exhibit crushing or shear strengths representative of the individual planks, rather than of the whole assembly, and the test results might be significantly higher or lower than the specified "minimum values" that refer to the behavior of the full size component. It is evident that erroneous information may be generated by ordinary coupon testing procedures, unless this size effect is accounted for.

Possible Alternative QA/QC Procedures

It may be possible to avoid the dilemma created by a conventional interpretation of the test results by expressing the stiffness or the strength requirements for balsa wood test samples not as fixed minimal values, but rather as lower-bound variables having a functional dependence on the specimen density. Such criteria could have the linear form of eqs. (1) and (2), or the exponential form of eqs. (3) and (4). The numerical constants should not represent the average behavior of a typical sample population as in Table II, however. Instead, these constants should be scaled lower, perhaps to coincide with a lower confidence limit (say, the 99.5 percent limit) that has been established by a sufficiently large baseline sample population. Alternatively, these constants perhaps could define a limit that is conservatively much lower than any sample behavior, analogous to the "allowable stresses," always lower than the yield or failure stresses that are commonly written into construction codes.

It should be recognized, however, that the above described density-dependent lower-bound criteria may not necessarily constitute valid "go, no-go" tests by which the mechanical strength of balsa wood laminates may be qualified as "acceptable" or "not acceptable." Suppose, for example, that a particular test specimen were to exhibit a low crushing strength relative to its own density, that is to say, it were to fail the density-dependent minimum strength criterion of eq. (1) or (3). This would by no means indicate that the structural component as a whole was deficient. It would need only one other plank laminated somewhere into the component whose crushing strength exceeded, by a like amount, the lower bound that is appropriate to its own density, for the component as a whole to meet the original design specifications. For fiber-reinforced laminates, this difficulty is usually eliminated by testing fiber bundles, or specially prepared uniaxial test laminates, rather than individual fibers. It may be desireable to develop analogous test procedures for balsa wood laminates.

SUMMARY AND CONCLUSION

It has been shown that balsa wood may be a preferred material for cryogenic applications, owing to its structural strength and good insulating properties. It has been pointed out, however, that balsa wood is unique among other more conventional materials of construction, because the strength of individual elements that comprise a laminate may vary considerably above or below the nominal strength assigned to the assembled component. This may create some difficulties when coupon tests are needed for QA/QC or for code compliance purposes. It has been suggested that some formulations that systematically relate the various strength and stiffness parameters to the specimen density may be a useful tool for establishing reliable QA/QC procedures for balsa wood, and that state-of-the-art "composite" mechanics could be employed to put such formulations rigorously on a scientific basis.

REFERENCES

1. R.R. Desai, SPI Reinf. Plast. Comps. Proc. 27th Annu. Tech. Conf., (1972), Sect. 7-A.

2. R.C. Ffooks, in: Advances in Cryogenic Engineering, Vol. 19, Plenum Press, New York (1974), p. 269.

3. Anon, Pipeline and Gas J. 199, 62 (1972).

4. D 143-52, Annual Book of ASTM Standards, Part 10, ASTM, Philadelphia, Pennsylvania (1975), p. 34.

5. R.E. Tatro and F.O. Bennet, Jr., in: Advances in Cryogenic Engineering, Vol. 20, Plenum Press, New York (1975), p. 315.

6. J. Navickas and R.A. Madsen, in: Advances in Cryogenic Engineering, Vol. 22, Plenum Press, New York (1977), p. 233.

7. Industrial Pipe, Tank, and Vessel Insulations, Product Bulletin No. 172-417-70, The Dow Chemical Company, Construction Materials Division, Midland, Michigan (1970).

8. R.W. Gerrish, in: Advances in Cryogenic Engineering, Vol. 22, Plenum Press, New York (1977), p. 242.

9. A. Lippay, Boat Construction and Maintenance, 21 (Nov. 1965).

10. Thermal Conductivity of Certified Kilndried Belcobalsa Structural Insulation, Product Data Sheet No. 5-C, Baltec Corporation, Cryogenics Division, Northvale, New Jersey (1962).

11. H. Hoerig, Z. Tech. Phys. 12, 369 (1931).

12. C.F. Jenkin, Report on the Materials of Construction Used in Aircraft and Aircraft Engines, H.M. Stationery Office, London (1920) - Quoted in 15.

13. H. Harrington, Philos. Mag. 6, 206 (1921) - Quoted in 11.

14. Forest Products Laboratory, Wood Handbook, No. 72, U.S. Department of Agriculture, Madison, Wisconsin (1955).

15. R.F.S. Hearmon and W.W. Barkas, Proc. Phys. Soc. 53, 674 (1941).

16. R.F.S. Hearmon, Proc. Phys. Soc. 55, 67 (1943).

17. C.A. Wiepking and D.V. Doyle, Strength and Related Properties of Balsa and Quipo Woods, Report No. 1511, U.S. Department of Agriculture, Forest Products Laboratory, Madison, Wisconsin (1944).

18. D.V. Doyle, J.T. Drow, and R.S. McBurnery, Elastic Properties of Wood - the Young's Modulus and Poisson's Ratio of Balsa and Quipo, Report No. 1528, U.S. Department of Agriculture, Forest Products Laboratory, Madison, Wisconsin (1945).

19. P.D. Soden and R.D. McLeish, J. Strain Anal. 11, 225 (1976).

20. G.E. Padawer, in: Advances in Cryogenic Engineering, Vol. 21, Plenum Press, New York (1977), p. 315.

21. C. Bernhardt, Preload Technology, Inc., Garden City, New York, private communication.

22. W. Voigt, Lehrbuch der Kristallphysik, 1st Edition, Teubner, Leipzig (1910), p. 574 - Quoted in 11.

23. A.T. Price, Philos. Trans. A. 228, 659 - Quoted in 15.

24. L.R. Calcote, The analysis of Laminated Composite Structures, Van Nostrand, New York (1969).

25. S.W. Tsai, J.C. Halpin, and N.J. Pagano, eds., Composite Materials Workshop, Technomic, Stamford, Connecticut (1968).

26. J.E. Ashton, J.C. Halpin, and P.H. Petit, Primer on Composite Materials: Analysis, Technomic, Stamford, Connecticut (1969).

27. J.E. Ashton and J.M. Whitney, Theory of Laminated Plates, Technomic, Stamford, Connecticut (1970).

CARBON-FIBRE-REINFORCED CARBON COMPOSITES: PROCESSING,
ROOM TEMPERATURE PROPERTIES, AND EXPANSION BEHAVIOUR AT
LOW TEMPERATURES

W. Fritz and W. Hüttner

Institut für Chemische Technik
Karlsruhe, West Germany

and

G. Hartwig

Institut für Technische Physik
Karlsruhe, West Germany

INTRODUCTION

The idea of carbon-fibre-reinforced carbon (C/C) composites
was born in the beginning of the last decade. Due to the outstand-
ing physical and chemical properties of carbon, such as high heat
of ablation, thermal shock resistance, strength improvement at
high temperatures and chemical inertness, C/C composites are used
as a high performance material for ablation and high temperature
application in sophisticated engineering.

As shown in the cross-sectional light micrograph of Fig. 1
(left side), carbon/carbon composites consist of a fibrous carbon
substrate in a carbonaceous matrix. The composite behaviour,
however, is very complex, although both constituents consist of the
same element. There exists a number of parameters that influence
the composite properties. One parameter group is dominated by the
carbon fibres like the type of carbon fibres selected for reinforcement (high modulus = type I or high tensile = type II), the fibre
volume fraction, and the fibre arrangement. A schematic of three
standard concepts of reinforcing fibre arrangements with continuous
carbon fibres, 1-D, 2-D, and 3-D, is illustrated in Fig. 2.[1] The
resulting mechanical properties of the composite correspond to

245

(a)   (b)

Fig. 1. Unidirectionally reinforced C/C composites with
         pitch as matrix precursor (~60V/o fibres).

         (a) light microscopy of a polished cross section.
         (b) free-standing unidirectionally reinforced
             C/C-composite bodies.

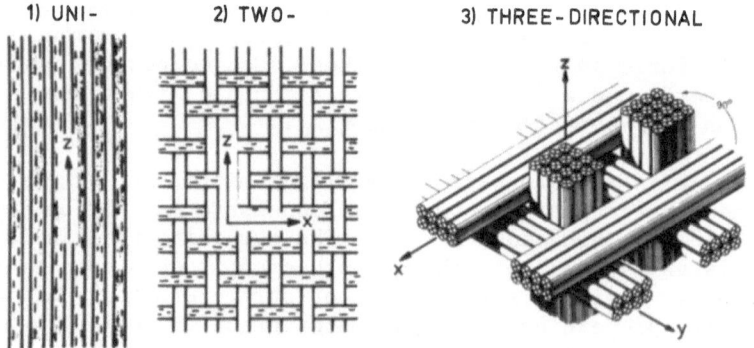

Fig. 2. Schematic of three standard concepts of fibre arrange-
         ments in C/C composites.[1]

         C/C-composite strength according to the rule of mixtures:

$$\sigma_{c(z)} \approx V_{f(z)} \cdot \sigma_f \qquad \sigma_{c(z,x)} \approx V_{f(z,x)} \cdot \sigma_f \qquad \sigma_{c(x,y,z)} \approx V_{f(x,y,z)} \cdot \sigma_f$$

σ = strength V_f = fibre volume fraction
c = composite   f   = fibre

the rule of mixtures.   3-D woven preforms are special fibre arrangements, fabricated from continuous carbon fibres in a special weaving
technique.   Several weave modifications to the basic 3-D orthogonal
design are available to form a highly isotropic woven structure.
Woven preforms with 5, 7, or 11 fibre directions are available.[2]

The second parameter group that influences the composite properties is dominated by the carbonaceous matrix.   The matrix structure can range from carbon to graphite and is influenced by the
selection of matrix precursors, the processing conditions, and the
temperature treatment of the C/C composite.

CARBON/CARBON COMPOSITE PROCESSING

Figure 3 illustrates the principal processing steps for
fabricating carbon-fibre-reinforced carbon composites.   The continuous C-fibre substrates, if not available as self-supporting frames
like 3-D fabrics, are impregnated with an organic matrix precursor
like pitch or resin by a wet-winding process, followed by molding,
pressing, and curing of the laminates to get the fibres fixed in
the selected arrangement and to regulate the fibre volume fraction.
Then the carbonization is accomplished.   Due to the weight loss and
volume shrinkage of the organic carbon matrix precursors during the
pyrolysis, a porous C/C-skeleton frame results, which has to be

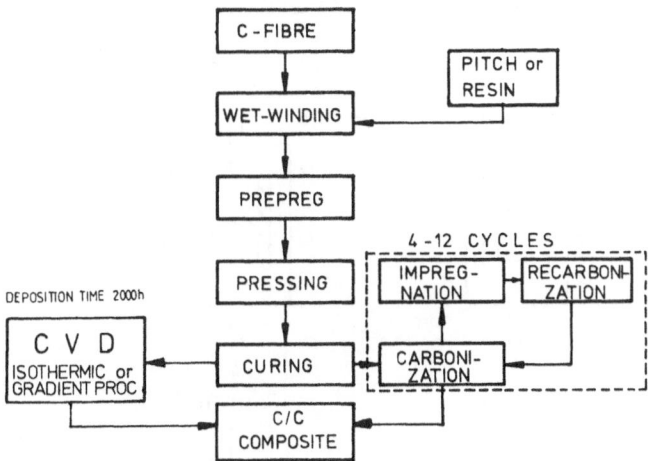

Fig. 3.   Flow chart of C/C-composite processing.

densified. For the densification of the porous composite, there
are two basic techniques: (1) the chemical vapour deposition (CVD)
of carbon from a hydrocarbon gas (e.g., methane or benzene) and
(2) multiple impregnations and recarbonizations with a liquid
organic resin or pitch.

CHEMICAL VAPOUR DEPOSITION PROCESS (CVD)

     The CVD technique is based on the diffusion of a carbon-
bearing gas into a porous substrate and the diffusion-controlled
reaction of the gas on the hot substrate into carbon and gaseous
by-products. A variety of parameters (such as temperature, pressure, flow rate, substrate density, and surface area) affect the
quality and structure of the deposited carbon and the deposition
presents a difficult and complex problem. The present deposition
conditions are based on empirical observations.[3] Various CVD techniques have been developed for the infiltration of porous substrates. The most extensively utilized are the isothermal and the
temperature gradient processes.[4,5,6] The isothermal process technique is illustrated schematically in Fig. 4a. The substrate is
heated by an induction-heated susceptor (graphite). If the hydrocarbon gas comes in contact with the hot substrate, carbon is
deposited and gaseous by-products (mainly H_2) are released. The
disadvantage of this technique is the overcrusting of the outer

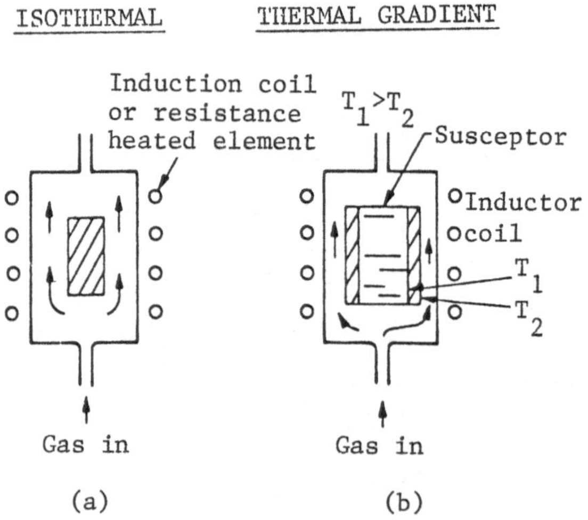

Fig. 4.   Schematic of chemical vapour deposition
          (CVD) apparatus.

          (a) isothermal deposition process.
          (b) thermal gradient deposition process.

surface by deposited carbon, which must be machined away to achieve
the maximal density. Multiple infiltration cycles are required.
In general, the infiltration lasts 2000 h.

The temperature gradient process, schematically shown in
Fig. 4b, allows continuous infiltration with shorter deposition
times (less than 1/2), but is limited to single-item infiltration,
and the complexity to achieve uniform deposits is very great. A
nonconducting sleeve serves as an insulator and channels the react-
ants and by-products. The sleeve is placed between the graphite
mandrel, which acts as susceptor, and the induction coil. The
porous substrate itself is not induction heated because of its low
density and, therefore, poor coupling. The inner side of the substrate, however, is hot, because it is in direct contact with the
mandrel. The outer zone of the substrate is exposed to the cool
environment, where the cooling effect of the high-velocity gas flowing over the substrate surface is the principal means of achieving
and controlling the thermal gradient. The deposition starts at
the mandrel and progresses radially through the substrate, as the
compacted parts of the substrate become inductively heated.

LIQUID IMPREGNATION TECHNIQUE

An alternative method to densify a porous C/C composite is
the impregnation with liquid organic carbon precursors and the
succeeding recarbonization in subsequent densification cycles.
The most critical problem in this process is the carbonization
behaviour of the matrix precursors and the interactions between
fibre and matrix during the carbonization.

The densification is performed in a combined vacuum-pressure
impregnation step, followed by recarbonization of the infiltrated
impregnant. The porous composite frame first is evacuated (400 Pa)
and then dipped into the liquid impregnant. The applied pressure
for the infiltration differs between 10 and 2000 bar (1 and 200 Pa),
depending on the precursor material. The composite to be infiltrated stays up to 20 h in the impregnant under pressure. Then
the recarbonization is followed by heating up to temperatures
between 1000°C and 2800°C. This densification cycle generally has
to be repeated 4 to 12 times, depending on the sample thickness and
the required final porosity of the composite.[7,8,9]

CARBON PRECURSOR MATERIALS AND CARBONIZATION BEHAVIOUR

The conversion of the organic precursor into carbon should
yield a high-coke residue and low-volume shrinkage during the
pyrolysis. Furthermore, no fibre or matrix damage during the
carbonization should occur. One has to distinguish between two

general types of precursor materials: (1) thermosetting resins
like phenolics and polyimides and (2) thermoplastic precursors like
pitches. The thermosetting precursors carbonize in a solid phase
and result in glass-like, nongraphitizable carbon, the thermoplastic precursors carbonize in a liquid phase and result in so-
called soft carbons, which are graphitizable. Some suitable matrix
precursors are compiled in Fig. 5. Phenolic novolacs (e.g., resin
A) are known to be easily handleable precursors with considerable
carbon yields of 60 to 65%, but high isotropic shrinkages (20%
linear).[10] Polyimides have nearly the same carbonization behaviour.[11,12] If pitches are used as matrix precursors, with high
pressure pyrolysis up to 550°C, the formation temperature of the
bulk mesophase, or chemical dehydrogenation at 250°C by addition
of elemental sulfur, has to be performed to achieve high carbon
yields.[13,14,15] Figure 6 illustrates the weight loss of coal tar
(CT) pitch versus temperature treatment, after application of
different gas pressures during the carbonization at 550°C or
600°C and when pitches are used, modified with different amounts
of elemental sulfur.[8,14,16] The highest coke residue of 90% is
achieved when the pyrolysis is performed applying a pressure of
1000 bar (100 MPa); 70% coke residue is received applying a pressure of 100 bar (10 MPa). In addition to the coke residue, the
microstructure of the resulting carbon is influenced by the pyrolysis pressure.[8] Synthetic pitches, like isomeric polyphenylenes,

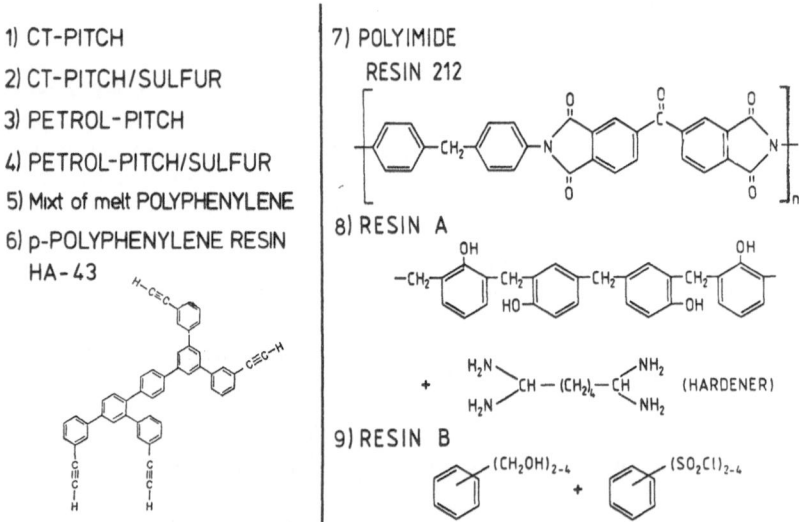

1) CT-PITCH
2) CT-PITCH/SULFUR
3) PETROL-PITCH
4) PETROL-PITCH/SULFUR
5) Mixt of melt POLYPHENYLENE
6) p-POLYPHENYLENE RESIN
   HA-43

7) POLYIMIDE
   RESIN 212

8) RESIN A

9) RESIN B

Fig. 5.   Compilation of some suitable carbon/carbon-
          composite matrix precursors.

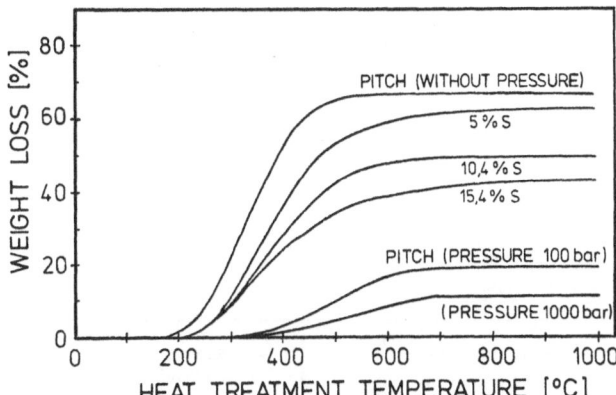

Fig. 6.   Influence of addition of elemental sulfur and
          pressure carbonization on the weight loss of
          CT pitch during the pyrolysis.

also result in high carbon yields, but the difficult handling
problems are not yet solved.[17]   The curable acetylene terminated
polyphenylene resin HA-43, however, is easily handleable, has a
low linear shrinkage of only 10%, and high coke residues of 85%.[18]
The linear shrinkage of the precursor materials can be correlated
with the corresponding coke yield.   Resins with high coke yields
show low linear shrinkage.   In the composites, however, the shrinkage parallel to the fibre axis is hindered, and the three-
directional, isotropic shrinkage of the precursors is changed into
a two-directional, cross-sectional shrinkage of the composite.
Figure 7 shows the cross-sectional shrinkage of C/C composites
during the primary carbonization; the shrinkage is dependent on
the type of C fibre and the matrix precursor.[19]   The cross-sectional
shrinkage of the composites is low if precursors with high-coke
residues are used and decreases further with decreasing concentration of active surface groups on the carbon fibre surface.   (Compare
T 300 90, a surface treated carbon fibre; T 300 99, a carbonized
type II C fibre; and M40, a graphitized type I C fibre.)   This
effect can be explained by the good adhesion between fibre and
matrix.   High cross-sectional shrinkage and good adhesion between
fibre and matrix causes pyrolysis cracks in the matrix perpendicular to the fibre axis (Fig. 8) and, in the worst cases, fibre
damage, which is due to the shear stress concentration initiated
by the matrix shrinkage (Fig. 9).[19]   The result of fibre damage
during processing on the mechanical properties of C/C composites

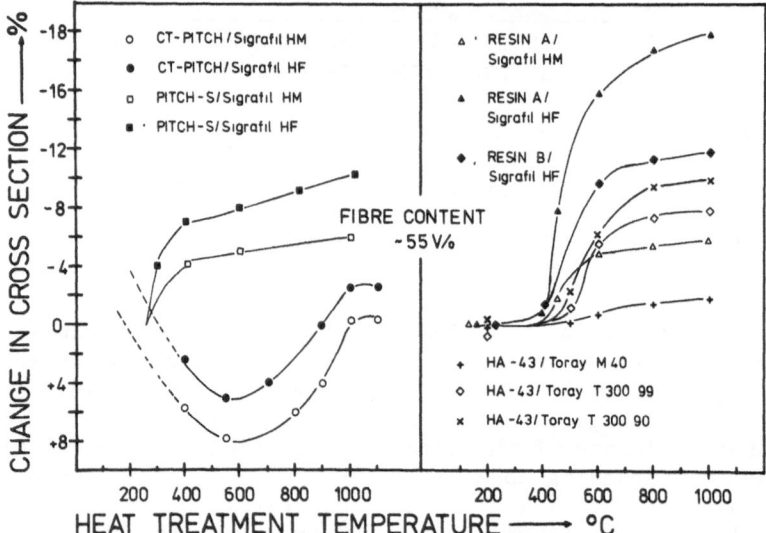

Fig. 7. Cross-sectional shrinkage behaviour during the primary
carbonization of "green composites," dependent on the type
of carbon fibre, carbon matrix precursor, and temperature.

Fig. 8. Photographs of C/C composites after the first carbonization with decreasing amounts of matrix pyrolysis
cracks, due to decreasing cross-sectional shrinkage
of the matrix precursor and decreasing adhesion
between fibre and matrix precursor.

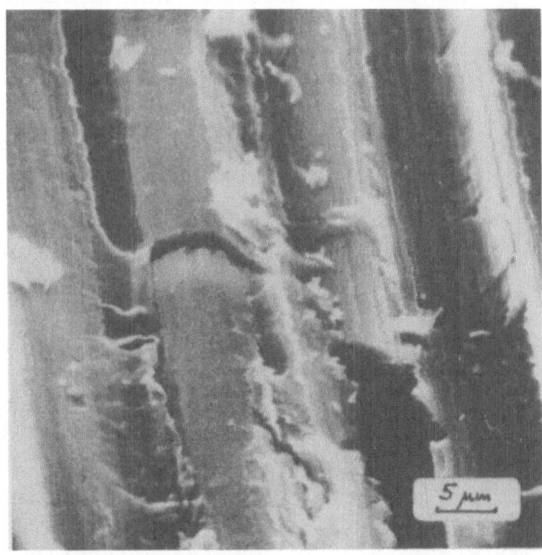

Fig. 9. SEM of damaged C fibres, caused by high shear stress concentration of the matrix on the fibre during the first
carbonization, due to good fibre/matrix adhesion and
hindered longitudinal shrinkage of the matrix precursor.

is shown in Fig. 10, where flexural strength data of the carbonized
composites reinforced with three different fibre types (Toray T
300 90, T 300 99, M 40 99) is plotted as function of the bulk
density.[19] The measuring points in brackets indicate the flexural
strength data before the carbonization; the measuring points without brackets indicate the steps of repeated impregnation-
recarbonization cycles. As one can see, fibre strength up to 100%
can be realized only if weak adhesion exists between fibre and carbon precursor in the green composite. The adhesion in the final
composite is realized mainly by a mechanical interlocking, which
is due to the repeated densification cycles. All process or material modifications that increase the adhesion in the green condition
decrease the strength of the C/C composites, although the flexural
strength of the green composites before carbonization is considerably increased (compare T 300 90 and T 300 99).

MECHANICAL PROPERTIES

     Table I compiles the mechanical properties of various C/C com-
posites at room temperature. After 4 to 6 densification cycles and
final heat treatments at 1000°C, unidirectional (1-D) reinforced

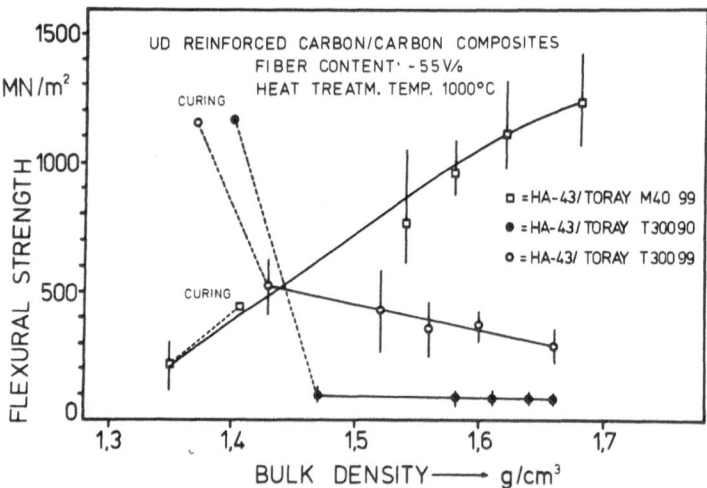

Fig. 10.   Flexural strength of C/C composites, reinforced with
           different C-fibre types as a function of bulk density.

Table I. Mechanical Properties of C/C Composites
          at Room Temperature.

CFRC	Flexural strength, MN/m^2	Young's modulus, GN/m^2	Interlaminar shear strength, MN/m^2
1-D C/C, ‖ Fibre content: 55 V/o	1200 - 1400	150 - 200	20 - 40
2-D C/C, ‖ 8 H/S cloth warp Fibre content: 35 V/o	300	60	20 - 40
3-D C/C, z cloth Fibre content: 50 V/o	250 - 300	50 - 150	50 - 80
3-D C/C Felt Fibre content: 35 V/o	170	15 - 20	20 - 30

composites achieve reproducible flexural strengths of 1200 to 1400 MN/m^2 and Young's moduli of 150 to 200 GN/m^2 parallel to the fibre axis, if reinforced with 55 V/o type I fibres. The strength perpendicular to the fibre axis is only 1/100 that perpendicular to the longitudinal axis.[14,15,19,20,21,22] The interlaminar shear strengths (ILSS) are 20 to 40 MN/m^2. The strengths of two-directional (2-D) carbon-cloth-reinforced composites depend on the type of weave pattern of the carbon cloth. If an 8-harness satin fabric is used, strengths of 300 MN/m^2 in the x direction (warp direction) can be realized. The strength in the y direction (fill direction) amounts to 150 MN/m^2.[23] If a square-weave carbon fabric is used for reinforcement, the strengths in the x and y directions are identical and amount to about 100 to 170 MN/m^2.[24]

Three-directional (3-D) orthogonal woven C/C composites, heat treated to 2200°C with a fibre volume fraction of 50 V/o in a fibre arrangement of 2_x, 2_y, 3_z, have flexural strengths of 250 to 300 MN/m^2 in the z direction.[25] The strength is strongly dependent on the weave pattern and the fibre content in the x, y, and z directions and can be tailored from isotropic to anisotropic, according to special requirements, by means of the kind of weave pattern.[26]

A different kind of 3-D fibre arrangement is the reinforcement of carbon with short fibres or carbon felt. Composites with iso-tropic properties result, having flexural strengths ranging from 100 to 170 MN/m^2.[27,28]

As illustrated in Fig. 11, the flexural strengths and Young's moduli of carbon/carbon composites include a wide range of data: flexural strengths range from 100 to 1400 MN/m^2 and Young's moduli from 20 to 200 GN/m^2. By selecting the fibre type, fibre volume fraction, fibre arrangement, matrix precursor, processing technique, number of densification cycles, and final heat treatment, one has the possibility to tailor the required properties of the C/C com-posite.

Further possibilities to vary the properties of C/C composites are given by so-called angle ply fibre arrangements. This is a fibre arrangement, where unidirectional C fibres are laminated in alternating angles from 0° to 90° bearing on the direction of the load attack.

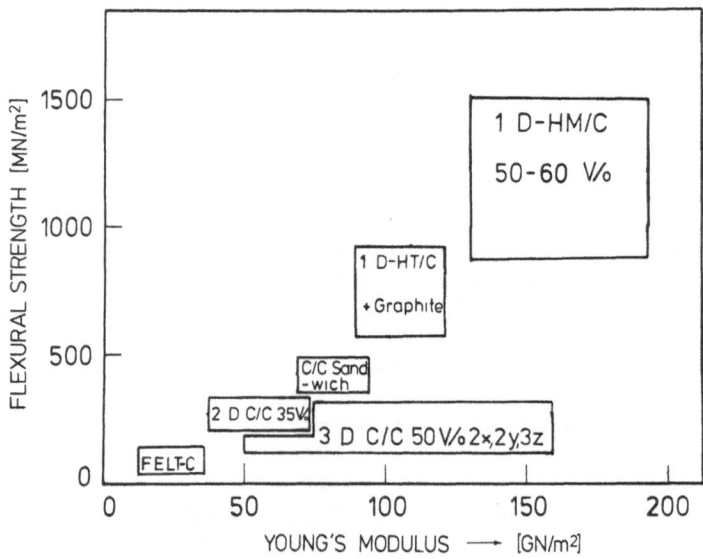

Fig. 11.   Strength and modulus range of C/C composites with different fibre arrangements.

FRACTURE BEHAVIOUR

Figure 12 shows schematically the stress-strain curves of 1-D, 2-D, and 3-D reinforced C/C composites at room temperature. 1-D composites exhibit brittle fracture behaviour, the 2-D composites fail in a "semi-brittle" manner by a continuous step drop in load.[7,12,20,23] The mode of failure of 3-D composites, however, is not of a brittle type. One observes strain rates up to 5%.[29] This nontypical fracture behaviour of 3-D composites is due to a continuous crac system inside the composite, as illustrated schematically in Fig. 13. This crack pattern depends on the weave pattern and originates during the processing of the carbon/carbon composite, as a result of the heating and cooling cycles. These cracks are able to annihilate fracture energy. If the cracks are closed at higher temperatures because of the thermal expansion of the material, the typical brittle fracture behaviour of C/C composites is found (see Fig. 13).[30,31]

THERMAL EXPANSION BEHAVIOUR

The thermal expansion of C/C composites is strongly influenced by the same parameters also responsible for the mechanical properties. The coefficients of thermal expansion as well as the linear thermal expansion are dependent on temperature, and differences occur between the data for thermal expansion behaviour of the graphite single crystal in the a, b direction and c direction.

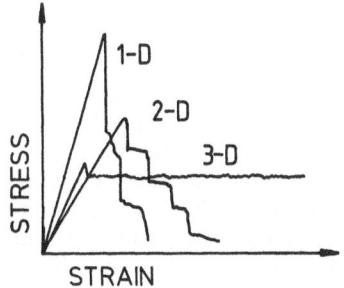

Fig. 12.   Schematic of stress/strain behaviour
           of 1-D, 2-D and 3-D C/C composites
           at room temperature.

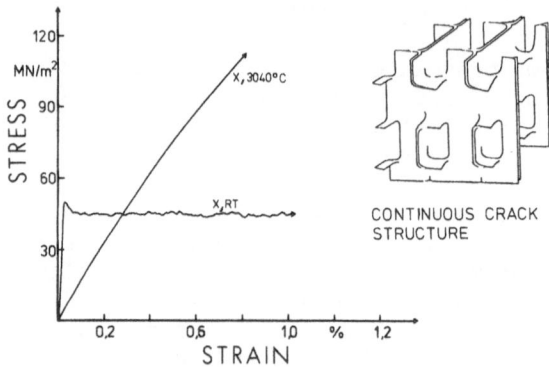

Fig. 13.   Different fracture behaviour of 3-D C/C
composites at room temperature and 3040°C,
induced by the crack structure of the
composite.

Parallel to the fibre axis, the thermal expansion behaviour is
controlled only by the thermal expansion of the C fibre, whereas
the thermal expansion perpendicular to the fibre axis is dominated
mainly by the carbonaceous matrix. In Fig. 14, the thermal expansion and the coefficient of thermal expansion of unidirectionally
reinforced C/C composites parallel and perpendicular to the fibre
axis is shown versus temperature. In Fig. 14a, curves are plotted
of composites with polyimide as the matrix precursor with different
final heat treatment temperatures of 1500°C, 2000°C, and 2700°C.
Figure 14b illustrates, in comparison with the polyimide-based
composites, the expansion behaviour of pitch-based composites reinforced with different types of graphitized C fibres.[32]

Up to 200 C, the polyimide-based C/C composites have negative
coefficients of thermal expansion parallel to the fibre axis (α =
$-1 \cdot 10^{-6} \cdot {}^\circ C^{-1}$. This expansion characteristic, generally typical of
high-performance carbon fibres, is due to the carbon matrix, which
has a partially preferred orientation structure and an expansion
behaviour similar to C fibres. A proof of the preferred orientation can be seen in the fact that the coefficient of thermal expansion at room temperature is more negative for composites with higher
final heat-treatment temperatures. Between the temperatures of 400
and 1400°C, the thermal expansion is linear, with $\alpha = 2 \cdot 10^{-6} \cdot {}^\circ C^{-1}$.

Perpendicular to the fibre axis, α increases from $-1 \cdot 10^{-6} \cdot {}^\circ C^{-1}$
to zero between room temperature and 400°C. Above 400°C, α is
$5.5 \cdot 10^{-6} \cdot {}^\circ C^{-1}$ and remains constant when the temperature is raised.

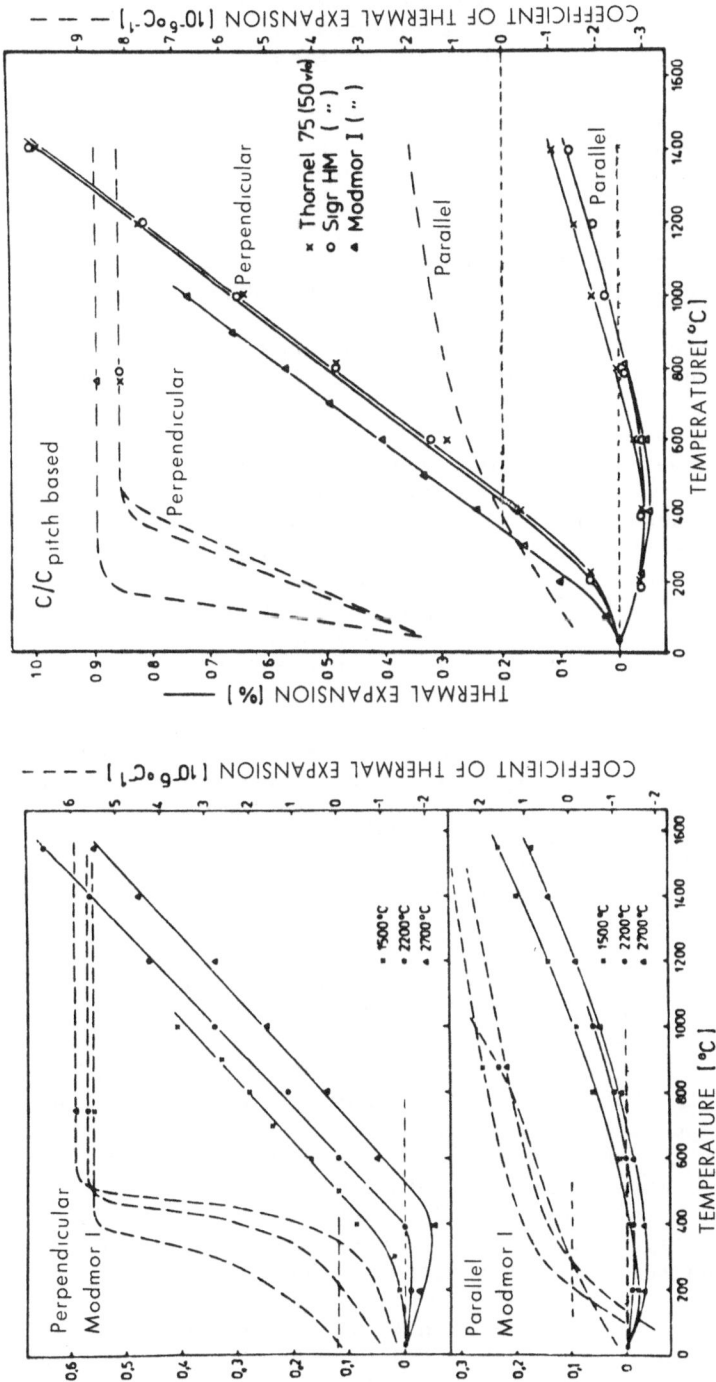

Fig. 14. Linear thermal expansion and coefficient of thermal expansion of 1-D reinforced C/C composites parallel and perpendicular to the fibre axis as a function of temperature.

(a) Polyimide-based carbon matrix, heat treated to 1500°C, 2000°C, and 2700°C.
(b) Pitch-based carbon matrix reinforced with three different types of HM fibres.

The value of $5.5 \cdot 10^{-6} \cdot {}^{\circ}C^{-1}$ is approximately the same as that calculated for the coefficient of thermal expansion of C fibres perpen-
dicular to the fibre axis, as well as for glassy carbon.[33,34]

In the case of C/C composites with CT pitch as matrix precursor, the coefficient of thermal expansion parallel to the fibre
axis is an average of $0.5 \cdot 10^{-6} \cdot {}^{\circ}C^{-1}$ lower than α of the polyimide-
based C/C composites, but increased to about $2.0 \cdot 10^{-6} \cdot {}^{\circ}C^{-1}$ perpendicular to the fibre axis. This behaviour is due to a higher degree
of preferred orientation of the matrix structure of the pitch-based
C/C composites compared with the polyimide-based matrix structure.

The influence of the different graphitized fibre types on the
thermal expansion behaviour is not significant.

Figure 15 shows measurements of linear thermal expansion and
coefficients of thermal expansion at temperatures between 4.2 K and
room temperature, which were preformed with unidirectionally reinforced C/C composites parallel to the fibre axis. The composites
are fabricated with two different carbon matrix precursors, CT pitch
and CT pitch modified by elemental sulfur, as well as two different
fibre types, Sigrafil HF (type II) and Sigrafil HM (type I). The
C/C composites consist of 50 V/o fibre volume fraction and are
densified four times with a final heat treatment of 1000°C. Details
of the measuring apparatus are described by Hartwig et al.[35]

No significant differences of the thermal expansion behaviour
were observed for samples reinforced with the same type of C fibre,
but different matrix precursors. Also, at low temperatures, the coefficient of thermal expansion is dominated by the fibres. α at
liquid He temperature is nearly zero. The composites, reinforced
with different fibre types, possess extremely different expansion
behaviour. The coefficient of thermal expansion of the high-modulus
(HM) fibre-reinforced composites is $-4 \cdot 10^{-6} \cdot {}^{\circ}C^{-1}$ at liquid HE temperature, increases to zero at 100 K, and decreases to
$-2.5 \cdot 10^{-6} \cdot {}^{\circ}C^{-1}$ at a temperature of 250 K, where a second point of
extension exists. α at room temperature is $-2 \cdot 10^{-6} \cdot {}^{\circ}C^{-1}$.

To examine the influence of other parameters on thermal expansion behaviour, the thermal expansion of 2-D carbon-cloth-reinforced
C/C composites with different porosity was measured at low temperatures. The C/C composites were reinforced with 35 V/o 8-harness
satin-carbon cloth (sigrafil GDS 8/30) and have porosites of 18 and
15 V/o, respectively. The matrix precursor was CT pitch, modified
with elemental sulfur and heat treated to 1000°C.

As one can recognize clearly in Fig. 16, the coefficient of
thermal expansion at liquid HE temperature for the 2-D composite
containing 15 V/o porosity is $-2 \cdot 10^{-6} \cdot {}^{\circ}C^{-1}$, whereas the specimen

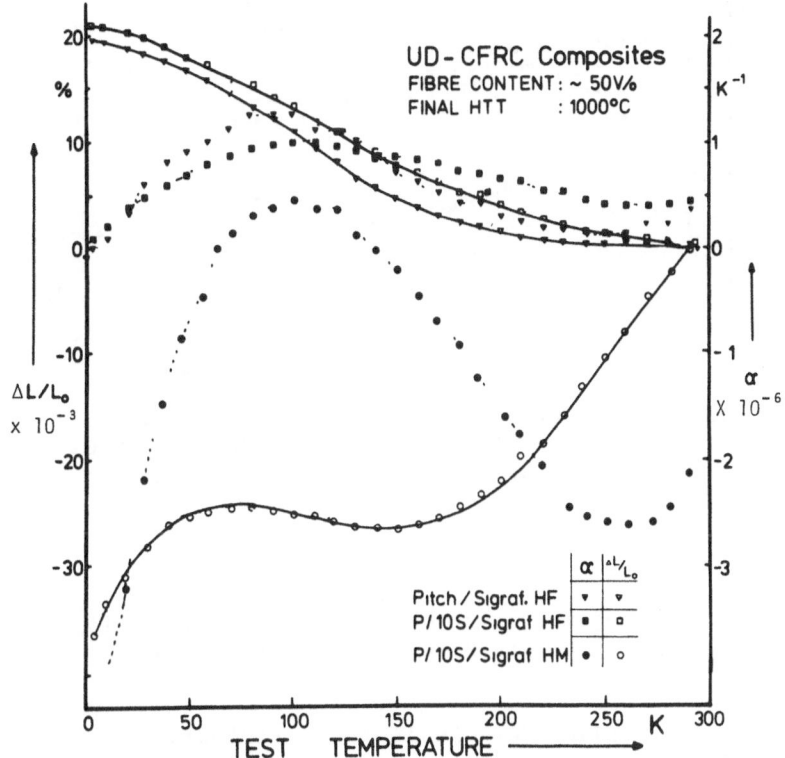

Fig. 15. Linear thermal expansion and coefficient of thermal
          expansion of 1-D reinforced C/C composites parallel to
          the fibre axis as a function of carbon matrix and carbon
          fibres at temperatures between 4.2 K and room tempera-
          ture.

with 18 V/o porosity has a positive coefficient of $0.25 \cdot 10^{-6} \cdot {}^{\circ}C^{-1}$.
This difference can be caused only by the carbon matrix.

     It must be pointed out that these measurements of the thermal
expansion behaviour of C/C composites at low temperatures shall
primarily demonstrate that it may be possible to tailor a high-
performance material with a required thermal expansion behaviour
for use as cryogenic material. For tailoring the required properties, there are a lot of parameters like those of C fibres: the

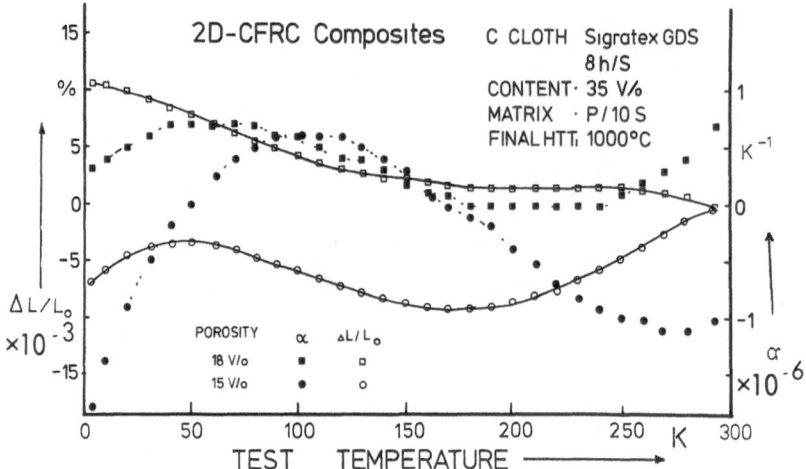

Fig. 16. Effect of porosity on thermal expansion
          behaviour of 2-D reinforced C/C composites
          at temperatures between 4.2 K and room
          temperature.

kind of fibre arrangement, on the one hand, and the type of carbonaceous matrix, carbon or graphite, on the other hand.

APPLICATIONS OF CARBON/CARBON COMPOSITES

C/C composites are mainly applied as high temperature and
ablation material due to their excellent high temperature properties. Multidirectionally reinforced C/C composites are used in
space technology mainly as heat shields and nose tips for reentry
vehicles.[36,37]

C/C composite brake discs are used in some military aircrafts
like F-15, F-16, YF-16 and also in the commercial supersonic airplane, the Concorde. Some European racing cars are also equipped
with carbon brakes.[38,39]

Furthermore, C/C composites are used as rocket nozzles for
solid propellants.[40,41] Hot pressing dies are commercially available. C/C composites with porosities up to 90% serve as self-
supporting insulation material for high temperature furnaces.[42,43]
A new field of application for C/C composites is given in bioengi-
neering. Due to the excellent biocompatibility of carbon, connected
with the possibility to tailor the mechanical properties to the
requirements of the biological and physiological surroundings, C/C
composites can be used as prosthetic material for joint replace-

ment[44] (Fig. 17a) and as bone plates[23] (Fig. 17b) or bone nails.[45]

In our minds, the suitability of C/C composites as armaments for magnets in fusion technology should be proved. First steps in this direction were done by investigating the thermal properties of carbon–fibre–reinforced carbon composites at low temperatures. Because of the chemical and physical properties of carbon, along with the various parameters that influence the properties of this composite material, it seems possible to tailor a material to the special requirements of that application.

(a)

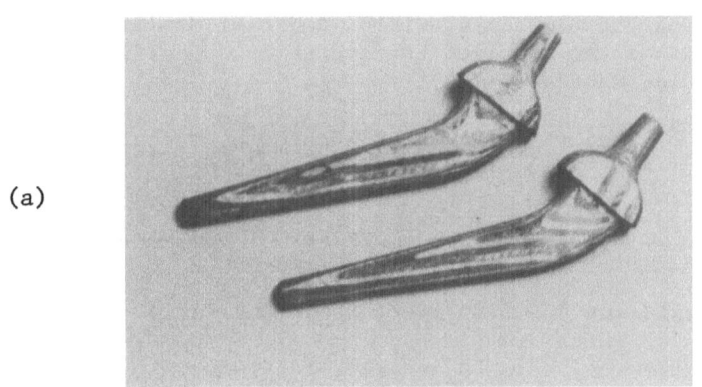

(b)

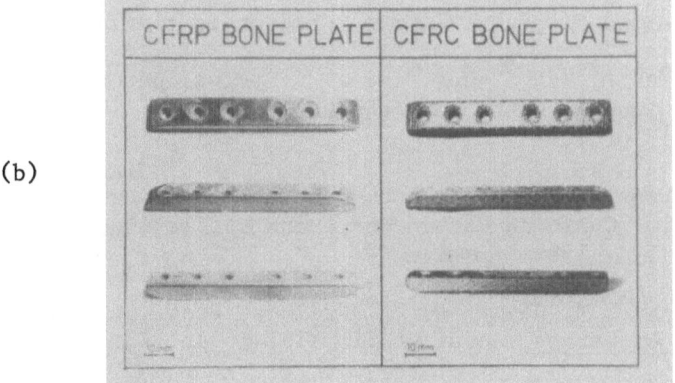

Fig. 17.   C/C composites for use as biomaterial:
　　　　(a) hip joint stems.[44]
　　　　(b) osteosynthesis plates.[23]

ACKNOWLEDGMENT

The authors wish to thank Prof. Dr. E. Fitzer, Director of the Institute für Chemische Technik, University of Karlsruhe, for his valuable advice and help, which was the supposition to present this lecture.

REFERENCES

1. E. Fitzer and M. Heym, invited plenary paper presented at the conference, Carbon and Graphite as Refractory Materials, Newcastle, England (1977).

2. W.L. Lachmann, J.A. Crawford, and L.E. McAllister, paper presented at the 2nd Int. Conference on Composite Materials, Toronto, Canada (1978).

3. R.J. Diefendorf, Proc. of Carbon Comp. Techn. Symp., 10th Annual ASME Symposium, New Mexico (1970).

4. W.V. Kotlensky, Proc. 16th Nat. Sampe Symp. and Exhib., Anaheim, California (1971).

5. H.M. Stoller and E.R. Frye, Sampe Quarterly, 3(3), 10 (1971).

6. H.O. Pierson, Sampe Quarterly, 14 (1968).

7. E. Fitzer and B. Terwiesch, Carbon, 11, 570 (1973).

8. J.S. Evangelidis, paper presented at 13th Biennial Conf. on Carbon and Graphite, Irvine, California (1977), p. 16.

9. G.W. Weber, K.R. Young, and A.J. Taylor, paper presented at 13th Biennial Conf. on Carbon and Graphite, Irvine, California (1977), p. 68.

10. E. Fitzer, K.H. Geigl, and M. Heym, paper presented at 13th Biennial Conf. on Carbon and Graphite, Irvine, California (1977), p. 168.

11. E. Fitzer, M. Heym, and K. Karlisch, Proc. 4th London Int. Conf. on Carbon, Session III, (1974), p. 172.

12. A. Bürger, E. Fitzer, M. Heym, and B. Terwiesch, Carbon, 13, 149 (1975).

13. E. Fitzer and A. Bürger, Chem-Ing.-Tech. 42, 1203 (1970).

14. E. Fitzer and B. Terwiesch, Carbon, 10, 383 (1972).

15. E. Fitzer, M. Heym, and B. Rhee, paper presented at 2nd Int. Kohlenstofftagung, Baden-Baden, Germany (1976), p. 508. [Compare High Temperatures - High Pressures 8, 307 (1976).]

16. E. Fitzer and H. Tillmanns, paper presented at the 12th Biennial Conference on Carbon, Pittsburgh, Pennsylvania (1975), p. 217.

17. E. Fitzer and F. Grieser, paper presented at 2nd Int. Kohlenstofftagung, Baden-Baden, Germany (1976), p. 513.

18. W. Bradshaw, P.C. Pinoli, and R.F. Karlak, Report No. CR 134 625, NASA (1974).

19. E. Fitzer, K.H. Geigl, and W. Hüttner, Proc. 5th London Int. Conf. on Carbon and Graphite, (1978) in print.

20. E. Fitzer and M. Heym, Z. Werkstofftechnik, 7, 269 (1976).

21. I. Hill, C.R. Thomas, and E.J. Walker, Proc. 2nd Int. Carbon Fibres Conf. London (1974).

22. W. Bradshaw and A.E. Vidoz, Ceram. Bull. 57, 193 (1978).

23. E. Fitzer, W. Hüttner, L.M. Manocha, and D. Wolter, Proc. 5th London Int. Conf. on Carbon and Graphite, (1978) in print.

24. D.W. Bauer and W.V. Kotlensky, paper presented at 23rd Pacific Coast Regional Meeting, ASCI, San Francisco, California (1970).

25. A. Levine, J.A. Roetling, E.R. Stover, and J.J. Gebhardt, paper presented at the 12th Biennial Conference on Carbon, Pittsburgh, Pennsylvania (1975).

26. C.R. Rowe, Proc. 19th Nat. Sampe Symp. and Exhib., Buena Park, California (1975), p. 359.

27. J.W. Warren and R.M. Williams, Proc. 4th Nat. Sampe Symp. and Exhib., Palo Alto, California (1972).

28. K. Karlisch, thesis, Institut für Chemische Technik, University of Karlsruhe, Karlsruhe, Germany (1977).

29. P.G. Rolincik, paper FC 26 presented at 11th Biennial Conference on Carbon, Gatlinburg, Tennessee (1973).

30. J. Jortner, paper presented at 13th Biennial Conference on Carbon, Irvine, California (1977), p. 443.

31. D.A. Eitmann, L.B. Greszczuk, J. Jortner, and C.R. Rowe, Proc. 19th Nat. Sampe Symp. and Exhib., Buena Park, California (1974), p. 346.

32. M. Heym, thesis, Institut für Chemische Technik, University of Karlsruhe, Karlsruhe, Germany (1973).

33. W.N. Reynolds, Proc. 3rd Int. Conf. on Ind. Carbon & Graphite, London (1970), p. 427.

34. B. Lersmacher and H. Lydtin, Chem.-Ing.-Tech. 42, 659 (1970).

35. G. Hartwig, A. Puck, and W. Weiss, Kunststoffe, Bd 64, 32 (1974).

36. T.R. Guess and C.W. Bert, paper presented at the Conference on Continuum Aspects of Graphite Design, Oak Ridge National Laboratory, Oak Ridge, Tennessee (1970).

37. H.W. Schmidt, paper FC 57A presented at 10th Biennial Conference on Carbon, Bethlehem, Pennsylvania (1971).

38. R. Fisher and J.V. Weaver, paper 98 presented at the 4th London International Conference on Carbon and Graphite (1974), p. 223.

39. Auto, Motor, Sport, Heft 19, 231 (1977).

40. G.W. Driggers and H.S. Starrett, paper presented at the 13th Biennial Conference on Carbon, Irvine, California (1977), p. 78.

41. R. Laramee, G. Lamere, and B. Precott, paper presented at the 13th Biennial Conference on Carbon, Irvine, California (1977), p. 74.

42. Product brochure of Fiber Materials, Inc., Biddeford, Maine (1977).

43. Product brochure FTR, of the Kureha Corporation, Tokyo, Japan (1977).

44. P. Rose, G. Gistinger, U. Gruber, F. Gerstenberger, C. Burri, and D. Wolter, paper presented at the 3rd Conference on Materials for Use in Medicine and Biology, Keele, England (1978).

45. G.W. Jenkins and J. Carvalho, Carbon, 15, 33 (1977).

DYNAMIC ELASTIC MODULUS AND INTERNAL FRICTION

IN FIBROUS COMPOSITES

H. M. Ledbetter

National Bureau of Standards
Boulder, Colorado, U.S.A.

INTRODUCTION

Solid-state elastic constants fill many needs. Engineering design calculations require them for estimating load-deflection and thermoelastic stress. Derived from fundamental interatomic forces, elastic constants index both cohesion and strength. They relate to other physical properties such as specific heat and thermal expansion, all of which help define a solid's equation of state.

A composite's elastic constants serve further purposes. Their accurate measurement permits tests of theories relating composite properties to constituent properties. Their tensor character permits detecting fiber-induced mechanical anisotropy. Elastic constants may also facilitate quality control[1] and incipient-failure detection.[2] For some composites, empirical relationships exist between elastic constants and important mechanical properties.[3]

Anisotropic internal friction in composites has several possible causes:

1. Fiber properties

2. Matrix properties

    a. Intrinsic (for example, "molecular engineering" in the case of resins)

    b. Extrinsic (for example, fillers)

3. Fiber-matrix interface

    a. Relative displacement

    b. Localized stress

  4. Cracks and voids.

Besides depending on direction, internal friction usually varies
with both deformation mode (flexure, torsion, etc.) and frequency.
The present study used an extensional, or Young's-modulus-type,
deformation, which has alternating uniaxial tensile and compressive
regions with corresponding transverse Poisson contractions or ex-
pansions. Internal friction not only characterizes composites,
but it may serve to detect incipient failure.[2] In engineering
design, higher internal friction improves dimensional stability by
minimizing stress-induced deflections.

    The presently reported study had two principal purposes:
(1) to apply to composites an experimental technique used previously
almost exclusively for noncomposites[4] for determining dynamic
Young's modulus and internal friction; (2) to compare dynamic moduli
with static moduli reported for the same materials by other workers.

    The experimental technique, using a three-component oscillator,
offers several possible advantages over static methods: relatively
small, simple-geometry specimens; a simple and relatively small experimental arrangement (reducing refrigeration costs during cooling
experiments, for example), readily obtained one-percent inaccuracies;
elastic-constant variations caused by changes in variables, such as
temperature, pressure, and magnetic field, easily measured to a few
parts in 10^5; and nearly simultaneous internal-friction measurements.
However, the technique does not yield accurate dilatation-type elastic constants, such as Poisson's ratio and the bulk modulus, compared with directly measured shear-type elastic constants such as
Young's modulus and the torsional modulus.

                              EXPERIMENT

                              Materials

    Table I summarizes studied-material properties. Schramm and
Kasen,[5-7] who studied both mechanical properties and static elastic
properties of these composites, gave further materials details.

                              Specimens

    The experiments used rods, either circular, square, or rec-
tangular, with typical thicknesses of 2 to 4 mm. Lengths varied
from 1 to 10 cm depending on material, fiber orientation, and chosen

Table I. Properties of Studied Materials

Material	Fiber	Matrix	Average ply thickness, mm	Mass density, g/cm^3	Fiber volume fraction
Boron-aluminum	5.6 mil boron	6061-F aluminum	0.17	2.53	0.472
Boron-epoxy	5.6 mil boron	2387 epoxy	0.17	1.97	0.522
Glass cloth-epoxy 1	1581 S-glass cloth	NASA resin 2	0.25	1.75	0.56
Glass cloth-epoxy 2	1581 S-glass cloth	E787 resin	0.25	1.84	0.556
Graphite-epoxy 1	AS graphite	NASA resin 2	0.19	1.59	0.628
Graphite-epoxy 2	HMS graphite	934 epoxy	0.13	1.60	0.625
Graphite-epoxy 3	GY70 graphite	934 epoxy	0.12	1.73	0.742
Kevlar 49-epoxy	Kevlar 49	NASA resin 2	0.23	1.16	0.624

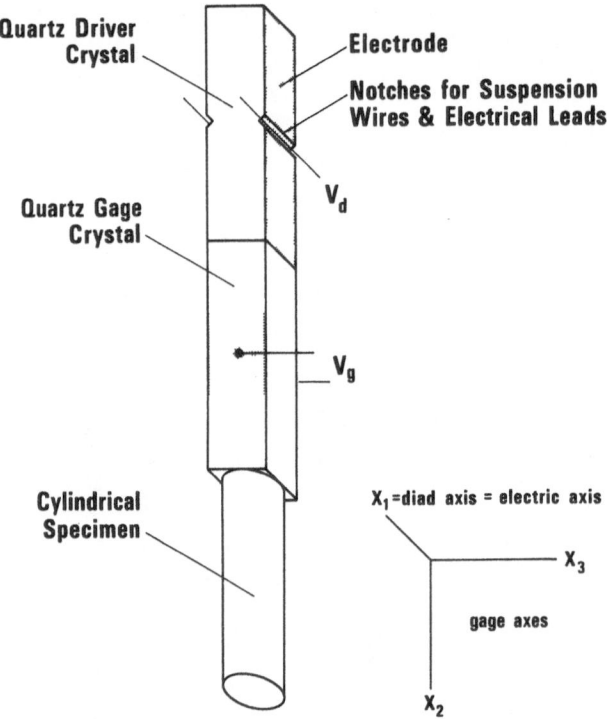

Fig. 1.   Marx composite oscillator for determining standing-wave
          (resonant) sound velocities.

resonance frequency. Two to fourteen specimens were studied for
each material and orientation.

Elastic-Constant Measurements

Determining Young's modulus, E, required measuring the sound
velocity, v, and the mass density, ρ, and using the relationship

$$E = \rho v^2 = \rho f_s^2 \lambda^2 \tag{1}$$

where f_s denotes specimen resonance frequency and λ denotes resonance wavelength. At half-wave resonance, $\lambda = 2l_s$ (where l_s denotes
specimen length), and

$$E = 4\rho f_s^2 l_s^2 \tag{2}$$

A Marx[8,9] oscillator (shown schematically in Fig. 1) yielded the
experimental resonance frequency according to

$$m_o f_o^2 = \sum_i m_i f_i^2 \tag{3}$$

where m denotes mass, o denotes the complete oscillator, and i denotes the oscillator components: quartz-crystal driver, quartz-
crystal gage, and specimen. It follows simply from Eq. (3) that

$$f_s^2 = f_o^2 + (f_o^2 - f_q^2) \, m_q/m_s \tag{4}$$

where subscript q denotes quartz crystals. The quartz-crystal
length fixes f_q. A maximum voltage V_g, corresponding to a maximum
strain amplitude at the gage-crystal midpoint, detects f_o. Sweep-
ing frequency near f_q reveals the voltage maximum. Extensional
elastic waves were launched along the rod axis by using $-18.5°$
X-cut quartz crystals. Multiplying the right side of Eq. (2) by
the well-known Rayleigh factor

$$R = \left[1 + \left(\frac{n \pi \nu d}{4 l_s} \right)^2 \right]^2 \tag{5}$$

introduced the necessary short-rod correction, where $n = 2l_s/\lambda$, ν
denotes Poisson's ratio, and d denotes either the diameter of
circular rods or $d^2 = \frac{2}{3} (a^2 + b^2)$ for a x b rectangular rods.
Detecting resonance at the first and third harmonics and extra-
polating to zero yielded the zero-frequency Young's moduli.

Internal-Friction Measurements

For a sharp, Lorentzian resonance peak (as shown schematically in Fig. 2) the composite oscillator's internal friction, Q_o^{-1}, is

$$Q_o^{-1} = (f_2 - f_1)/f_o \tag{6}$$

the resonance-peak width at 0.707 of maximum in a strain (proportional to V_g)-versus-frequency plot. Equation (6) holds for $\phi = Q^{-1}$ up to 0.1 or 0.2, where ϕ relates the complex, real, and imaginary elastic moduli as shown in Fig. 3. Experimentally, one observes M^*, the modulus in phase with strain. But one usually seeks M_{real}, the usually more relevant storage modulus. For zero internal friction, $M_{real} = M^*$.

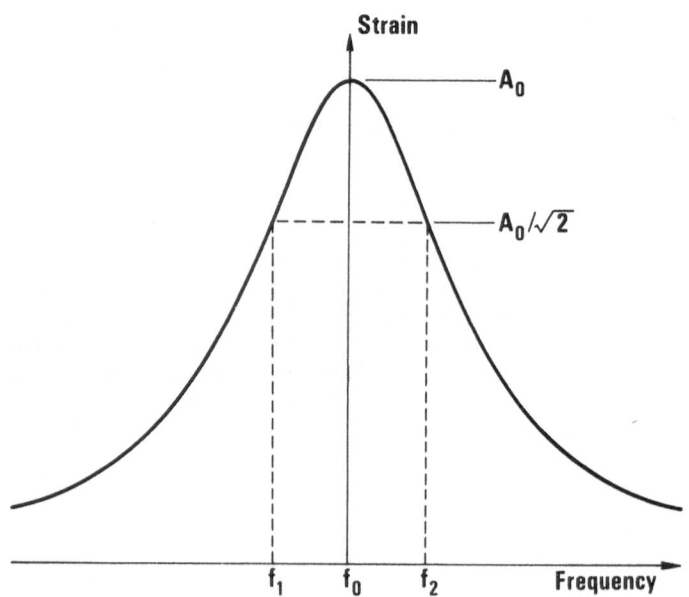

Fig. 2.   Forced-vibration Lorentzian-shaped resonance peak. Adapted from Nowick and Berry.[10]

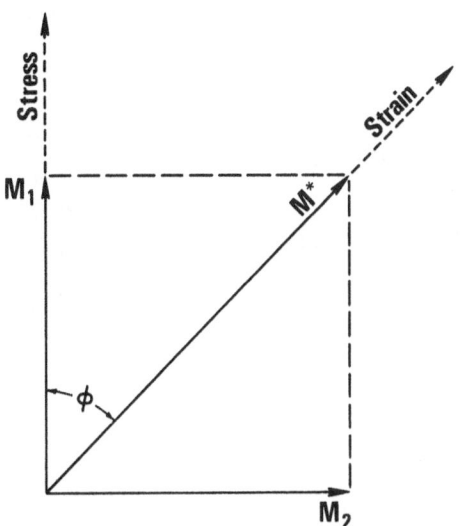

Fig. 3. Complex-plane diagram showing vector relationships
among stress, strain, M* = complex modulus, M_1 =
M_{real} = storage modulus, M_2 = $M_{imaginary}$ = loss
modulus, and ϕ = loss angle = internal friction.
Adapted from Nowick and Berry.[10]

Similar to Eq. (3),

$$m_o \, Q_o^{-1} = \sum_i m_i Q_i^{-1} \tag{7}$$

Thus, a relationship similar to Eq. (4) yields the specimen's
internal friction

$$Q_s^{-1} = Q_o^{-1} + (Q_o^{-1} - Q_q^{-1}) \, m_q/m_s \tag{8}$$

RESULTS

Table II contains room-temperature internal-friction values
and Young's moduli, the latter both at the quartz-crystal resonant
frequency (30 to 90 kHz) and at zero frequency. Linear

Table II. Dynamic Young's Modulus and Internal Friction of Several Fibrous Composites at Room Temperature

Material	Fiber orientation, degrees	Static Young's modulus,* GN/m²		Dynamic Young's modulus, GN/m²		Internal friction, Q⁻¹, 10^{-4}	First-harmonic frequency, kHz
		Tension	Compression	f = fres	f = 0		
Boron-aluminum	0	200 ±1	257 ±51	226.2 ±0.2	226	5.6 ±0.4	50
	90	161 ±18	148 ±40	139.2 ±0.2	138.7	17.0 ±0.5	80
Boron-epoxy	0	231 ±4	212 ±42	226 ±4	226	17.6 ±1.4	55
	90	17 ±1	19 ±4	22.7 ±0.2		400.8 ±20.6	35
	0,±45,0	120 ±2	102 ±1	121.4 ±0.8		79.9 ±3.3	70
Glass cloth-epoxy 1	woof	26 ±2	24 ±1	30.6 ±0.6		62.1 ±8.2	40
Glass cloth-epoxy 2	woof	28 ±1	27 ±3	29.6 ±0.2		64.9 ±7.5	40
Graphite-epoxy 1	0	118 ±7	129 ±7	133.5†	134.7	11.1	90
	90	7.8 ±1	9.7 ±1	10.0	10.3	164.2	45
	±45	15.6 ±2					
Graphite-epoxy 2	0	323 ±12	368 ±22	300.5 ±0.4		17.9 ±0.4	65
	90	6.7 ±0.1	7.8 ±1.0	7.3 ±0.5		258.8 ±1.8	45
	±45	21 ±0.5					
	0,±45,90						
Graphite-epoxy 3	0	186 ±7		177 ±4		15.0 ±3.8	50
	90	8.6 ±0.4		8.5		218.0 ±11.7	30
	0,±45,90			70 ±1		49.8 ±2.1	40
Kevlar 49-epoxy	0	71 ±1		66.1 ±0.5	67.1	114.8 ±6.1	60

* All static values are from Schramm and Kasen.[5-7]   † Uncertainty statements are not made for this material because only one specimen of each orientation was tested.

extrapolation of the first-harmonic ($\lambda = 2l_s$) data and the third-harmonic ($\lambda = 2l_s/3$) yielded the zero-frequency value. Internal-friction values apply to the resonant frequency. Uncertainties in the table represent one standard deviation. Figure 4 shows typical resonance peaks in strain versus frequency for quartz, boron-aluminum, and boron-epoxy. For clarity, these graphs show only a fraction of the data points. Figure 5 shows Young's modulus versus temperature for three graphite-epoxy composites.

DISCUSSION

Several observations follow from the results in Table II:

1. Young's modulus can be measured more accurately dynamically than statically, as shown by the average dynamic-case uncertainty of 3 percent, and the average static-case uncertainties of 6 percent in tension and 12 percent in compression.

2. Despite higher uncertainties in the compressive values, they correspond as closely to dynamic moduli as do tensile values.

3. As a whole, static elastic moduli scatter around dynamic values with an average standard deviation of 15 percent, indicating no bias toward higher or lower values.

4. Static transverse elastic moduli, which reflect the matrix material, tend to be lower than the dynamic modulus by about 6 percent. They scatter less around the dynamic value than do the static longitudinal moduli.

5. Experimental uncertainties in dynamic values relate to the material, not to specimen orientation. The same material shows similar uncertainties for longitudinal, transverse, and crossply specimen orientations.

6. Young's-modulus anisotropy, $E(0)/E(90)$, varied from near unity to near 100, and it varied considerably in the three graphite-reinforced epoxies.

7. Internal-friction anisotropy varied from about 3 to 20. Internal friction was always higher in the transverse direction, which tends to sample the matrix. Despite their wide elastic-modulus-anisotropy variation, the graphite-reinforced epoxies varied little in internal-friction anisotropy.

8. Higher Young's modulus usually corresponds to lower internal friction, and vice versa.

9. Longitudinal graphite-epoxy results indicate a possible exception to modulus/internal-friction reciprocity. The

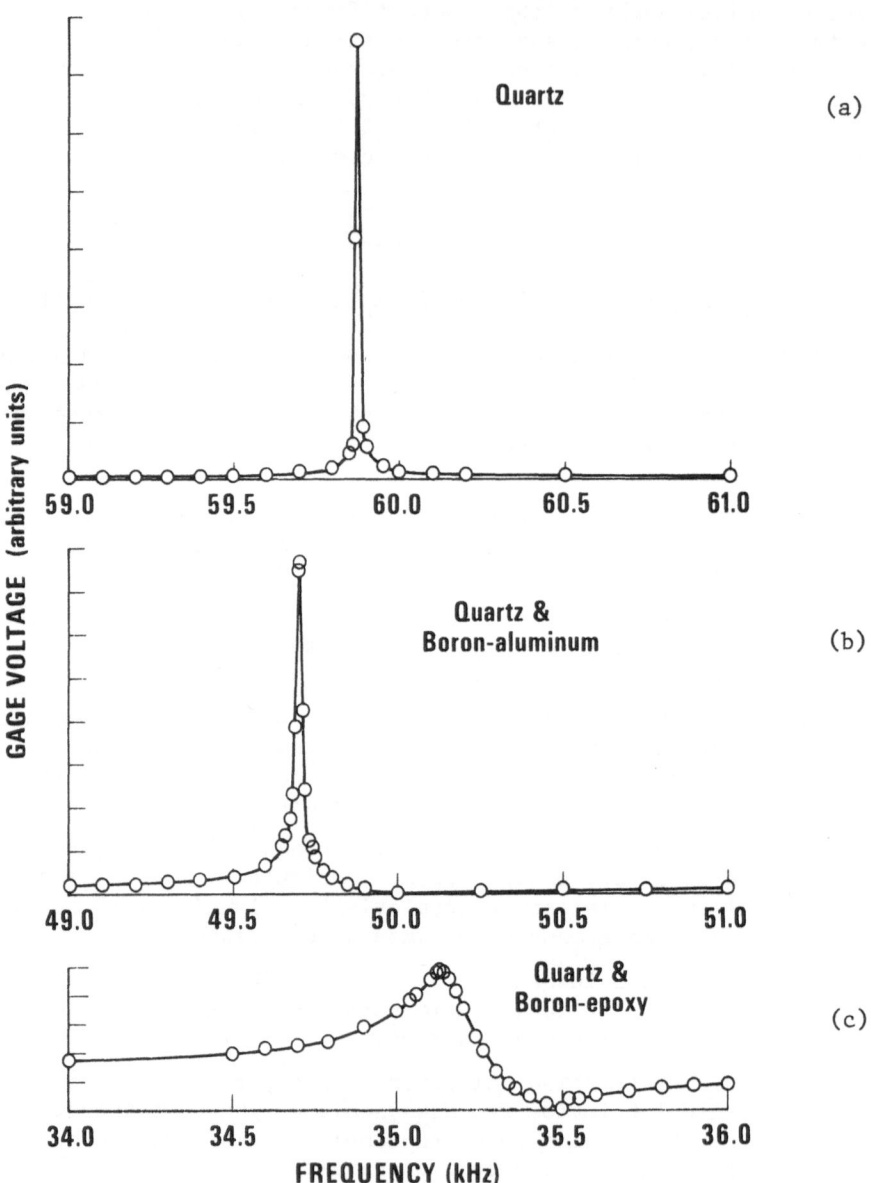

Fig. 4. Strain-amplitude-versus-frequency resonance peaks for
         (a) quartz, (b) quartz plus boron-aluminum, and (c)
         quartz plus boron-epoxy.

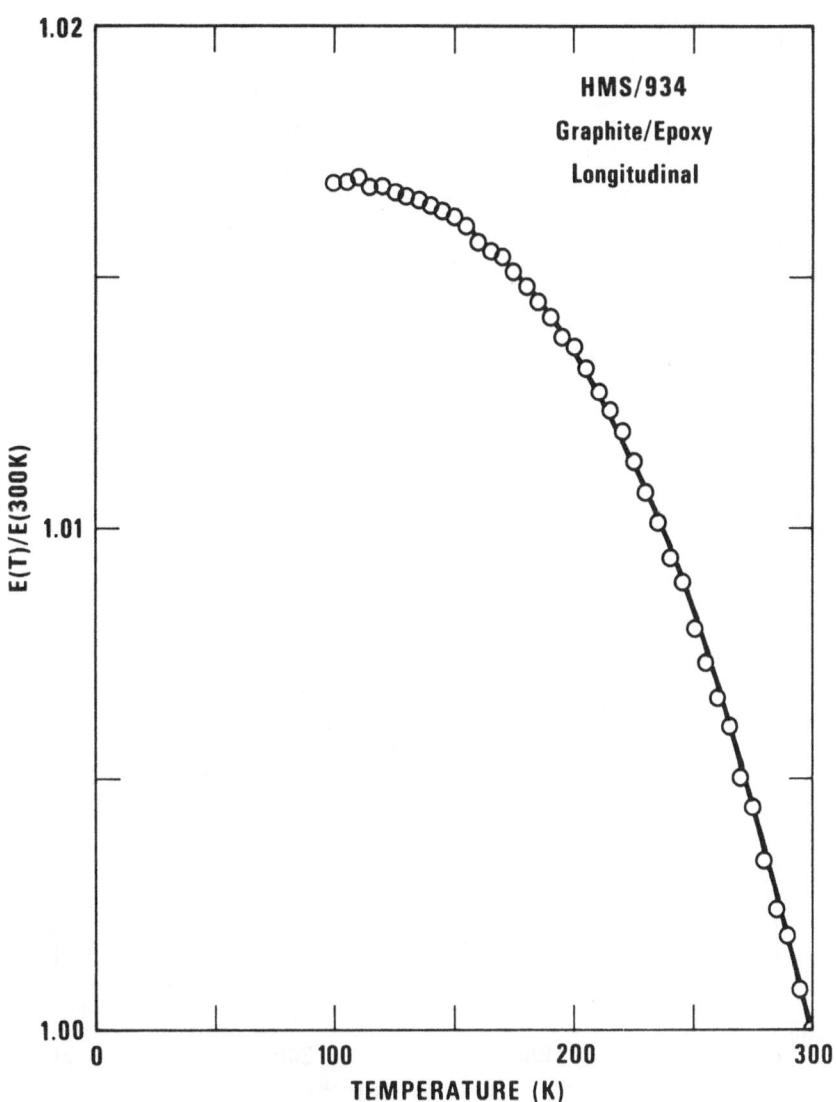

Fig. 5. Young's modulus versus temperature for a uniaxial
         graphite-epoxy composite in a longitudinal direction.

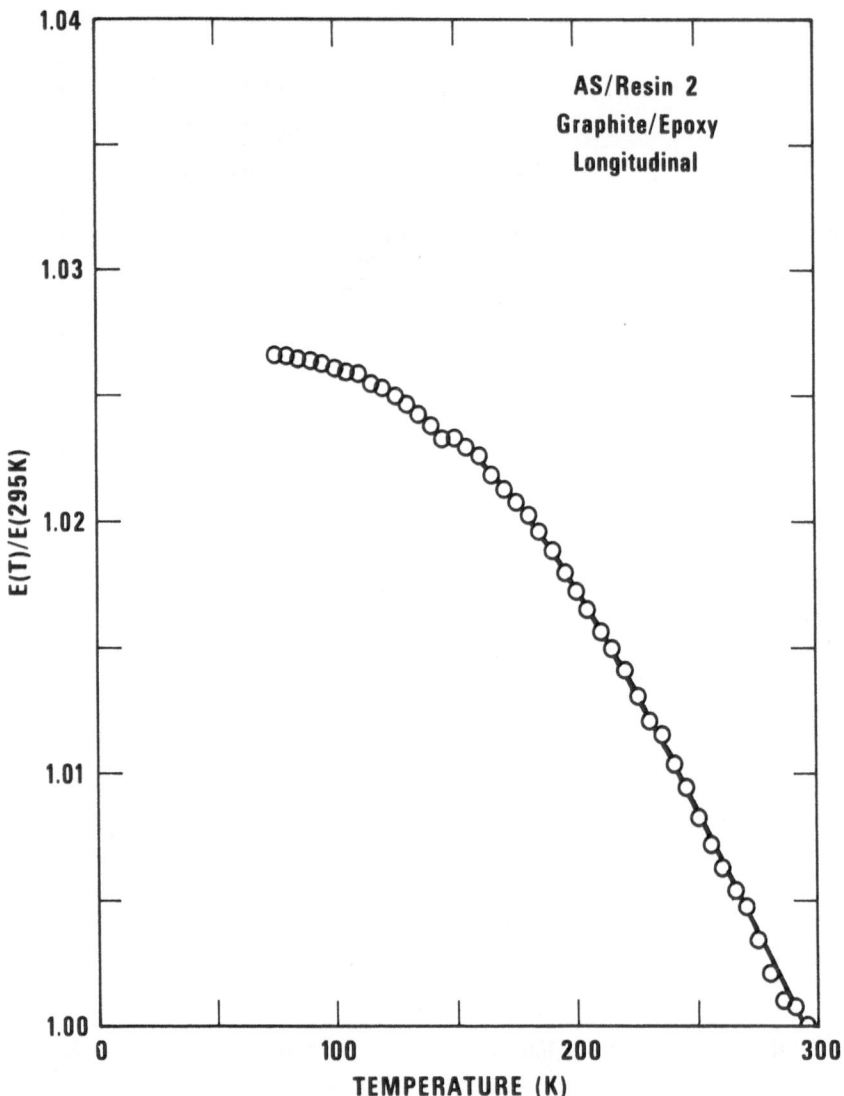

Fig. 6. Young's modulus versus temperature for a uniaxial
graphite-epoxy composite in a longitudinal direction.

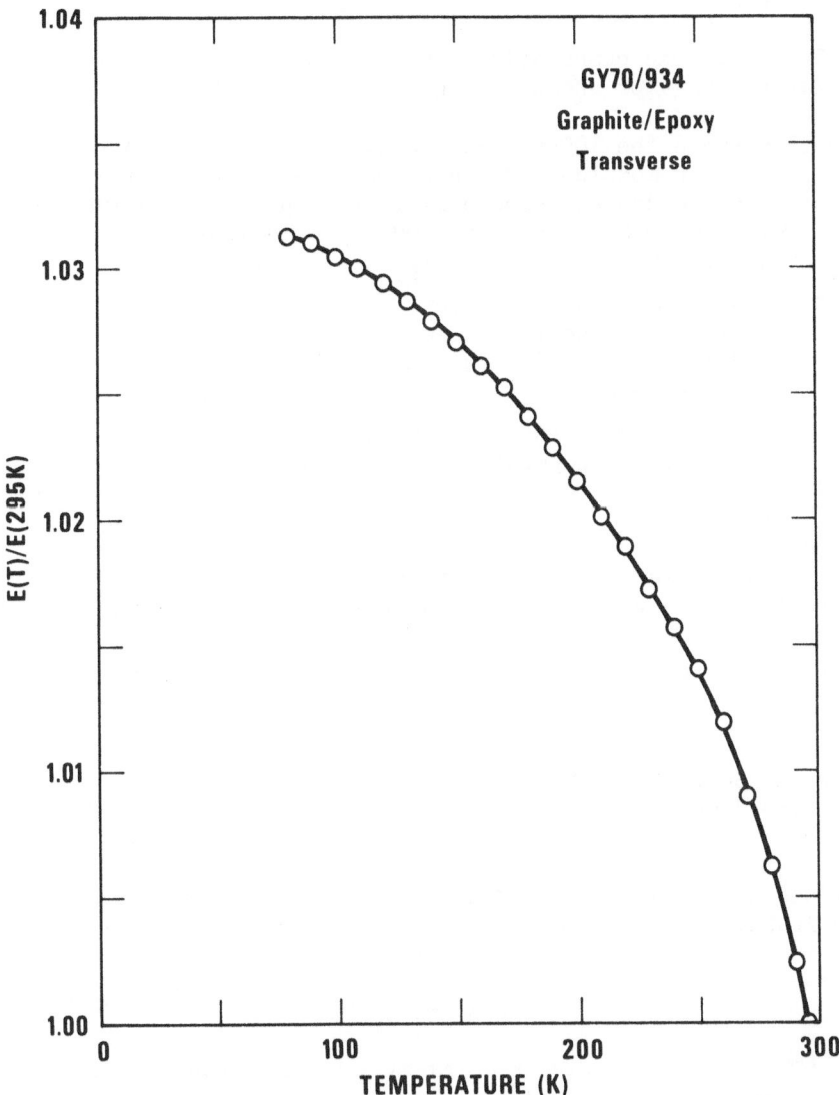

Fig. 7. Young's modulus versus temperature for a uniaxial
graphite-epoxy composite in a transverse direction.

results indicate that higher modulus may correspond to higher damping. To confirm this, further experimental studies should eliminate the frequency effects and strain-amplitude effects that perhaps occurred in the present study.

The following observations emerge from the Young's-modulus/temperature behavior shown in Figs. 5 to 7 for the graphite epoxies:

1. Despite the 100-percent variation in room-temperature Young's modulus, all three materials, for the particular modes measured, show only 2- to 3-percent modulus increases during cooling to below 100 K. In comparison, most metals and alloys show 5- to 15-percent increases.

2. Longitudinal and transverse moduli change in magnitude by a similar percentage. Considering the high Young's-modulus anisotropy, further studies may negate this surprising result.

3. Modulus/temperature curves show regular behavior: linearity at high temperatures, lower absolute slope at lower temperatures, and approaching zero slope near zero temperature.

Internal friction (results omitted here) decreases with lower temperatures to a value at 76 K about 1/2 to 1/3 its room-temperature value.

ACKNOWLEDGMENTS

R. E. Schramm and M. B. Kasen furnished materials for study and provided unpublished data. M. Austin and G. Maerz made the measurements.

This study was supported in part by the U.S. Air Force Aerospace Propulsion Laboratory.

REFERENCES

1. K. Thomas, D.E. Meyer, E.C. Fleet, and M. Abrahams, J. Phys. D. (Appl. Phys.) 6, 1336 (1973).

2. L.E. Nielsen, J. Compos. Mater. 9, 149 (1975).

3. A. Vary and K.J. Bowles, NASA Tech. Memo. TMX-73646, NASA-Lewis Research Center, Cleveland, Ohio (1972).

4. W.H. Robinson, S.H. Carpenter, and J.L. Tallon, J. Appl. Phys. 45, 1975 (1974). References 4-12 therein.

5. R.E. Schramm and M.B. Kasen, Mater. Sci. Eng. 30, 197 (1977).

6. R.E. Schramm and M.B. Kasen, in: Advances in Cryogenic Engineering, Vol. 24, Plenum Press, New York (1978), p. 271.

7. R.E. Schramm and M.B. Kasen, NBS Thermophysical Properties Division, Boulder, Colorado, personal communication (1978).

8. J. Marx, Rev. Sci. Instrum. 22, 503 (1951).

9. W.H. Robinson and A. Edgar, IEEE Trans. Son. Ultrason. SU-21, 98 (1974).

10. A.S. Nowick and B.S. Berry, Anelastic Relaxation in Crystalline Solids, Academic, New York (1972), p. 18.

COMPRESSIVE FATIGUE TESTS ON A UNIDIRECTIONAL GLASS/POLYESTER

COMPOSITE AT CRYOGENIC TEMPERATURES

E. L. Stone, L. O. El-Marazki, and W. C. Young

University of Wisconsin
Madison, Wisconsin, U.S.A.

INTRODUCTION

Superconductive magnetic energy storage (SMES) units are large potential users of fiberglass reinforced polymer matrix composites. A 1000 MWh unit would require 1.6 Gg of material.[1] The superconducting magnet requires low thermal conductance structural supports to transmit the magnetic loads from the 1.8 K conductors to the room temperature bedrock. In particular, it needs support struts with a high strength to thermal conductivity ratio, but low elastic modulus and a low strength to weight ratio are quite acceptable. This is unlike many aerospace applications. The large mass of material needed for struts necessitates the use of easily obtainable low-cost materials. This generally means that commercially fabricated composites will be used. The total capital cost must be balanced against operating cost.

At the University of Wisconsin (U.W.), a program has been under way to obtain cryogenic engineering data on commercial composites with emphasis on the strut requirements of SMES. The major loads of a SMES unit are carried by the struts in compression. Therefore, the first measurements were of the ultimate compressive strengths of several commercial "high pressure industrial laminates." The results at room temperature, 77 K, and 4.2 K have been previously reported.[2,3] It was found that unidirectional glass/polyester had high strengths, comparable with glass/epoxies, but was one-half to one-third as expensive. The glass/polyesters were therefore of considerable interest and chosen for further testing.

A SMES unit operating for 50 years at 1 cycle a day undergoes 18 250 cycles. There is, however, no cryogenic fatigue data available on unidirectional glass/polyester. The existing cyclic compressive fatigue data is on unidirectional glass in an epoxy matrix at room temperature.[4,5] At cryogenic temperatures, there have been 2 cyclic tensile fatigue studies made on a unidirectional glass/epoxy.[6,7] Therefore, it was important that compressive cyclic fatigue tests be carried out on the unidirectional glass/polyester.

The U.W. measurements are believed to be the first 77 and 4.2 K compressive fatigue data reported on unidirectional fiberglass reinforced polyester.

FATIGUE TESTING

One aim of the fatigue testing is to simulate the loading schedule of a SMES unit. This schedule is cyclic, once a day, varying smoothly from peak load to approximately one-quarter peak load. The cyclic life to failure is of interest, not life to a certain percentage change in modulus, since slight changes in modulus are acceptable. To simulate this SMES loading, a 1 Hz haversine waveform is imposed upon a constant baseload to give a ratio of minimum stress to maximum stress, R, of 4.0.

The glass/polyester composite chosen for fatigue testing has been previously tested for ultimate compressive strength and designated specimen A.[2,3] The ultimate compressive strength of two rods of specimen type A, nominally the same material, is presented in Table I. For the fatigue tests, rod 5 is used. It has the lower of the two ultimate compressive strengths.

The apparatus used for compressive fatigue testing has been previously described.[2,3] Load control of a closed-loop electrohydraulic test machine is used. The test specimens are right circular cylinders 1.27 cm diameter by 3.81 cm long. End caps are used to prevent premature failure due to the ends splitting and brooming. Before testing at 77 K, the specimens are cooled in the vapor above a liquid nitrogen bath for about 10 min, then immersed for 30 more min. Before testing at 4.2 K, the specimens are cooled for 2 h in the apparatus to 120 to 150 K. Then liquid helium is transferred and the samples are cooled further by the helium vapor for about 15 min before being covered by the liquid. The specimens equilibrate in liquid helium for 30 min before testing. At least five specimens at each of four different stress levels are tested at room temperature. At 77 K, five specimens at each of three different stress levels are run. Three specimens at one stress level are run at 4.2 K.

Table I. Average Ultimate Compressive Strength of Glass/Polyester

		Testing temperature					
		R.T.		77 K		4.2 K	
		Rod 1	Rod 5	Rod 1	Rod 5	Rod 1	Rod 5
σ_{Cu},	GN/m^2	0.73	0.66	1.38	1.18	1.35	–
	ksi	106	96	200	171	196	–
Number of specimens, type A		6	5	5	5	3	–
C.V., %		6	12	9	5	10	–

R.T. = room temperature

σ_{Cu} = average ultimate compressive strength

Specimen type A: pultruded rod, polyester resin, unidirectional
glass fiber from the Glastic Corporation, Cleve-
land, Ohio. "HIR" round rod.

C.V. = coefficient of variation = $\dfrac{\text{standard deviation}}{\text{average value}}$

RESULTS AND DISCUSSION

    As in the specimen A ultimate strength testing, almost all of
the room temperature and most of the cryogenic temperature fatigue
failures result in a uniform mushrooming of the specimen near one
end. The other failure mode is an approximately 45° fracture line
through the middle of the specimen. This is indicative of compressive failure in shear. There is no systematic difference in fatigue
life between these two failure modes.

    The room temperature and cryogenic compressive fatigue data
are presented in Table II. The peak stress versus median fatigue
life is plotted in Fig. 1 for the three temperatures. The 4.2 K
fatigue life point is similar to the 77 K data. The peak stresses
normalized by the ultimate compressive strengths are plotted in
Fig. 2 versus median fatigue life. The room temperature normalized
stresses are slightly higher and change more with fatigue than at
77 K. The difference in fatigue dependence is small, on the order
of the experimental uncertainty. At both room temperature and 77 K,

Table II. Cyclic Compressive Fatigue Data for Glass/Polyester, Specimen A, Rod 5

Testing temperature	Peak stress	Peak stress / Ultimate strength	Life (cycles)	Median life*
~295 K	595 MN/m^2 86.2 ksi	0.90	1.5 112 312 16 280 42 696 157 746[†]	2 300
	562 MN/m^2 81.4 ksi	0.85	347 3 235 5 748 6 721 10 873 101 742	6 200
	529 MN/m^2 76.6 ksi	0.80	8 754 15 462 89 331 100 519 256 404[†]	89 331
	462 MN/m^2 67.1 ksi	0.70	10 957 24 367 88 836 159 176 291 376[†] 434 253[†]	119 000
77 K	1060 MN/m^2 153.6 ksi	0.90	1.5 185 194 5 082 10 952	194
	942 MN/m^2 136.6 ksi	0.80	850 2 334 10 567 12 778 15 543	10 567

Table II   (Continued)

Testing temperature	Peak stress	Peak stress / Ultimate strength	Life (cycles)	Median life*
77 K	824 MN/m 119.5 ksi	0.70	18 503 31 217 95 448 104 028 162 499	95 448
4.2 K	942 MN/m 136.6 ksi	~0.8	597 2 858 17 000[†]	2 858

*Median life for even number of samples is calculated using the average of log life of the two middlemost values.

[†]Test stopped, run out.

fatigue failure at 100 000 cycles occurs at a stress level of 70 to 75% of the respective ultimate compressive strengths.

The fatigue life data for peak stresses of 70% and 80% of ultimate compressive strength have spreads of ±1 order of magnitude, whereas at 90% of ultimate, the spreads are ±2 orders of magnitude. One reason for this large scatter is believed to be sample variability. This sample variability shows up in the ultimate compressive strength measurements (see Table I). As can be seen in Fig. 2, a variation of ±10% in the ultimate compressive strength would be expected to result in a change in the fatigue life of ±1 order of magnitude. This is seen in the fatigue life data for peak stress levels of 70% and 80% of ultimate strength. Why the scatter is so much larger at peak stresses of 90% of the ultimate strength is not known.

A comparison can be made between these room temperature results and the fatigue resistance of S glass/epoxy reported in the literature. The room temperature strengths in the present study are intermediate between the compressive loading results reported by Cornish et al.[4] and those by Hofer and Olsen,[5] as can be seen in Fig. 3. The specimens in Cornish's study had a higher glass content than those in the study by Hofer and Olson. The slope of the fatigue line in the present study is only slightly steeper than both the S glass/epoxy curves.

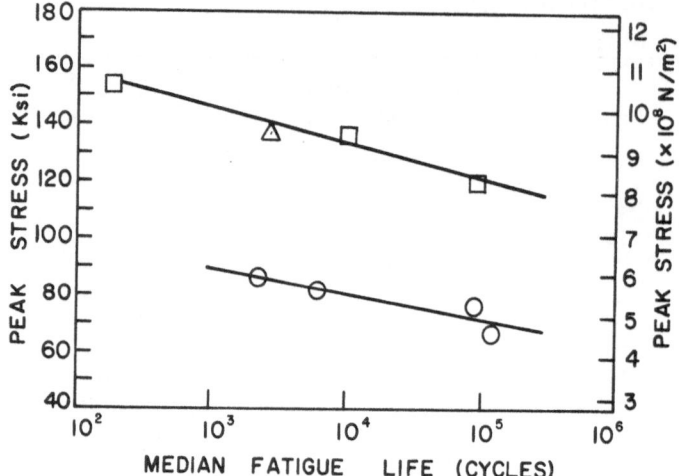

Fig. 1. Peak compressive stress versus median fatigue
         life for glass/polyester, specimen A, rod 5.
         0 - room temperature, □ - 77 K, Δ - 4.2 K.

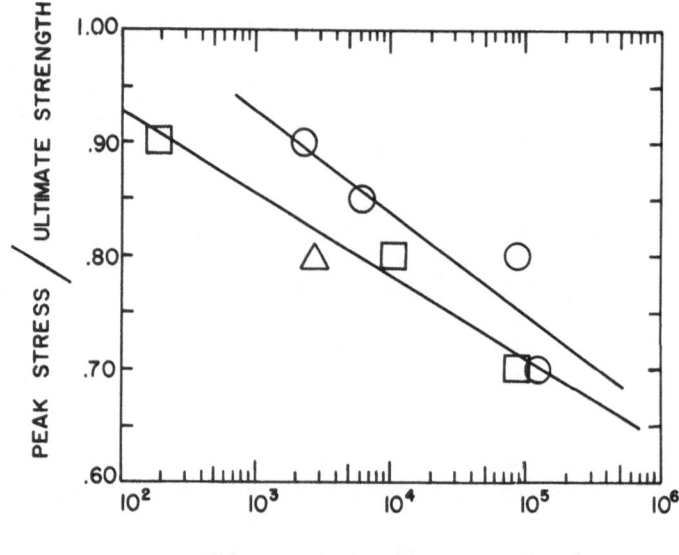

Fig. 2. Peak compressive stress normalized by ultimate
         compressive strength at given temperatures. 0 -
         room temperature, □ - 77 K, Δ - 4.2 K. 4.2 K
         point normalized to 77 K strength.

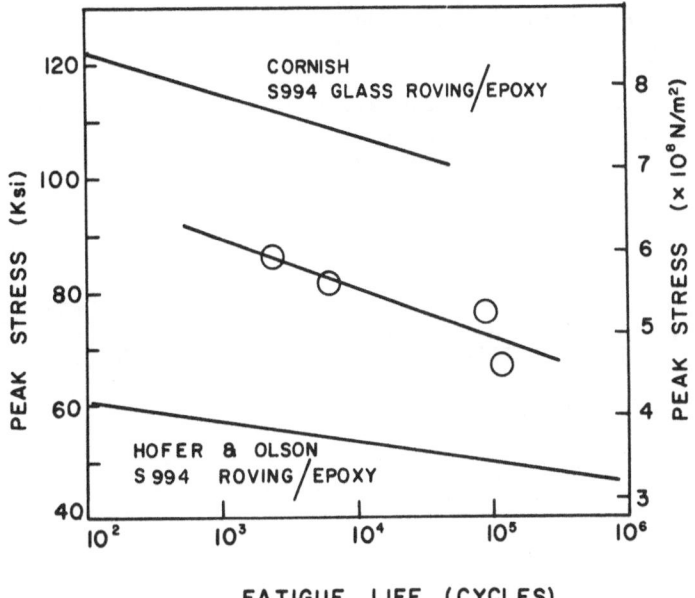

Fig. 3. Peak compressive stress versus fatigue life.
0 - room temperature data of present study.

The glass/polyester composites will probably cost less than one half as much as epoxy composites. Therefore, glass/polyester is being considered for struts in superconducting energy storage designs.

ACKNOWLEDGMENT

This work is supported by the U.S. Department of Energy.

REFERENCES

1. R.W. Boom et al., Wisconsin Superconductive Energy Storage Project Report, Vol. I (1974), Vol. II (1976), Annual Report (1977), University of Wisconsin, Madison, Wisconsin.

2. E. L. Stone and W.C. Young, in: Advances in Cryogenic Engineering, Vol. 24, Plenum Press, New York (1978), p. 279.

3. E.L. Stone and W.C. Young, in: Proceedings 7th Symposium on Engineering Problems of Fusion Research II, Knoxville, Tennessee, IEEE Publ. No. 77CH1267-4-NPS (1977).

4. R.H. Cornish, H.R. Nelson, and J.W. Dally, Proc. SPI 19, 9E (1964).

5. K.E.Hofer, Jr. and E.M. Olson, An Investigation of the Fatigue and Creep Properties of Glass Reinforced Plastics for Primary Aircraft Structures, Final Report (N67-32662) or (AD652415), (1967).

6. R.L. Tobler and D.T. Read, J. Compos. Mater. 10, 32 (1976).

7. D.W. Chamberlain, B.R. Lloyd, R.L. Tennant, Determination of the Performance of Plastic Laminates at Cryogenic Temperatures, Report No. ASD-TDR-62-794, Part 2 or N64-24212 (1964).

POWDER-FILLED EPOXY RESIN COMPOSITES OF

ADJUSTABLE THERMAL CONTRACTION

K. Ishibashi*, M. Wake, M. Kobayashi, and A. Katase*

National Laboratory for High Energy Physics
Ibaraki, Japan

*Kyushu University
Fukuoka-shi, Japan

INTRODUCTION

For the construction of superconducting dipole magnets for a synchrotron accelerator, it is necessary to use a number of spacers for separating a winding block to others. If the "cos θ" winding developed at Brookhaven National Laboratory[1] is chosen, then the number of spacers needed will be more than 10 000 for a synchrotron. Moreover, these spacers, wedges of special shape, need high dimensional precision to obtain permissible homogeneity of the magnetic field and therefore cannot be individually machined. From the viewpoint of mass production, a molding method may be the best way to obtain the above mentioned accuracy.

Nevertheless, the cured epoxy resins without fillers have a thermal contraction about four times higher than those of typical metals between room temperature (293 K) and liquid helium temperature (4.2 K). It has been reported,[2] however, that strain energy stored by differential thermal contraction between a superconducting cable and cured epoxy while cooling down leads a magnet to have premature quench and to experience training effect. To adjust the thermal contraction of epoxy to that of the winding material, therefore, it is necessary to add some fillers with a lower thermal contraction than metals to the epoxy resin. If we do not choose both amounts and kinds of fillers carefully, however, they would reduce not only the good molding characteristics but also the mechanical strength of the composite. In finding so-called good composites that have low thermal contraction comparable with metals, easy molding properties, and enough mechanical strength, it is very difficult to determine both the amounts and kinds of fillers as well as the combination of epoxy resin and its curing agent. A composite, named XD-580, developed at Rutherford Laboratory in England, partially satisfies these requirements on thermal contraction and mechanical strength. Nevertheless, this mixed pate before curing is too stiff and its pot life is too short, i.e., about half an hour. It is, therefore, somewhat difficult to mix completely the epoxy resin with curing agent and also to mold them into the spacers. If the molds we intend to make are relatively larger or longer, moreover, this short pot life may make their mechanical strength comparatively weaker, because short pot life leads to inhomogeneous curing owing to temperature distribution induced by quick reaction. In this present work, therefore, we have attempted to develop a composite that has better molding characteristics, relatively low thermal contraction, and sufficient mechanical strength, even for molding of larger or longer spacers.

THEORETICAL CONSIDERATION FOR THE COMPOSITE

For the development of the composite, the thermal contraction may be the most important problem. Following Hartwig,[3,4] we assumed a simple spherical model, as illustrated in an inset of Fig. 1. If we can solve the problem with the condition that only one kind of filler exists, the outward surface of matrix is completely free, and the stored stress energy resulting from contraction within this sphere becomes minimum, the thermal contraction can be written[3,4]

$$\alpha_c = \alpha_m - \frac{3}{2} \frac{(1-\nu_m)(\alpha_m-\alpha_f)\phi^+}{(1-2\nu_m)\phi^+ + 1 + \nu_m + (2-4\nu_f)(1-\phi^+)E_m/E_f} , \qquad (1)$$

where α is normalized thermal contraction between room temperature and liquid helium temperature, i.e., $(L_{293} - L_{4.2})/L_{293}$; ν is Poisson's ratio; ϕ^+ is the effective volume ratio of filler to matrix; E is the elastic modulus; and the subscripts c, m, and f mean composite, matrix, and filler, respectively.

As is well known, spheres of the same size occupy geometrically at most 74% of the total volume in which the spheres are contained. Considering this fact, Hartwig[3,4] has introduced a relation of $\phi = 0.74\phi^+$, where ϕ is actual volume ratio of filler to matrix. A chain line in Fig. 1 shows the results calculated using this relation and choosing the typical values for properties of matrix and filler. In this figure, the filler is a silica flour, and the experiment was carried out by Hartwig.[3] Nevertheless, the chain line is far below the experimental results. This disagreement

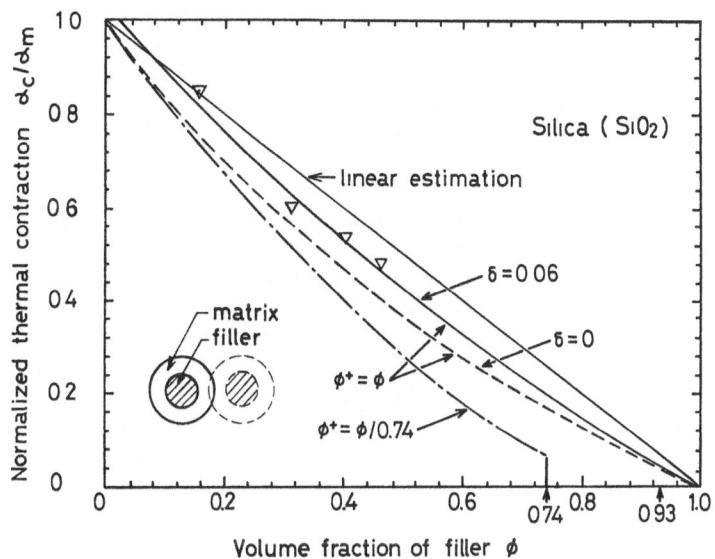

Fig. 1.   Thermal contraction as a function of filler concen-
          tration. The experimental data has been quoted from
          Reference 3. The inset illustrates the sphere model.

can be expected, because the theory overestimates the effect of
filler owing to the relation $\phi^+ = \phi/0.74$.

        Therefore, it seems reasonable to convert this relation to
$\phi^+ = \phi$. A dashed line in Fig. 1 shows the calculated results using
this relation. Agreement with the experimental ones is appreciably
improved. Nevertheless, the value expected by the theory is still
lower than the experimental one, probably because the free surface
on the matrix is assumed. In fact, there may exist a compressive
force among cells, as shown in Fig. 1. To include this effect,
therefore, we will assume a simple relation, $P_s = \delta P_b$, where P_s
and P_b are the pressure on the outer surface of the matrix and the
one on the boundary between the filler and matrix, respectively,
and δ is the adjustable parameter. Using this assumption, the
thermal contraction may be written as

$$\alpha_c = \alpha_m - \frac{(\alpha_m-\alpha_f)\,[3\phi(1-\delta)(1-\nu_m) - 2(1-\phi)\delta(1-2\nu_m)]}{2(1-\phi)(2-\nu_f)E_m/E_f + (1-\delta)[2(1-2\nu_m)\phi+1+\nu_m] - 2\delta(1-\phi)(1-2\nu_m)}$$

$$(2)$$

The solid line in Fig. 1 shows the calculated results with $\delta = 0.06$.
Good agreement with the experiment is obtained, although a small
discrepancy appears only where ϕ is near zero. This discrepancy
may arise because we used too simple an assumption of $P_s = \delta P_b$,
i.e., P_b and, consequently, P_s increase significantly with decreasing ϕ, as shown in Fig. 2. If the strength of the matrix is not
sufficient, the composite, even with a smaller filler concentration, will show degraded mechanical properties.

Fig. 1 also indicates that the lower limit of thermal contrac-
tion, corresponding to $\phi = 0.74$, may be 2.8×10^{-3}, using a typical
matrix thermal contraction of 1.4×10^{-2}, if only one type of filler is used. To attain a lower contraction than this, therefore,
it is necessary to use different sizes of fillers, i.e., one normal
size and the other much smaller. If we use two different sizes of
fillers, the smaller filler will enter among the larger ones in
the composite, and the maximum volume fraction of filler may become
0.93, meaning the filler percentage is given by $0.74 + (1 - 0.74)$
$\times 0.74$ and its corresponding thermal contraction is as low as
0.42×10^{-3}, as indicated in Fig. 1. This value, of course, shows
only the possibility, and it is quite difficult to attain without
any mechanical defect.

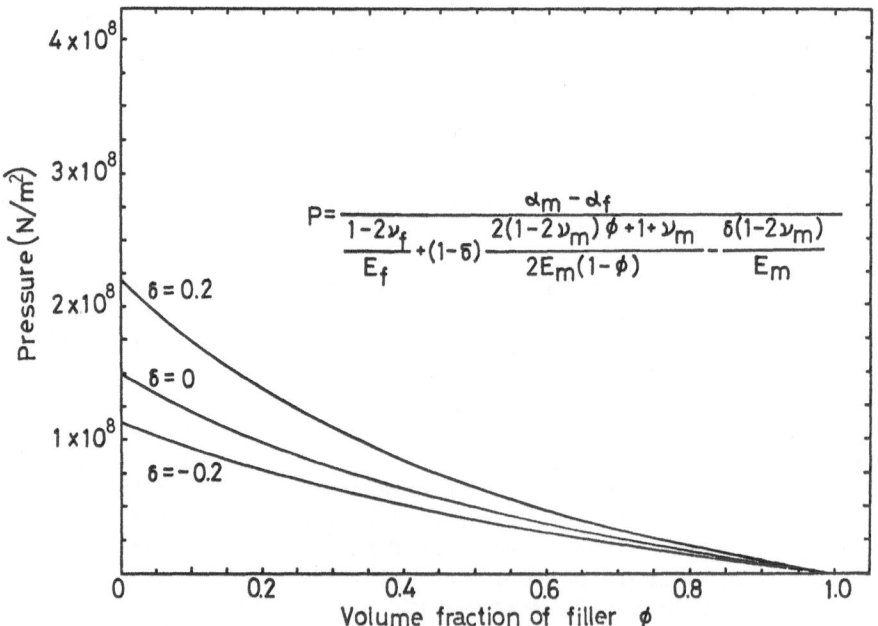

Fig. 2. Compressive force on the boundary between the filler
         and matrix. The calculated results are plotted as
         a function of ϕ and δ.

This figure, however, can give no information about the strength of the composite and ease of molding. Empirically speaking, the mixed pate of matrix and silica flour, of which the volume fraction is above 0.5, is poor in molding works, because it is hard enough and besides, it does not stick to other materials. Even if it can be cured, the strength of the composite is extremely low, and it cannot be used for any purpose. To obtain enough strength and easy molding characteristics besides desirable thermal contraction, therefore, it is important to choose the proper filler sizes and also the kinds of fillers.

TEST SAMPLES OF THE COMPOSITE AND THEIR PROPERTIES

Considering above mentioned facts, we have prepared test samples as shown in Table I. The combination of bisphenol-A and chainaliphatic amine, used in samples K-1 to K-5, is popular for use at low temperatures, and its strength has been confirmed by thermal cycle tests, where iron bolts have been imbedded in these

Table I. Specification of Test Samples.

Sample	Epoxy resin	Curing agent	Filler	Pot life
K-1	Bisphenol-A*	Cycloaliphatic amine	Calcium carbonate	5 h
K-2	Bisphenol-A*	+		
K-3	Bisphenol-A*	chainalipatic amine	Talc	
K-4	Bisphenol-A*		Asbestos	
K-5	Bisphenol-A	Cycloaliphatic amine	Coloring agent	5 h
K-6	Bisphenol-A*	Modified amine		3~4 h
K-7	Bisphenol-A*	+		
K-8	Bisphenol-A	polyamide		
K-9	Bisphenol-A			
K-10	Bisphenol-A			
XD-580			Silica?	0.5 h

* Resin of different molecular weight was also added.

composites and the pieces have been immersed many times in liquid
nitrogen to see whether cracks are generated or not. This thermal
cycle test has given us the knowledge that cycloaliphatic amine
and also a mixture of modified amine and polyamide are useful as
curing agents, and, consequently, they have been adopted to samples
K-1 to K-5 and K-6 to K-10, respectively. Some kinds of curing
agent are poisonous, but this cycloaliphatic amine is harmless to
the human body and, consequently, is sometimes useful to mass production of things by molding methods. As is seen from Table I, a
mixture of different kinds of epoxy resins or curing agents is
used so as to have a good affinity with filler.

Filler has consisted of calcium carbonate, talc, asbestos,
and coloring agent. Calcium carbonate has been used mainly to decrease thermal contraction and, moreover, chosen because of its
good affinity with epoxy resin and curing agents. Talc has been
used for ease of molding and asbestos to increase the strength of
the composite. Particle sizes and the mixing ratio of these components have been determined empirically to have good properties.
Particle sizes and amounts of filler components for K-5 and K-8
are the same as K-1, K-7 and K-9 are the same as K-2, K-10 is the
same as K-3, and K-6 is the same as K-4.

Resultant pot life of samples K-1 to K-5 is about 5 h at room
temperature, and that of K-6 to K-8, about 3.5 h, considerably
longer than the 0.5 h for XD-580. These composites would therefore not show any degradation in mechanical strength as a result
of their homogeneous curing, even for molding of large blocks.
It takes usually a few times longer than the pot life to cure the
composite completely.

Measurement for thermal contraction has been carried out using
the apparatus made of quartz and invar illustrated in Fig. 3. The
measuring process is as follows: after liquid helium is filled in
a cryostat, it is evaporated by a heater; then the sample is gradually warmed up to room temperature, taking about 30 h. During
this temperature rise, both sample elongation and temperature are
recorded continuously by another heater wound around a copper block.
Figure 4 shows the measured results for thermal contraction plotted
as a function of temperature in K. In Fig. 4a, the thermal contractions for various metals are indicated. K-1 and K-5 have rather
lower thermal contractions, near to that of iron. K-2 may be suitable for spacers of cold iron yoke magnets, considering the lower
thermal contraction of the iron yoke, and K-3 may be suitable for
spacers of the warm iron yoke magnets. Thermal contractions of
samples K-6 to K-10 in Fig. 4b are considerably higher than those
corresponding to the filler in Fig. 4a, because of the difference
in curing agents, as shown in Table I.

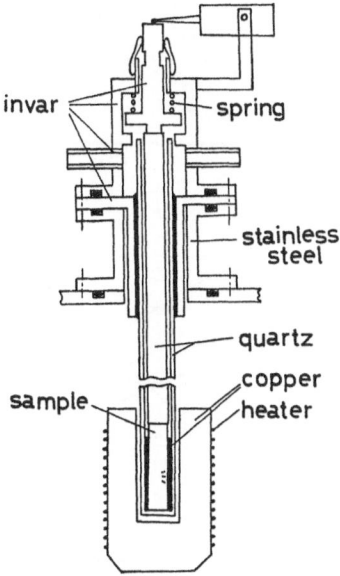

Fig. 3.   Apparatus for thermal
contraction measurement.

        Measurements for mechanical properties have been carried out
at both room temperature and liquid nitrogen temperature. These
are the tensile, bending, compressive, elastic, and Charpy impact
strength measurements, according to Japanese Industrial Standards
(JIS).   Table II shows the results measured for these properties.
A temperature of -160°C has been attained by blowing cold nitrogen
gas. Five test pieces have been prepared for each test and three
figures, except the highest and lowest ones, have been averaged.
These values sometimes have curious behavior from sample to sample;
for example, the tensile strength at 23°C for sample K-2 is appreciably lower than those of K-1 and K-3, although the filler composition for K-2 is between K-1 and K-3.   Nevertheless, the measured
strengths are of the same figure as XD-580 and may be suitable for
normal use.

                        RESULTS AND DISCUSSION

        The mechanical properties of our composites have relatively
high strength. Nevertheless, asbestos added to improve the
mechanical properties may not be so effective as expected. To
obtain mechanical properties more suitable for the molding, however,
it may be better to add the glass fiber of short length instead of

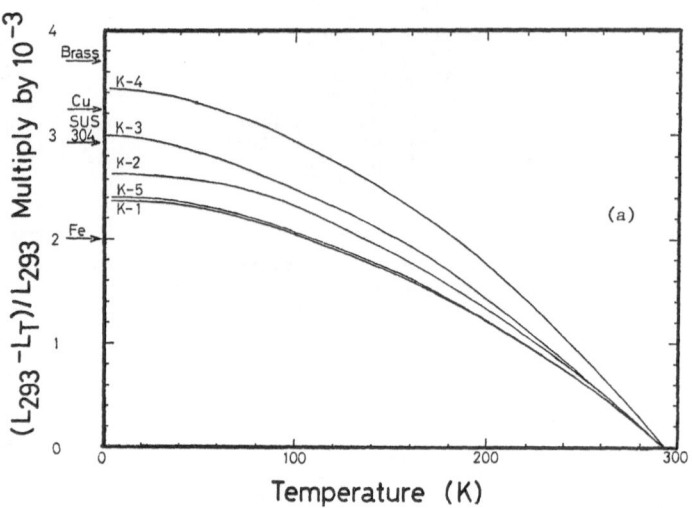

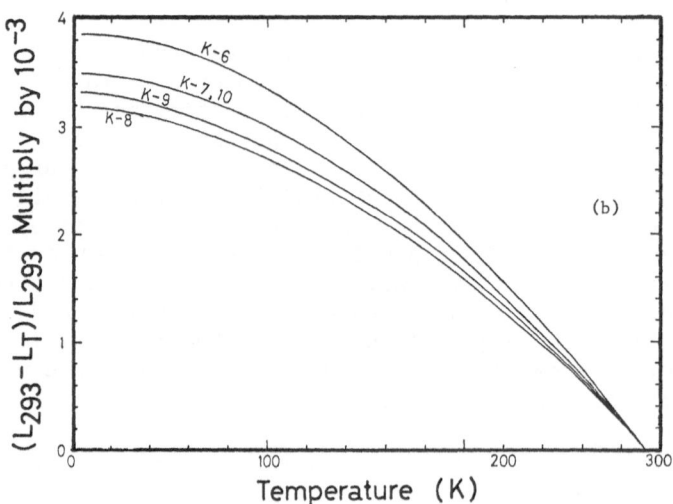

Fig. 4.   Thermal contraction for test samples.

Table II. Mechanical Properties of Test Samples.

Sample	Tensile strength, kg/mm²		Bending strength, kg/mm²		Compressive strength, kg/mm²		Elastic modulus kg/mm²		Charpy impact strength, kg/cm²
	Temp. 23°C	Temp. -160°C	Temp. 23°C	Temp. -160°C	Temp. 23°C	Temp. -160°C	Temp. 23°C	Temp. -160°C	Temp. -160°C
K-1	2.2	6.7	6.0	9.2	–	–	1400	2550	–
K-2	1.5	3.3	5.0	8.3	8.7	19.7	1760	1900	3.28
K-3	2.6	3.7	5.9	9.7	10.6	28.9	1480	2230	3.39
K-4	2.2	2.7	6.0	9.4	10.8	31.9	1320	2040	3.37
K-5	2.7	2.9	6.2	8.8	10.5	27.8	1680	2560	3.93
K-6	2.1	3.9	6.3	10.9	11.4	26.8	1210	2030	4.09
K-7	1.4	5.5	6.0	9.7	11.6	26.6	1350	2150	3.50
K-8	2.4	–	5.5	8.6	–	–	1380	2320	3.75
K-9	2.2	–	6.2	10.3	–	–	1470	2340	3.47
K-10	2.7	–	6.3	10.0	–	–	1320	2190	3.50
XD-580	3.1	6.4	5.9	9.7	10.4	24.8	1470	2170	4.0

asbestos, because it improves the strength in the longitudinal
direction.

As already mentioned, we have succeeded in developing a composite that has, for spacers of dipole magnets, proper thermal
contraction, easy molding characterisitics, and relatively high
mechanical strength. Also, now it is possible to make an epoxy
resin composite of thermal contraction adjustable to any metals;
for example, iron, stainless steel, and brass.

ACKNOWLEDGMENTS

We are very grateful to Professors M. Morimoto and H. Ishimoto
for helpful discussions and N. Kudo for the fabrication of some
apparatus.

REFERENCES

1.   A.D. McIntruff et al., Isabelle Ring Magnets, BNL 21708,
     Brookhaven National Laboratory, Upton, New York (1977).

2.   M.N. Wilson, Stabilization of Superconductors for Use in
     Magnets, Applied Superconductivity Conference, Rutherford
     Laboratory, Chilton, Didcot, Oxon, England (1976).

3.   G. Hartwig, A. Puck, and W. Weiss, Mater. Sci. Eng. 22, 26
     (1976).

4.   G. Hartwig, in: Advances in Cryogenic Engineering, Vol. 24
     Plenum Press, New York (1978), p. 17.

EFFECT OF STRAIN ON EPOXY-IMPREGNATED

SUPERCONDUCTING COMPOSITES

J. W. Ekin, R. E. Schramm, and A. F. Clark

National Bureau of Standards
Boulder, Colorado, U.S.A.

INTRODUCTION

Flux-jump stabilized superconducting magnets are usually impregnated with epoxy to prevent wire movement and enhance stability. Hoop stress experienced by the windings when the magnet is energized would otherwise cause wire movement, leading to localized heating and, quite probably, thermal runaway.

Unfortunately, the epoxy itself can act as a source of localized heating and initiate thermal runaway. This occurs when the epoxy is stressed to the point of fracture by a combination of magnetic hoop stress and stress introduced by differential thermal contraction between the epoxy and superconducting wire during cooldown. Usually, failures take place only locally in regions of stress concentration. The local density of released energy, however, is usually more than adequate to initiate thermal runaway. Thus, epoxy impregnation of flux-jump stabilized magnets would be expected to contribute to their stable operation only at moderate stress levels. At some point of strain, localized failure of the epoxy is expected to degrade magnet performance. The purpose of this study is to see if this point can be predicted from the mechanical properties of the epoxy. A further purpose is to determine if the epoxy properties have any correlation with the common problem of magnet "training," wherein a flux-jump stabilized magnet must be quenched a number of times before it reaches the final design field.

This study is divided into two parts: first, a study of three types of epoxies commonly used in the construction of potted magnets; second, a preliminary study of superconducting rings potted with these epoxies.

EPOXY TESTS

Specimen Preparation and Tensile-Test Apparatus

Mixing of epoxy components was according to manufacturers'*
directions; in some cases, outgassing the mixture at about 600 Pa
(6×10^{-3} atm) for 5 to 10 minutes followed. Curing was in a sili-
cone-rubber mold and yielded a flat, tapered specimen, accord ng
to ASTM D 638, Type I;[1] the gage length was 5 cm long and 1.3 cm
wide with a thickness of about 0.5 cm. After the prescribed cure,
commercial metal-foil strain gauges were attached directly to the
specimen surface (the strain-gage-cement cure temperature was
well below that of the test epoxy). Specimen conditioning prior
to testing was at least 40 hours at 50% relative humidity and 23°C
(ASTM D 618, procedure A).[2]

The specimen grip system was 100 mesh stainless steel screen
wrapped around 2.5 cm of each end before clamping between two stainless steel plates (0.6 cm thick) cross-serrated with 10 lines/cm.
After assembly in an alignment jig, three pairs of 10-32 screws,
torqued to 2.3 to 4.5 N-m (20 to 40 in-lbs), were used to clamp the
grip onto the specimen. Holes in the steel plates allowed a load
pin attachment to clevises on each pull-rod. The test frame design[3] made it possible to immerse the entire system in a dewar of
liquid helium.

During the test, a two-pen X-Y plotter recorded strain gage
output as a function of load. Our procedure was to load cycle the
specimen (strain rate, $\dot{\varepsilon} = 3 \times 10^{-4}$ s^{-1}) to 25 to 35% of its ultimate strength four times with high sensitivity settings on the
plotter; these data yielded the elastic parameters. We decreased
the plotter sensitivity on the final loading cycle to record the
fracture parameters.

Epoxy Results

Table I shows the mechanical property results for three types
of epoxies at both room temperature and at liquid helium temperatures.

---

*Epoxy A, Armstrong A12, manufactured by Armstrong Products Company,
Warsaw, Indiana; Epoxy B, Stycast 2850FT, manufactured by Emerson
and Cuming, Canton, Massachusetts; Epoxy C, Araldite 6004, manufactured by Ciba Products Company, Summit, New Jersey, mixed with
Lindride 16 hardener (manufactured by Lindau Chemicals, Inc.,
Columbia, South Carolina) in the ratio 100 parts 6004 to 85 parts
hardener and cured for 8 h at 80°C. Trade names are used to define
the materials that were studied. Their use implies no endorsement
of particular products by NBS.

Table I. Epoxy Properties.

Epoxy	Temperature, K	Young's modulus, GPa	Ultimate strength, MPa	Ultimate strain, %
A	4	8 (CV = 17%)*	30 (CV = 14%)	0.4 (CV = 27%)
B	4	24 (CV = 4%)	80 (CV = 11%)	0.32(CV = 7%)
C	4	8 (CV = 13%)	60 (CV = 21%)	0.8 (CV = 12%)
A	295	3	25	0.6
B	295	13	40	0.32
C	295	4	34	0.8

*CV = coefficient of variation among test specimens.

The flexibilized epoxy A has a low modulus and high ultimate strain
at room temperature, but a much higher modulus and lower ultimate
strain at 4 K, about 0.4%. The filled epoxy B has the highest
modulus, but a relatively low ultimate strain, 0.32% at 4 K. The
low-viscosity epoxy C has a low modulus at 4 K, but the highest
ultimate strain, about 0.8% at 4 K.

Note the substantial change in properties of the flexibilized
epoxy A on cooling to 4 K. At liquid helium temperature, its
elongation is no better than the filled epoxy B. By contrast, the
low-viscosity epoxy C has, by far, the greatest elongation at 4 K.
The comparative differences in elongation at 4 K are the most
important property for the purposes of this study.

COMPOSITE RING TESTS

Ring Specimens and Test Apparatus

Several superconducting composite rings were fabricated using
each of these epoxies. Approximately 150 turns of copper-stabilized
NbTi (1.8:1 copper-superconductor ratio) were wound into rings 15 cm
in diameter, 1 cm thick by 1 cm high. Fiberglass cloth was used
to separate each layer of superconducting wire. A cross-sectional
view of the final ring structure is shown in Fig. 1. For epoxies
A and B, the rings were potted as they were wound, using a wet
layup technique. Rings made with the low-viscosity epoxy C were
vacuum impregnated after winding.

To simulate the hoop stress that would be generated in an
actual magnet solenoid, a split-D fixture was used to apply stress
to the ring, as shown schematically in Fig. 2. With this arrangement, the stress is concentrated in the gap region between the two
D's and, unfortunately, a significant amount of bending strain is
also introduced. To better simulate the actual hoop stress in a
magnet, a multifingered-cone and magnet arrangement is now being
developed. Nevertheless, tests with the simplified apparatus have
served to give comparative results among the different epoxy types,
although the relative strain values indicated below should not be
taken to represent absolute hoop strain. Strain was calculated
from measurements of D separation using an extensometer. These
calculated values were also correlated with strain measured directly
with gages attached to various sections of the ring. The strain
values given in this paper represent the maximum combined hoop
strain and bending strain occurring in the ring.

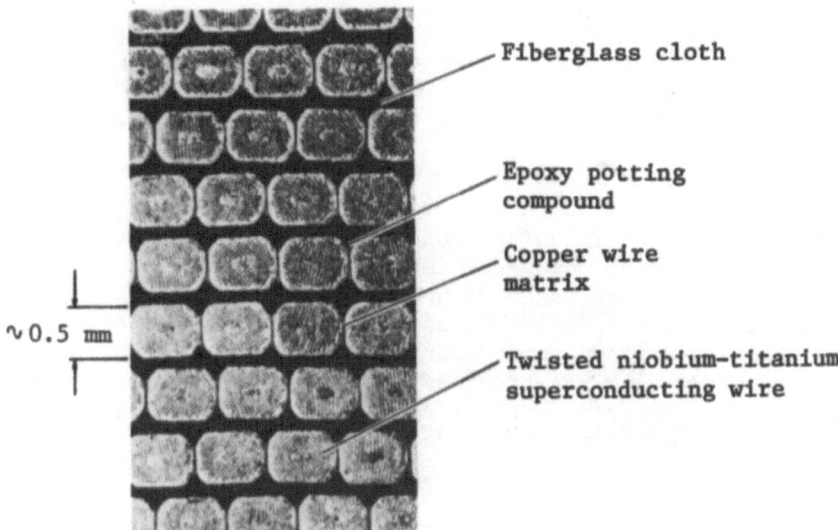

Fig. 1. Cross-sectional view of a superconducting composite ring showing its components.

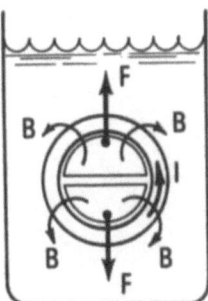

Fig. 2. Schematic of the ring test setup showing the configuration of force F applied to the D's, current I applied to the superconducting ring, and the resulting self-field B, calculated to be about 3 T at maximum current.

Composite Ring Results

While under strain, each ring was energized with current until it quenched. At constant strain, the quench current kept increasing until it reached an ultimate value, i.e., the ring "trained." The number of quenches it took to teach the ultimate current is shown in Fig. 3.

What causes training? We suggest that a dominant cause of the training process in potted magnets may be associated with localized fracture of the epoxy impregnant, as follows: When the coil is energized, the magnetic hoop stress on the coil increases. This charging process continues until a failure of the epoxy occurs in the region of stress concentration sufficiently large to initiate thermal runaway. However, the net result is that the stress concentration in the epoxy is relieved, and the next time the coil is energized, it can withstand a higher hoop stress, allowing the coil to go to a higher current. This process continues until the highest levels of stress concentration are all relieved, or the superconducting wire reaches its short-specimen critical current.

As may be seen from Fig. 3, all three epoxies showed little or no training initially, but all trained to varying degrees at high strain. The training in rings impregnated with epoxies A

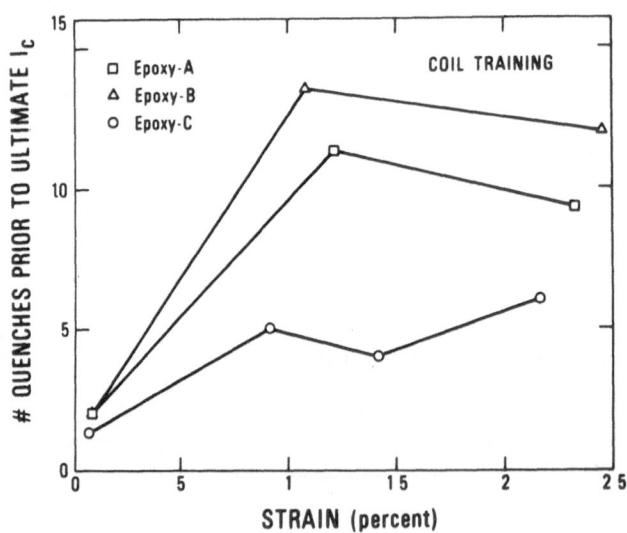

Fig. 3. Training behavior of each ring type as
a function of applied strain.

and B was more severe, in correlation with their relatively low
ultimate strain. In contrast, the ring made with epoxy C showed
significantly less training, consistent with its much higher strain
tolerance and the training mechanism described above.

The ultimate critical current reached after training is plotted for each ring type in Fig. 4. The critical current degradation
shown in Fig. 4 is mainly a function of the localized strain in the
superconducing wire rather than the epoxy. Recently, it was shown
in a separate study on nonpotted NbTi wire specimens[4] that the
critical current of NbTi will begin to degrade significantly at
about 1% strain. This effect is aggravated, however, by cracks in
the epoxy, which lead to localized strain in the superconductor.
It is seen in Fig. 4 that the critical-current degradation is most
pronounced in the composite rings potted with epoxies A and B,
again correlating with the lower strain tolerance of these epoxies
compared with epoxy C.

The method of epoxy impregnation may also be playing a role
in both these results. Rings potted with epoxies A and B were

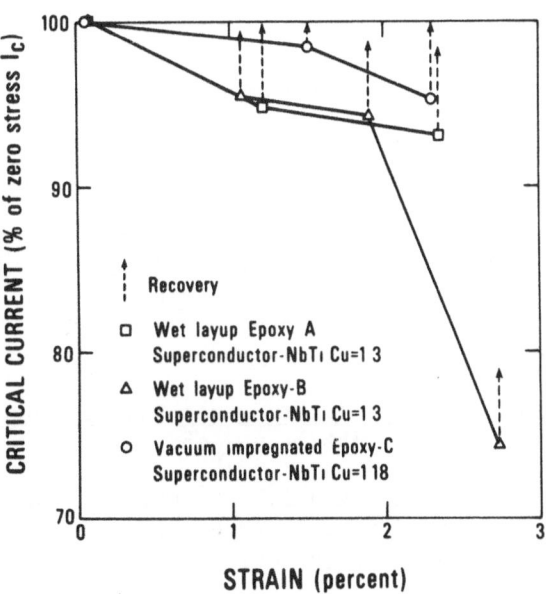

Fig. 4.   Critical-current degradation of each ring
          type as a function of applied strain.

fabricated using a wet layup technique. The ring made with epoxy
C was made using a vacuum-impregnated technique. Voids introduced by the wet layup technique may well serve as crack initia-
tion sites, leading to even lower strain tolerance of epoxies A
and B.

CONCLUSIONS

These preliminary data indicate that the training process in
potted superconducting magnets is associated with a succession of
stress-concentration relief in the epoxy impregnant. The elongation of the epoxy at 4 K and the method of impregnation appear to
be key factors controlling training.

ACKNOWLEDGMENTS

We are grateful to American Magnetics, Inc. and to M. J. Super-
czynski and F. E. McDonald of the Naval Ship Research and Development Center for fabricating the superconduction composite rings
tested in this study. We wish to particularly thank M. J. Super-
czynski for his suggestions and helpful discussions relating to
these results.

This work was supported by the U.S. Naval Ship Research and
Development Center.

REFERENCES

1. "Tensile Properties of Plastics," Annual Book of ASTM Stand-
    ards, Part 35, ASTM D638-76, American Society for Testing and
    Materials, Philadelphia, Pennsylvania (1977), p. 215.

2. "Conditioning Plastics and Electrical Insulating Materials for
    Testing," Annual Book of ASTM Standards, Part 35, ASTM D618-61,
    American Society for Testing and Materials, Philadelphia,
    Pennsylvania (1977), p. 198.

3. R.P. Reed, in: Advances in Cryogenic Engineering, Vol. 7,
    Plenum press, New York (1962), p. 448.

4. J.W. Ekin, F.R. Fickett, and A.F. Clark, in: Advances in
    Cryogenic Engineering, Vol. 22 (Proc. Int. Cryogenic Materials
    Conf., Kingston, Ontario, Canada, 1975), Plenum Press, New
    York (1977), p. 449.

THERMOPHYSICAL PROPERTIES OF COMPOSITE MATERIALS BASED ON HIGH-MOLECULAR COMPOUNDS WITH FIBROUS FILLER BETWEEN 10 AND 400 K

L. L. Vasiliev, L. S. Domorod, and S. A. Tanaeva

The Luikov Heat and Mass Transfer Institute
Minsk, USSR

Low-temperature thermal physics is a comparatively young field of science closely connected with low-temperature physics, cryogenics, and rocket engineering. The knowledge of thermophysical properties of materials is necessary to design powerful superconducting solenoids, cryogenic power lines, thermal insulation of liquefied gas transport tanks, etc. In the majority of cases, thermophysical properties are of interest when they are measured under the effect of conditions approaching operational conditions.

Presently in cryogenics, composite polymer materials are widely used. These are glass and carbon plastics based on epoxy and phenolformaldehyde resins. But there are scanty data on their thermophysical properties, and they refer, as a rule, to positive temperatures.

It is difficult to calculate thermal conductivity of oriented filled polymers, all the more to ascertain the temperature dependence of thermal conductivity (λ), thermal diffusivity (a), and specific heat (c). The calculation formulae cannot allow for such phenomena as the glass-transition of polymers, the possible lamination of polymer films due to the great discrepancy between the coefficients of linear expansion of the binder and filler, the effect of multiple thermal loading, etc. Therefore, most valuable are the experimental data on thermophysical properties of composite polymers in a wide temperature range (between 10 and 400 K).

Thermal conductivity, thermal diffusivity, and specific heat have been measured during nonsteady-state sample heating using the installation and methods described in Ref. 1. Test samples were square plates, and thermal properties were measured in two directions, i.e., with a heat flux along and across the filling fibre. As the measurements have shown, thermal properties of filled polymers depend considerably on filler orientation. Thermal conductivity and specific heat of glass plastics with formaldehyde and epoxy binder increase with increasing temperature, whereas thermal diffusivity falls in inverse proportion with temperature. The direction of the heat flux and orientation of the filler are responsible for the conductance and thermal diffusion in a given direction. Specific heat does not practically depend on the heat flux direction, since it characterizes the scalar value, i.e., energy accumulation.

The properties of filled polymers are greatly affected by solid fillers, such as glass and carbon fibres. This effect is due to wedging pressure and intermolecular forces between the polymer and adjacent phases. A specific structure of these layers contributes to specific differences in their properties.

In the flow direction, the molecules are chemically bound, which ensures high strength and large values of thermophysical properties. Across the flow, natural physical forces are acting, and, in this case, these quantities are lower. When affected by these forces and the orientation effect of the filler surface, polymer molecules may distribute more or less in order. Thus, anisotropy of filled polymer properties occurs because of a different heat transfer mechanism between the molecules of the filler and polymer binder and heat transfer in each of the components.

The authors have studied the thermophysical properties of glass, carbon, and glass-carbon fibres. As has been noted already (Fig. 1), thermal properties of glass fibres increase monotonically with growing temperature,[2] whereas carbon and glass-carbon fibres have some specific features. Thus, the behavior of carbon-fibre thermal diffusivity along and across the fibres is anomalous, which is a maximum between 50 and 200 K (Fig. 2). On the one hand, this anomaly is explained[3] by formation of new overmolecular structures in the binder causing anisotropy of polymer properties and, on the other, by high content (up to 70%) of graphite.

Of interest is the study of the properties of glass-carbon fibres on the basis of epoxy resin EDT-10 with modifying additives as well as of the filler composed of carbon and glass fibre ropes in a temperature range between 10 and 400 K. In Fig. 3, specific heat (J/kg·K), thermal conductivity (W/m·K) and thermal diffusivity (m^2/s) of glass-carbon fibres along (‖) and across (⊥) the fibres are plotted against temperature. Temperature dependences $\lambda_{\parallel}(T)$, $\lambda_{\perp}(T)$, $c_{\parallel}(T)$, and $c_{\perp}(T)$ increase monotonically; temperature dependence $a_{\perp}(t)$ is a hyperbola; and the curve $a_{\parallel}(T)$ has a maximum.

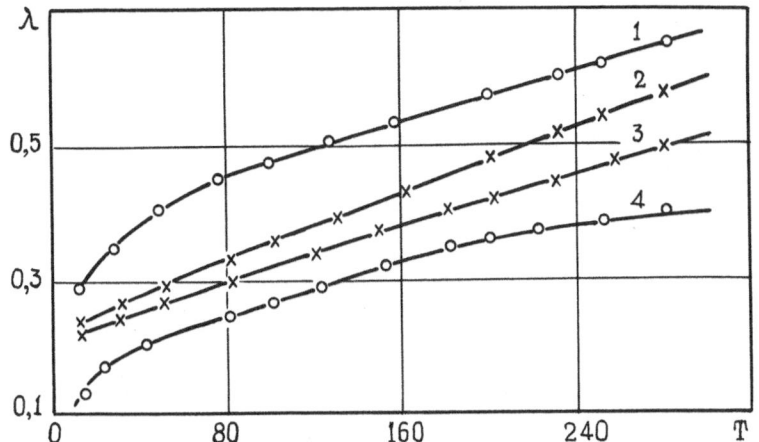

Fig. 1.   Thermal conductivity (λ, W/m·K) of glass–carbon fibres
          EDT-10 on the basis of epoxy binder and AG-4C on the
          basis of phenoloformaldehyde binder.

          1 - EDT-10 along the fibres; 2 - AG-4C along the fibres;
          3 - AG-4C across the fibres; 4 - EDT-10 across the fibres.

The shape of $\lambda_{\parallel}$(T) and $\lambda_{\perp}$(T) qualitatively agrees with the
phenomenological Debye representation

$$\lambda = c_V u \bar{l}/3 \qquad\qquad (1)$$

where c_V is the specific heat, u is the sound speed, and $\bar{l}$ is the
mean-free-path length of phonons. At very low temperatures, the
mean-free-path length of phonons does not change, in general, and
depends on crystallite sizes. λ(T), therefore, depends mainly on
c(T). With increasing temperature, $\lambda_{\parallel}$(T) of the composite increases
sharply. This is because $\lambda_{\parallel}$ of carbon fibres greatly depends on
temperature and has a maximum at a temperature about 100 K, and,
for some types of carbon fibres, $\lambda_{\parallel}$ sharply increases.

Since glass-carbon fibre is antisotropic material, then abso-
lute values of $\lambda_{\parallel}$(T) and $\lambda_{\perp}$(T) greatly differ from one another.
The numerical ratio of $\lambda_{\parallel}$ and $\lambda_{\perp}$ (degree of anisotropy) increases
from a factor of 2 to a factor of 7 in the test range of tempera-
tures. It must be noted that anisotropy of glass-carbon fibre
properties has increased markedly against that of carbon and glass
fibres. Thus, anisotropy of glass fibre EDT-10 ranges from 1.75

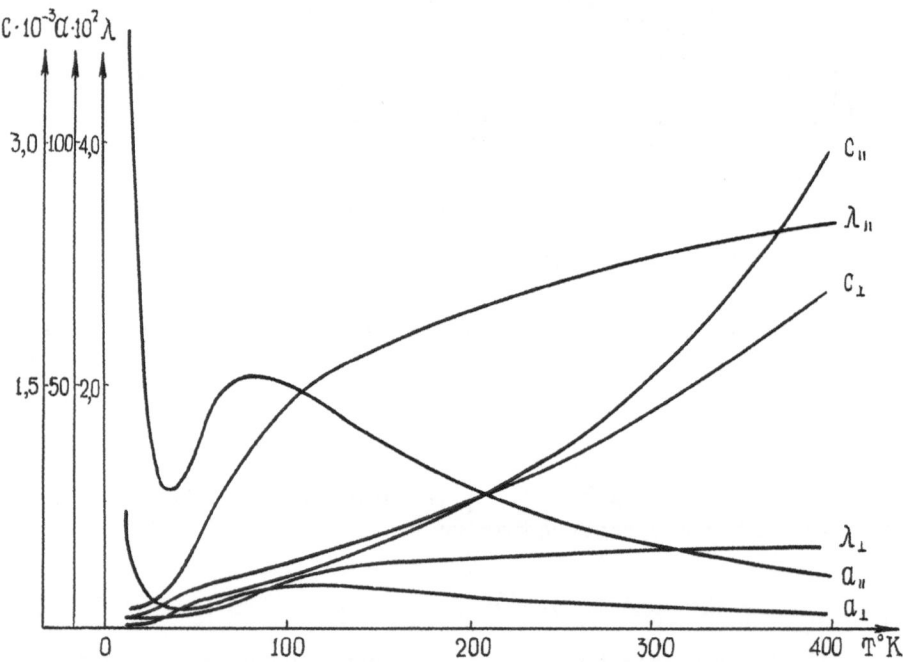

Fig. 2. Temperature dependences of λ (W/m²·K), a (m²/s), and c
(J/kg·K) of carbon fibres along and across the fibres.

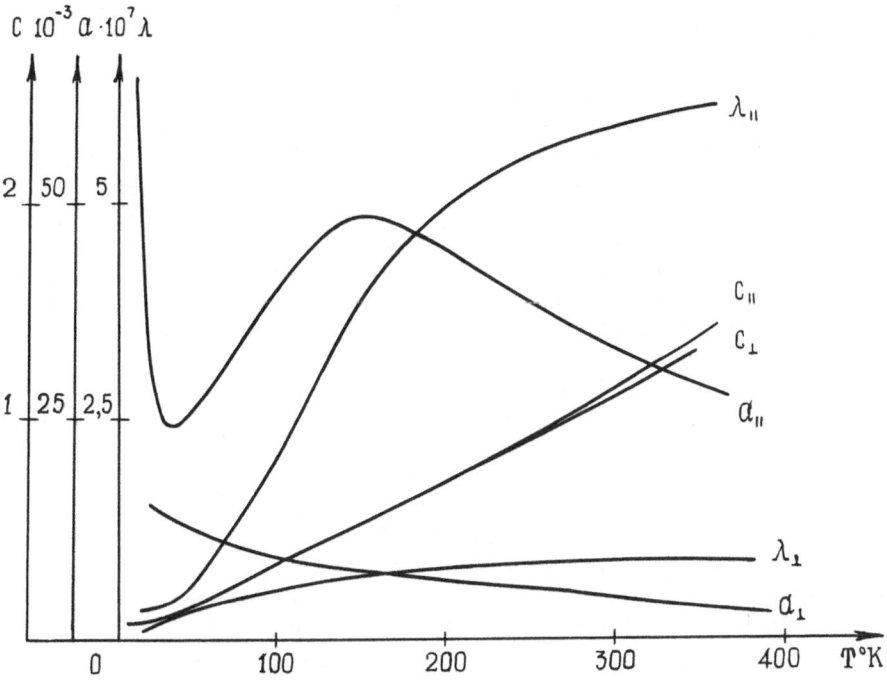

Fig. 3.   Temperature dependences of λ (W/m·K), a (m^2/s), and c (J/kg·K) of glass-carbon fibres along and across the fibres.

to 2, whereas the degree of anisotropy of carbon fibre on the basis of resin EDT-10 varies from 2 to 5.

The absolute value of $\lambda_{\perp GCF}$ is little greater than $\lambda_{\perp CF}$ within the whole temperature range. At T > 110 K, $\lambda_{\parallel GCF}$ is much higher than $\lambda_{\parallel CF}$, whereas at T < 110 K, $\lambda_{\parallel GCF}$ is smaller than $\lambda_{\parallel CF}(\lambda_{GF} \ll \lambda_{CF})$. In the range of nitrogen-helium temperatures, the dependence $\lambda_{\parallel GCF}(T)$ is similar to $\lambda_{\parallel CF}$, and $\lambda_{\perp GCF}(T)$ is similar to $\lambda_{\perp GF}(T)$. This may be explained by analysing the material structure. The composite properties are considerably affected by fillers. The intermolecular forces between polymer and glass and carbon fibres are different. In the longitudinal direction, the binding of glass and carbon fibres is covalent. In the direction normal to fibres, the binding between phases is by van der Waals' forces that are more weak than covalent forces. Therefore, heat transfer along the fibres is more intensive than across them. Anisotropy is known to be the higher, the more distinct is the orientation of the polymer binder along the filler surface. But, as in the test sample, when ropes of glass and carbon fibres are used, then it may be supposed that in the binder layers between glass and carbon ropes new overmolecular structures appear, which are stretched in the direction of reinforcing and are of a higher order than those in carbon fibres. This is proved by Ref. 2, where it is shown that a macrolattice of the epoxy binder may form on the filler fibre surface, and by Ref. 3, showing the effect of the epoxy binder on λ of carbon fibres.

The behavior of the temperature dependence $a_{\parallel GCF}(T)$ is anomalous. This anomaly shows itself in the maximum between 50 and 100 K. The maximum is achieved at T = 140 K. $a_{\parallel GCF}(T)$ changes in the same way as $a_{\parallel CF}(T)$, while $a_{\perp GCF}(T)$ varies similarly to $a_{\perp CF}(T)$. The absolute values of $a_{\parallel GCF}$ and $a_{\perp GCF}$, on the average are greater than the appropriate values for carbon and glass fibres. As the phenomenological Debye representations may give the following expressions for

$$a \approx u\bar{\ell}/3 \qquad\qquad (2)$$

then the temperature dependences $a_{\parallel GCF}(T)$ and $a_{\perp GCF}(T)$ are explained by a peculiar temperature dependence of the mean free path of composite phonons that results from the correlation of the free path lengths of the composite components.

It may be supposed that in the test material affected by residual and thermal stresses, which are due to sign-different coefficients of linear expansion for carbon fibre (negative) and for glass fibre and binder (positive), two simultaneous processes take place in the microvolumes around each monofibre. These are formations of overmolecular microfibril structures of higher order and overmolecular binder film structures.

In the composite, the free path length of phonons will be proportional to T^n, n being dependent on the direction of heat flux and temperature range. At low temperature, the free path length is commensurable with linear dimensions of crystallites. With increasing temperature, scattering in phonons prevails over overlapping, and the dependences $a_{\parallel GCF}(T)$ and $a_{\perp GCF}(T)$ (above 200 K) have, therefore, the form of an inverse proportion. The difference in $c_{\parallel GCF}(T)$ and $c_{\perp GCF}(T)$ is within the experimental accuracy.

Thermal properties of glass-carbon fibres depend on the anisotropy of the binder film properties that is caused by thermal stresses, whose value and direction depend on the temperature range and filler composition and which is of a variable nature.

Thermal properties that characterize GCF along the fibres are of the form typical for carbon fibres, whereas across the fibres they are similar to glass fibres.

REFERENCES

1. L.L. Vasiliev and S.A. Tanaeva, Thermophysical Properties of Porous Materials, Nauka i Tekhnika, Minsk (1971).

2. L.L. Vasiliev, L.S. Domorod, and S.A. Tanaeva, in: Superconductivity, Vol. 5, Atomizdat, Moscow (1977), p. 59.

3. L.S. Domorod, L.E. Evseeva, and S.A. Tanaeva, Inzh.-Fiz. Zh. 32, 111 (1977).

CRYOGENIC APPLICATIONS OF COMPOSITE TECHNOLOGY

IN THE U.S.A.

M. B. Kasen

National Bureau of Standards
Boulder, Colorado, U.S.A.

INTRODUCTION

It is reasonable to expect that, by the turn of the century, a substantial portion of the metallic materials currently used for structural purposes will be replaced by nonmetallic composites and aggregates. The impetus for the change will be primarily economic, as alloying elements become scarcer and more costly. However, improved product performance and lower life-cycle costs will also stimulate the change. Cryogenic technology will make widespread use of these new materials, as the high ratios of strength/conductivity (thermal and electrical) and high specific strengths and moduli of composites are particulary advantageous for low-temperature applications.

The term composite, as used in this paper, refers to materials having overall properties that are some average of the properties of several distinct components, one of which is contiguous and forms a matrix having a noncoherent interface with the reinforcing elements. Filamentary-reinforced materials, such as fiberglass-epoxy, and aggregates, such as concrete, fall within this definition. Composite laminates are built up from distinct plies of reinforcement in contrast to filament-would composites, which do not exhibit a laminated structure.

The selection of a composite material for cryogenic use would be made among the various composite categories listed on Fig. 1. The present discussion, therefore, will be organized around this format. Aggregates and filamentary-reinforced composites will be discussed separately, as their applications are distinctly different.

Within the general category of filamentary-reinforced composites, each subcategory has combinations of properties, costs, and fabrication flexibility that strongly influence selection for specific applications. The subcategories will be discussed in the order in which they appear in Fig. 1.

FILAMENTARY-REINFORCED COMPOSITES

The difference between high- and low-pressure laminates is primarily in the manner in which they are fabricated, which directly affects their applications. The pressure required to produce high-pressure, industrial laminates ($\sim 8\,MPa$) dictates that they be produced as bulk mill products (sheets, plates, tubes, and rods) from which desired shapes may be produced by machining. Such laminates are often used as if they were isotropic materials; however, they are orthotropic, usually reinforced with bidirectional fabric. The products are often referred to as Micarta, Textolite, Spauldite, etc., which are the tradenames assigned by various fabricators to their industrial laminate products.* More specific designations assigned by the National Electrical Manufacturers Association (NEMA)

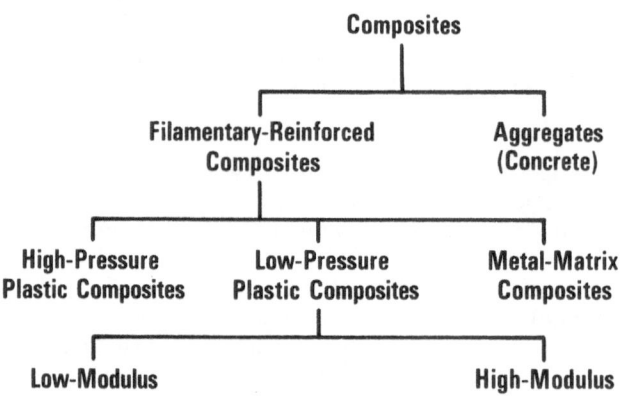

Fig. 1. Composite system flow chart.

---

*The use of tradenames in this publication does not imply approval, endorsement, or recommendation of specific commercial products by NBS.

are listed in Table I, which summarizes the types of industrial
laminates most frequently used for cryogenic service in the United
States. All high-pressure industrial laminates are reinforced with
low-modulus fibers. Close control over manufacturing variables
results in products of high quality at costs somewhat lower than
that of low-pressure laminates.

Low-pressure (0 to 2.8 MPa) laminating methods are used to
fabricate parts to their final shape and dimensions, providing great
flexibility in orienting the reinforcing fibers to "tailor make"
properties to meet service requirements. Additionally, low-pressure
laminates may be reinforced with high-modulus fibers, such as graphite or boron, to produce lightweight structures having strengths
and stiffnesses competitive with metal structures. Metal-matrix
laminates are always reinforced with a high-modulus fiber.

The author has recently reviewed the existing data on the
effect of cryogenic temperatures on filamentary-reinforced composites.[1-3] A systematic acquisition of data on specific contemporary composite systems has also begun.[4]

Applications of High-Pressure Reinforced Plastics

Glass-reinforced products of the G-10 and G-11 types are of
higher quality and have better mechanical properties than the other
types listed in Table I. They are, therefore, the choice for insulation and mechanical supports in demanding applications, such
as superconducting magnets. The less expensive cellulose-reinforced
laminates, such as types C or LE, are used in less critical applications, where lower mechanical properties and larger property variability are acceptable.

Table I. High-Pressure Industrial Laminates
Most Frequently Used for Cryogenic Service.

Designation	Description
G-10	Glass-epoxy
G-11	Glass-epoxy (high temperature)
LE	Cellulose (fine-weave)-epoxy
C	Cellulose (medium-weave)-epoxy
G-5	Glass-melamine
Permali	Phenolic-impregnated beechwood

A typical application of a G-10 type product as a coil form
and electrical insulator in a magnet is illustrated in Fig. 2.
This magnet, constructed at the Lawrence Livermore Laboratories,[5]
utilizes relatively thin laminate sections; however, other magnet
designs use laminates many centimeters in thickness. Figure 3
illustrates an alternative design in which a glass-epoxy industrial
laminate is used as an insulator and support for superconducting
windings in a magnet.[6] Figure 4 illustrates the principle employed in using a G-10 type laminate in combination with a metal
structure to produce a thermal standoff, which remains unchanged
in length upon cooling and in the presence of a thermal gradient
while supporting a compressive load with minimal thermal loss.
This principle, used by the National Accelerator Laboratory[7] to
support superconducting magnets through the walls of a dewar,
achieves dimensional stability because the thermal contraction of
the aluminum intermediate tube equals the sum of the contractions
of the outer and inner G-10 composite tubes.

NEMA G-10 and G-11 tubular and sheet products are also used in
commercial production of nonmetallic dewars, most often for the

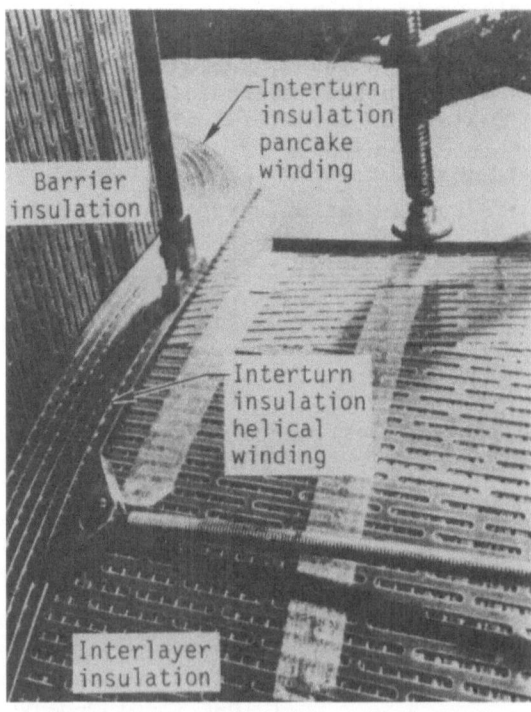

Fig. 2. Typical application of high-pressure, glass-epoxy
laminates for insulation in a superconducting magnet.[5]

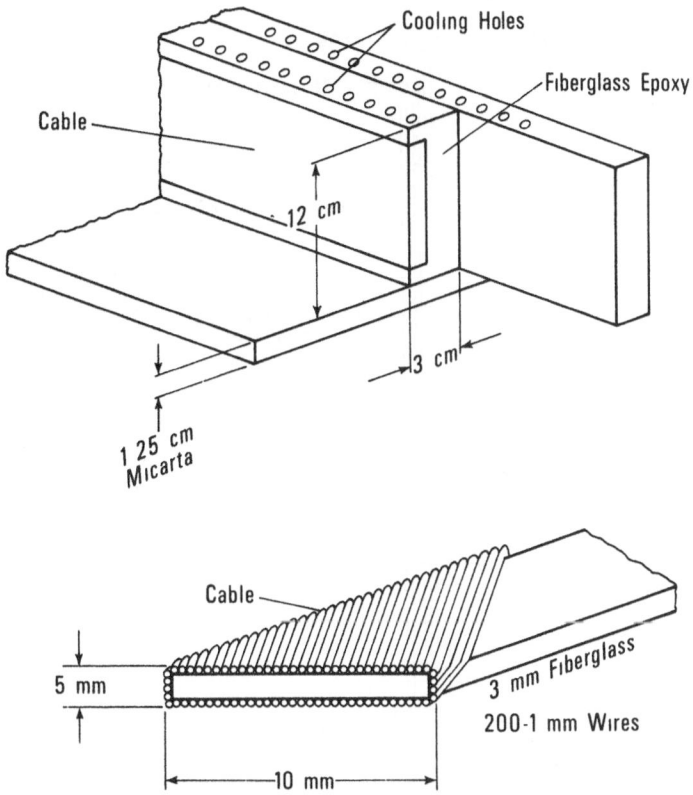

Fig. 3. Proposed use of glass-epoxy industrial laminates for direct support of superconducting windings in a magnet.

inner wall that is in contact with liquid helium. Practical experience has shown that helium permeability is not a problem below about 150 K; however, leakage through the laminate wall along pre-existing flaws may occur. Internal joints between sections of the preimpregnated tape from which the laminates are fabricated are likely sources of such leaks. Laminates free of internal joints may be obtained at extra cost for the most demanding leak-free service. Occasionally, it is desirable to reduce the wall thickness of a laminate by machining to achieve lower thermal losses; for example, in the necks of liquid helium dewars. In such cases, it has been found beneficial to specify a laminate reinforced with a very lightweight cloth, which increases the number of plies remaining in the wall and provides increased protection against the introduction of damaging flaws during machining. Matrix cracking

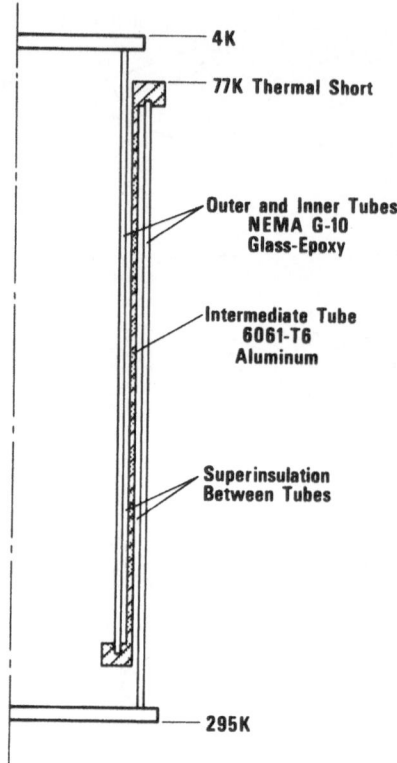

Fig. 4. A dimensionally stable, load-bearing cryogenic
         thermal standoff.[7]

due to thermal sock has not proven to be a problem, although care
must be taken to limit the resin content in bonded joints.

The current interest in constructing very large superconducting
magnets for prototype magnetic fusion energy (MFE) reactors and
for magnetohydrodynamic (MHD) energy conversion devices has created
the need for better mechanical, thermal, and electrical characterization of industrial laminates at cryogenic temperatures. Developing such a data base is complicated by the highly proprietary nature
of commercial laminate production. The NEMA specifications establish electrical performance criteria, while giving manufacturers
considerable leeway in selecting specific components and manufacturing methods to meet the criteria. Thus, it is found that NEMA-grade
laminates produced by different manufacturers have significantly
different mechanical and physical properties at cryogenic temperatures. To rectify this situation, NBS has been working with the
NEMA organization to provide the cryogenic industry with G-10 and

G-11 type laminates for which the components and manufacturing procedures are closely controlled. These products, designated G-10CR and G-11CR, are now commercially available in the United States and are undergoing intensive mechanical, thermal, electrical, and radiation-resistance characterization from 295 to 4 K. It is expected that these laminates will become the standard industrial laminates for a wide variety of low-temperature applications in the United States.

Applications of Low-Pressure Reinforced Plastics

Low-Modulus Composites. Several studies conducted in 1962 and 1964[8,9] concluded that epoxies were the best matrix materials for low-pressure glass-reinforced laminates intended for use at cryogenic temperatures, although polyesters were a close second. Today, epoxies are the almost universal choice. Some effort has been expended on developing special epoxy systems for cryogenic use,[10,11] usually with the intent of increasing the resin flexibility at low temperatures. These efforts have only been marginally successful because of the very low strengths of flexibilized systems at room temperature and the much higher thermal contraction of flexibilized systems compared with fully reacted systems. It is found that the high state of residual stress existing in a flexibilized-epoxy composite at 4 K limits the effective failure strain of the composite to a level commensurate with that of a laminate using a fully reacted system. It is the current United States practice to use one of the conventional epoxy systems that experience has proven to be satisfactory in past applications.[1-3]

Both the lime-alumina-borosilicate type E glass and the silica-alumina-magnesia type S glass are used for reinforcement. The less expensive E glass is preferred where electrical or thermal insulation properties take precedence over strength, whereas the more expensive S glass reinforcement is preferred where strength and modulus requirements dominate.

Low-pressure, glass-epoxy laminates are used primarily for the fabrication of electrical insulation, thermal standoffs, and nonmetallic dewars when the shape of the structure does not lend itself to fabrication from high-pressure laminates or where the required mechanical properties can only be met by a specific fiber orientation. The ability to work with preimpregnated and partially cured (B-staged) tape products offers some advantages--for example, spirally wrapping superconductors with tape prior to magnet winding has been proposed as a way of providing both electrical insulation and helium-cooling passages in the magnet. Proper choice of resin system will permit in situ curing at temperatures that will not damage the superconducting elements. In the case of small magnets

of potted construction, glass cloth is frequently wrapped between
sequential layers of the windings to become an integral part of the
structure after potting. This technique is currently being used in
the construction of prototype superconducting motors and generators.

Glass-epoxy structures are used to support and to achieve
thermal isolation of magnets in several superconducting motors
under construction in the United States. Figure 5 illustrates one
approach--a conical support made of S-glass-reinforced epoxy in
which the truncated end is at 295 K and the cone base is at 4 K.
Thermal shorts are provided by two copper rings embedded in the
cone during manufacture, the rings being cooled by helium vapor.
The magnet system using this type of support is now fully operational and has met the design heat leak requirements. Alternatively,
magnets may be supported in the dewar vacuum space by a series of
racetrack-shaped, filament-wound S-glass straps. The uniaxial
filaments provide a better ratio of strength to thermal conductivity
than the conical support does, but the need to maintain tension on
the straps increases the design complexity. Strap supports have
proven functional; they are used in one design of a superconducting
motor and are used to support a superfluid helium dewar in the
infrared astronomical satellite (IRAS). Tests have shown fatigue
performance at 4 K to be superior to that at 295 K.[12]

Fig. 5.   Conical thermal standoff and structural support for
          for the magnet in a 2.238 MW (3000 hp) super-
          conducting motor. Cone base diameter is 30.8 cm.

The ability to form complex shapes with low-pressure laminates has been utilized in fabricating helium dewars to the optimum shape for specific applications. A typical example is the cryostat assembly of Fig. 6, which is intended for use as an airborne geophysical exploration device.[13] Experience has shown that leakage problems are minimized when a wet layup is used with glass-cloth reinforcement of the epoxy. Attempts to fabricate dewars from preimpregnated tape with an autoclave cure have been less successful, probably because of the many internal joints that provide potential leak paths.

Studies are currently underway to select materials and fabrication methods for construction of glass-epoxy fan blades for the National Transonic Facility wind tunnel at NASA-Langley. Having a span of about 68 cm and an airfoil section of about 0.6 m^2, they

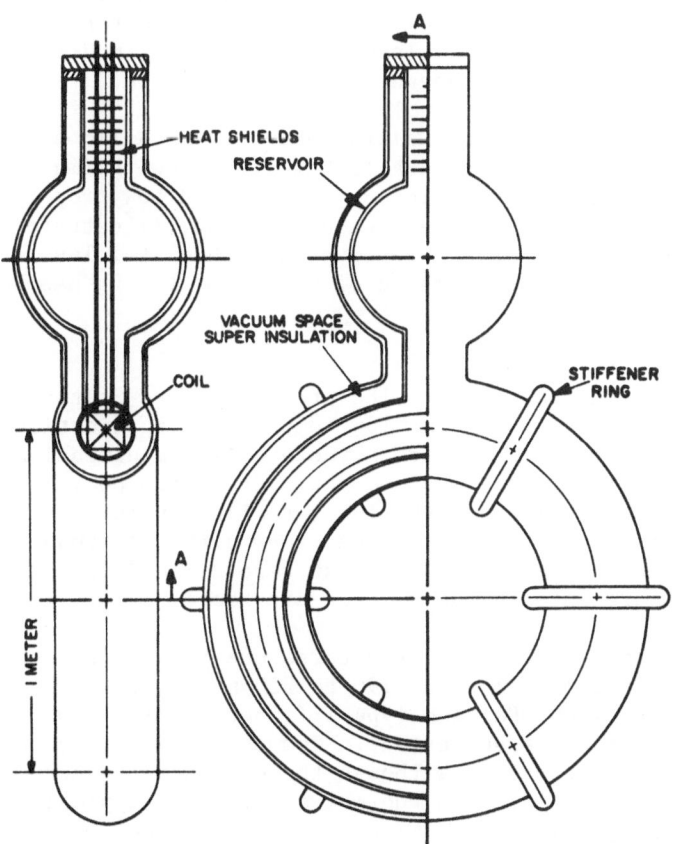

Fig. 6. Nonmetallic cryostat for electromagnetic geophysical exploration.[13]

will operate in dry air and in gaseous nitrogen at temperatures
from 77 to 353 K. Such blades have proven superior to metal blades
in conventional wind tunnels[14] and are expected to provide better
fatigue life, lower cost, and a higher strength/weight ratio than
could be provided by metal blades.

The reinforcement of cryogenic pressure vessels with glass
fibers has been studied for over a decade in the United States.[15-20]
Primary interest has been to reduce weight in weight-limited spacecraft structures. Many vessels have been produced; however, sufficient problems remained to preclude using such vessels in mission
spacecraft. Interest lay dormant during the Skylab space program,
where excess payload reduced the incentives for saving weight.
Interest in lightweight, efficient cryogenic pressure vessels has
now increased in anticipation of the need for very large vessels
operating at high pressures in space--refuelling stations for LH_2
and LO_2 are a case in point. Discussions with individuals know-
ledgeable about this technology suggest that recent progress in
improving vessel design will lead to a successful modern generation
of composite-reinforced cryogenic pressure vessels.

The principle of glass-reinforcement of metallic structures has
also been applied to produced a lightweight, low-thermal-flux tubing
intended to increase the efficiency of LH_2 and LO_2 propellant feed
lines in the Space Tug.[21] In the development, illustrated in
Fig. 7, glass is used as an overwrap on a thin stainless steel
liner. Satisfactory performance of prototype tubing has been demonstrated.

High-Modulus Composites. The primary United States interest
in high-modulus composites for low temperature applications is for
spacecraft structures which are structurally and dimensionally
stable while maintaining light weight. Graphite-epoxy is the
almost universal choice, as the low (or slightly negative) coefficient of thermal contraction in the fiber direction enables structures to be built that have an almost zero dimensional change from
room temperature into the deep cryogenic range.

Most current spacecraft structures use the lower-cost Type III
graphite fibers (fiber modulus 190 to 220 GPa), although Type II
(fiber modulus $\sim$ 260 GPa) and Type I (fiber modulus 310 to 520 GPa)
have been used to a lesser extent. Commercial resin systems appear
to perform satisfactorily, although there is some concern that
microcracking resulting from thermal excursions (typically 480 to
100 K) in crossply laminates might adversely affect the dimensional
stability of sensitive structures. Contamination of the surfaces
of mirrors from products outgassing or volatilizing from the matrix
can be a problem in some spacecraft; however, properly cured commercial resin systems appear to be satisfactory for most applications.

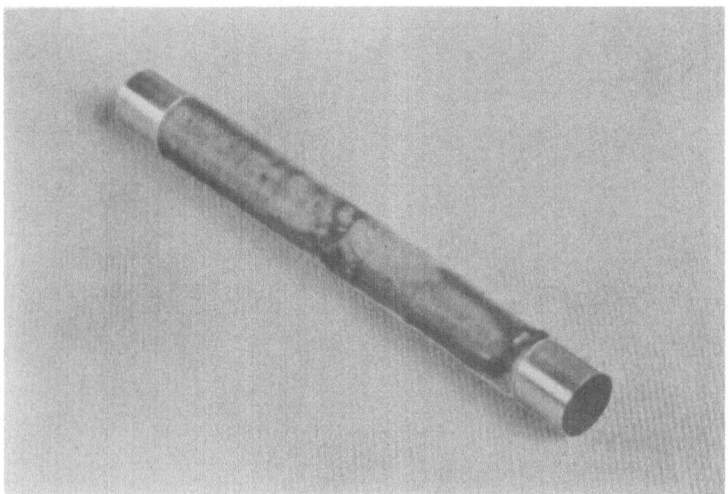

Fig. 7. Low thermal loss, glass-reinforced, stainless
steel cryogenic propellant feed line.[21]

Specific applications of advanced composites in spacecraft are
too numerous to be comprehensively reviewed in the present paper.
The several examples discussed here will serve to illustrate how
the material selection and structural design are related to operational constraints.

Graphite-epoxy nondeployable, high-gain reflector structures
are currently flying on Voyagers I and II and on the Viking spacecraft.[22] Requirements for extremely accurate and dimensionally
stable parabolic dish structures were met by using graphite-epoxy
face skins on an aluminum honeycomb core.   The Voyager craft also
carries a dimensionally stable, frequency-selective subreflector
made of metallized polyarimid epoxy. This vehicle also uses about
61 m of graphite-epoxy composite in various tubular truss structures.
Uniaxial HM-S/3501 graphite fibers are aligned along the tubes with
an outer and inner 90° wrap of Kevlar 49/3501 (polyaramid) fiber
for increased structural integrity and toughness. Graphite-epoxy
truss structures have also been proposed as supports for large
space telescopes. The structure illustrated in Fig. 8,[23] approximately 7.22 m (23.7 ft) high and 3.35 m (11 ft) in diameter, successfully met stiffness and thermal distortion requirements.

Similar structures for other aerospace venicles are in various
stages of design, fabrication, and testing by United States firms.
Lack of experience with large, high-modulus composite structures
usually necessitates several iterations before development of the
proper structure to meet specific flight requirements. For example,

Fig. 8.   Prototype graphite-epoxy truss structure
          for Space Tug.[23]

several alternate layup schemes were tried before it was determined
that a T300/934 graphite-epoxy face sheet used for a door in the
Space Shuttle would satisfactorily resist cracking and buckling
due to thermal stresses only if the laminate contained two uniaxial
layers plus a balanced-weave fabric layer at ±45°.[24] Uniaxial
layers crossplied at ±45° resulted in interlaminar cracking. It is
the result of this type of experience that knowledge of proper design for cryogenic service is being accumulated.

    Prototype T300/5208 graphite-epoxy antenna booms have been
constructed for the USAF Project to Study Spacecraft Charging at
High Altitudes (SCATHA). Construction is similar to the Voyager
trusses, except that a combination of graphite roving and glass

cloth is used for the inner and outer wraps. Supporting I beams having an aluminum honeycomb web and graphite-epoxy channel coverings and flange structures have also been produced for this vehicle.

The Space Tug intended to be the payload for the Space Shuttle is weight limited, stimulating NASA to sponsor development of an experimental graphite-epoxy body structure for the vehicle.[25] The prototype is built up of panels in which T300/934 graphite-epoxy, crossplied at ±45°, is used for skins and in a uniaxial layup with surface layers of fiberglass in stiffener flanges. Woven broadgoods are used in shear-web regions and at attachment points. Although the prototype structure met the design goals, a decision has been made to fly a structure made of metal in the initial Space Tug and to schedule composite structures for future vehicles. In contrast to optical or microwave structures, in which dimensional stability requirements make composite structure mandatory, the structural materials used in the Space Tug structure are selected on a cost/weight ratio, justifying the use of graphite-epoxy construction only when spacecraft sophistication has progressed to where the saving of weight becomes a significant economic factor.

At the present time, there are no high-modulus composites used structurally in superconducting magnets or in their support systems. It is worth noting, however, that the electrical conductor in a superconducting magnet shown in Fig. 9 is a metal-matrix composite, differing from structural metal-matrix composites only in its end function. Superconducting composites are severely strain limited; for example, Nb_3Sn superconducting elements severely degrade at tensile strains above about 0.2%.[26] This strain limitation has not been a major problem in the relatively small magnets that have been constructed in the past. However, electromagnetic stresses scale with magnet size, creating serious problems in the very large magnets presently being constructed for prototype MFE and MHD devices. It is possible to contain these stresses with massive stainless steel structures. However, there are circumstances where it may be necessary to accomplish the objective with high-modulus composites; for example, in the poloidal coils of Tokamak reactors, where eddy current losses from varying magnetic fields may place an unacceptable burden on the refrigeration system. Alternatively, it may be desirable to build very lightweight magnets for mobile superconducting motors and generators. Although these applications lie somewhere in the future, work is presently underway in the author's laboratory to establish the required cryogenic data base on graphite, boron, and polyaramid-reinforced laminates.

A number of programs in the United States have studied the relative merits of reinforcing cryogenic pressure vessels with high-modulus fibers.[27-29] Primary interest was in obtaining increased fatigue life of the liner by limiting liner strains. A typical vessel fabricated with graphite is shown in Fig. 10; others have

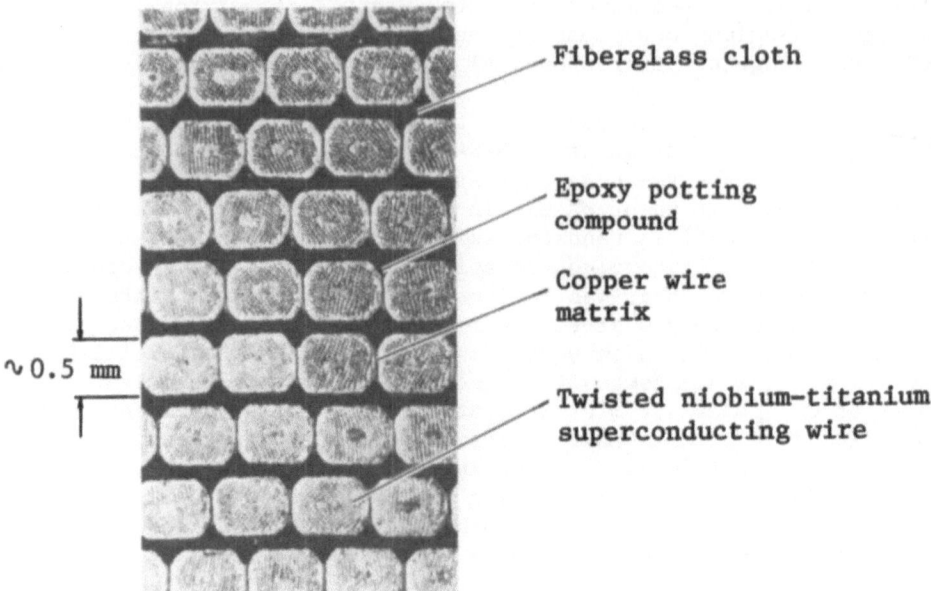

Fiberglass cloth

Epoxy potting
compound

Copper wire
matrix

∿0.5 mm

Twisted niobium-titanium
superconducting wire

Fig. 9.   Typical cross-section through an NbTi superconducting coil
          composite.   Each bundle contains 180 NbTi filaments.

been fabricated with boron and polyaramid reinforcement.   However,
it does not appear that the additional costs are justified, except
possibly for polyaramid, where both increased modulus and lower
weight provide advantages over glass reinforcement.   Vessels reinforced with the Kevlar 49 polyaramid fiber are currently being
fabricated and tested for aerospace use in the United States.

Metal-Matrix Laminates

    Superconducting composites aside, there is little interest in
the use of metal-matrix laminates for cryogenic structures in the
United States.   The primary reason is high thermal conductivity,
which is normally undesirable.   Studies have shown that uniaxial
boron-reinforced aluminum composites are superior to annealed stainless steel in specific strength and modulus, but inferior in this
regard to polymer-matrix composites reinforced with either graphite
or boron.[4]   It is possible that the high electrical conductivity of
boron-aluminum will make it attractive for specialized applications,
such as eddy-current shields in superconducting motors and generators.   It has also been suggested that metal-matrix composites
would be viable replacements for polymeric composites in weight-

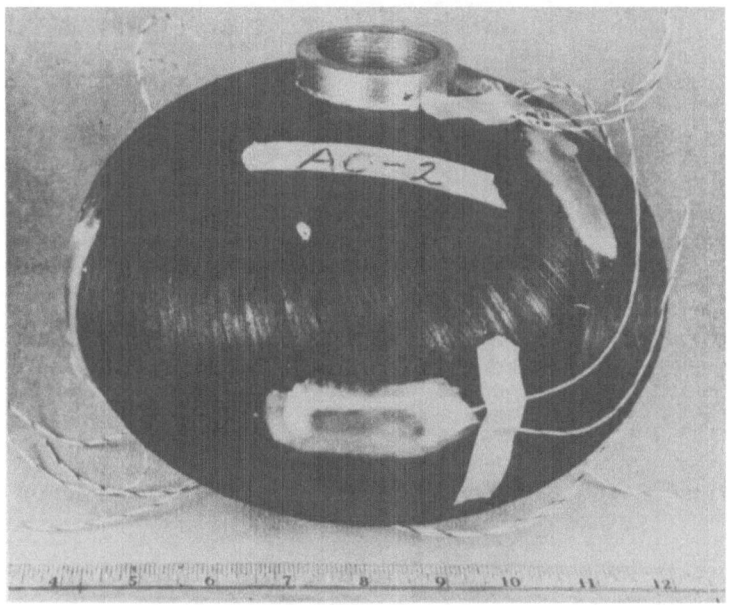

Fig. 10.   Oblate spheroid pressure vessel reinforced with filament-wound graphite-epoxy.[29]

limited space structures containing mirrors, which are in danger of contamination by products outgassing from the plastic matrix.

AGGREGATES

Conventional sand-and-gravel concrete has been successfully used in liquefied natural gas (LNG) storage tanks since 1962. Primary attractions are low cost, low thermal diffusivity, high compression, and buckling strengths. The ability of concrete of reasonable thickness to withstand external accident or sabotage is also an important factor.

Figure 11 illustrates the construction of a typical concrete LNG storage tank.[30] An inner shell of reinforced concrete is in contact with the LNG at 100 K and sustains the hydrostatic pressure. Inner liners of aluminum or 9% Ni steel are sometimes used. An outer shell of carbon steel forms a vapor-tight container for insulating materials placed between the shells. An outer ring of concrete forms a secondary barrier in some designs. Construction is primarily with precast sections, which are reinforced longitudinally during manufacture and circumferentially after tank erection.

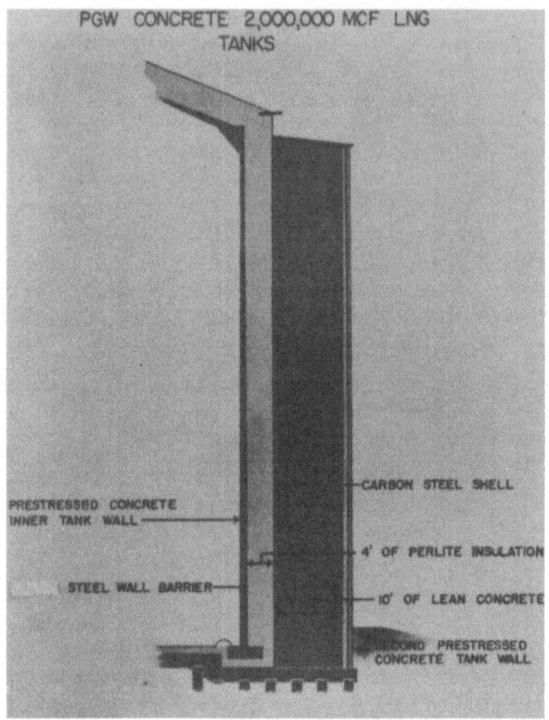

Fig. 11.   Cross-section through a typical
prestressed-concrete liquefied
natural gas storage tank.[30]

     Considering the catastrophic consequences of a major LNG spill
in the populated areas where many such tanks exist, there is a re-
markable lack of information in the literature on the effect of
LNG temperatures (and contact with LNG) on the mechanical and thermal properties of concrete. The low-temperature strength of concrete is known to depend on a number of factors including cement
content, water-to-cement ratio, type of aggregate, concrete age,
and air content. The Young's modulus of moist concrete probably
increases on cooling, while compressive and splitting strengths
increase substantially.[31]   Compressive strengths of moist concrete
apparently pass through a maximum, concomitant with development
of a minimum in thermal conductivity. Both phenomena are believed
related to freezing of evaporable moisture over a temperature range
that is determined by the dissolved impurity content. Limited data
suggest that improved insulating properties can be obtained at some
loss of strength when cellular concretes are used.[32]

     The lack of data correlating the many variables in concrete
aggregate properties with the fitness-for-service of LNG structures

suggests that present structures are designed so conservatively as to be insensitive to existing property variability. This approach may well be justified, considering the low cost of concrete and the low priority given to weight or volume savings in shore-based installations. Recently, however, there has been interest in exploiting reinforced concrete for ocean transport of LNG, which would require that it be qualified by the U.S. Coast Guard for such service. Weight and volume constraints are important in such applications, and structures will be subjected to tensile and flexural loads as well as compression. Programs are currently being formulated in the United States to generate the required cryogenic data base for cellular, polymer, and sulphur concretes, as well as for conventional sand-and-gravel types.

It is reasonable to expect concrete to be used structurally at temperatures down to 4 K in the future. It has already been proposed as a viable material for major portions of the Tokamak-type MFE reactors that are expected to become realities by the turn of the century.

MISCELLANEOUS COMPOSITES

Thus far, we have considered only those composite materials that are useful at both room temperature and at cryogenic temperatures. An alternative approach is to take advantage of low temperatures to create functional structural materials useful only at low temperatures. This concept seems strange until it is realized that the arctic peoples discovered long ago that moss-containing ice was a much tougher structural material than plain ice. The properties of reinforced ice have been studied;[33] however, the interest has remained in the area of climatic cold. Relatively recently there is an awakening to the potential of reinforced ice as a very inexpensive, versatile structural material for very large cryogenic installations.

SUMMARY

Composites are attractive for cryogenic structures where conventional metals are either too expensive or lack the combination of properties required for a given application. Cost saving is primarily obtained in very large structures typified by land storage facilities for LNG, where concrete aggregates have proven to be effective. Advantageous combinations of strength, modulus, and thermal and electrical conductivity provided by glass-epoxy and bonded-wood laminates have stimulated the use of such materials in superconducting magnets, nonmetallic dewars and pressure vessels, and in large vessels for transportation of LNG. The high specific moduli and specific strengths of the advanced composite laminates

have made them cost effective in some weight-limited aerospace
applications. However, the largest cryogenic application of
advanced composites is the use of graphite-epoxy for large, dimensionally stable, structurally stable, lightweight space structures,
such as are required for mirrors and antenna reflectors.

The primary factors inhibiting the wider use of composites
for cryogenic structures in the United States are the cost of the
materials, an inadequate data base, and lack of practical design
and operating experience. It is reasonable to expect that the use
of composites will expand as these problems become solved.

ACKNOWLEDGMENT

The author would like to thank the many individuals in United
States companies and Federal Agencies who contributed information
to this paper. In particular, the author wishes to thank Mr.
D. E. Spond, Martin Marietta Corporation, Mr. D. E. Skoumal,
Boeing Company, and Mr. J. J. Closner, Preload Technology, Inc.,
for providing documents and illustrative material of assistance
in both the conference presentation and in this paper.

The work was performed under the auspices of the Division of
Magnetic Fusion Energy, U.S. Department of Energy, and the Air
Force Aero Propulsion Laboratory, Wright-Patterson Air Force Base,
Ohio. Contribution of the National Bureau of Standards, not
subject to copyright.

REFERENCES

1. M.B. Kasen, in: Composite Reliability, ASTM STP 580, American
    Society for Testing and Materials, Philadelphia, Pennsylvania
    (1975), p. 586.

2. M.B. Kasen, Cryogenics 15, 327 (1975).

3. M.B. Kasen, Cryogenics 15, 701 (1975).

4. R.E. Schramm and M.B. Kasen, Mater. Sci. Eng. 30, 197 (1977).

5. A.R. Harvey and J.A. Rinde, in: Proc. 6th Symposium on Engi-
    neering Problems of Fusion Research, San Diego, California
    (1975), p. 606.

6. W.M. Stacey et al., Tokamak Experimental Power Reactor Concep-
    tual Design, Vol. 1, ANL/CTR-76-3, Argonne National Laboratory,
    Argonne, Illinois (1976).

7. J.R. Heim, A Low Heat Leak Temperature Stabilized Support, TM-334A, National Accelerator Laboratory, Batavia, Illinois (1971).

8. N.O. Brink, Determination of the Performance of Plastic Laminates under Cryogenic Temperatures, ASD-TDR-62-794, Air Force Systems Command, Wright-Patterson Air Force Base, Ohio (1962).

9. D.W. Chamberlain, B.R. Lloyd, and R.L. Tennant, Determination of the Performance of Plastic Laminates at Cryogenic Temperatures, ASD-TDR, 62-794, Part II, Air Force Systems Command, Wright-Patterson Air Force Base, Ohio (1964).

10. L.M. Soffer and R. Molho, Cryogenic Resins for Glass Filament-Wound Composites, NASA CR-72114 (final), National Aeronautics and Space Administration, Lewis Research Center, Cleveland, Ohio (1967).

11. A. Lewis and G.E. Bush, Improved Cryogenic Resin/Glass-Filament-Wound Composites, NASA CR-72163 (final), National Aeronautics and Space Administration, Lewis Research Center, Cleveland, Ohio (1967).

12. R.L. Tobler and D.T. Read, J. Compos. Mater. 10, 32 (1976).

13. W.V. Vogen, B. Clawson, and H.F. Morrison, in: Advances in Cryogenic Engineering, Vol. 21, Plenum Press, New York (1975), p. 51.

14. C.P. Young, NASA-Langley Research Center, Langley, Virginia, personal communication.

15. M.P. Hanson, H.T. Richards, and R.O. Hickel, Preliminary Investigation of Filament-Wound Glass-Reinforced Plastics and Liners for Cryogenic Pressure Vessels, NASA TN D-2741, National Aeronautics and Space Administration, Lewis Research Center, Cleveland, Ohio (1965).

16. M.P. Hanson, Static and Dynamic Fatigue Behavior of Glass Filament-Wound Pressure Vessels at Ambient and Cryogenic Temperatures, NASA TN D-5807, National Aeronautics and Space Administration, Lewis Research Center, Cleveland, Ohio (1970).

17. E.E. Morris, J. Mater. 4, 970 (1969).

18. M.R. Sanger, M. Molho, and W.W. Howard, Exploratory Evaluation of Filament-Wound Composites for Tankage of Rocket Oxidizers and Fuels, AFML-TR-65-381, Air Force Materials Laboratory, Wright-Patterson Air Force Base, Ohio (1966).

19. C.B. Shriver, Design and Fabrication of an Internally Insula-
     ted Filament Wound Liquid Hydrogen Propellant Tank, NASA CR-
     127, National Aeronautics and Space Administration, Washington,
     D.C. (1964).

20. J.M. Toth, W.C. Sherman, and D.J. Soltysiak, Investigation of
     Structural Properties of Fiber-Glass Filament-Wound Pressure
     Vessels at Cryogenic Temperatures, NASA CR-54393, National
     Aeronautics and Space Administration, Lewis Research Center,
     Cleveland, Ohio (1965).

21. C.A. Hall, D.J. Laintz, and J.M. Phillips, Composite Propul-
     sion Feedlines for Cryogenic Space Vehicles, NASA CR-121137,
     National Aeronautics and Space Administration, Lewis Research
     Center, Cleveland, Ohio (1973).

22. G.K.H. Dharan, "Mechanical and Thermal Behavior Characteriza-
     tion of Composite Materials for Communications Spacecraft,"
     paper presented at the Second International Conference on
     Composite Materials, Toronto, Canada (April 1978). To be
     published.

23. S. Oken and D.E. Skoumal, Design, Fabrication, and Test of a
     Graphite/Epoxy Metering Truss, Report No. D180-19335-1
     (final), Boeing Aerospace Company, Seattle, Washington. Pre-
     pared for NASA, George C. Marshall Space Flight Center,
     Huntsville, Alabama (1975).

24. J.P. Cupp, Rockwell International, Tulsa, Oklahoma, personal
     communication.

25. J.R. Lager, "Spacecraft Utilization of Fibrous Composites,"
     paper presented at ASCE Spring Convention and Exhibit, Pitts-
     burgh, Pennsylvania (April 1978). To be published.

26. J.W. Ekin, in: Advances in Cryogenic Engineering, Vol. 24,
     Plenum Press, New York (1978), p. 306.

27. R.J. Alfring, E.E. Morris, and R.E. Landes, Cycle-Testing of
     Boron Filament-Wound Tanks, NASA CR-72899, National Aero-
     nautics and Space Administration, Lewis Research Center,
     Cleveland, Ohio (1971).

28. M.P. Hanson, Glass-, Boron-, and Graphite-Filament-Wound Resin
     Composites and Liners for Cryogenic Pressure Vessels, NASA
     TN D-4412, National Aeronautics and Space Administration,
     Lewis Research Center, Cleveland, Ohio (1967).

29. A. Feldman and J.J. Damico, Graphite Filament Wound Pressure
     Vessels, NASA CR-120951, National Aeronautics and Space
     Administration, Lewis Research Center, Cleveland, Ohio (1972).

30. J.J. Closner, Preload Technology, Garden City, New York,
     personal communication.

31. G.E. Monfore and A.E. Lentz, J. PCA Res. Dev. Lab. 4, 333
     (1962).

32. T.G. Richard, J.A. Dubogai, T.D. Gerhardt, and W.C. Young,
     IEEE Trans. Magn. MAG-11, 500 (1975).

33. R.G. Stanley and P.G. Glockner, in: Materials Engineering in
     the Arctic, American Society for Metals, Metals Park, Ohio
     (1977), p. 29.

PROPERTIES OF PLASTIC TAPES FOR CRYOGENIC POWER CABLE INSULATION

A. C. Muller

Brookhaven National Laboratory
Upton, New York, U.S.A.

INTRODUCTION

A superconducting ac power transmission cable is under development at Brookhaven National Laboratory (BNL). This project was undertaken in 1972 in response to growing national power requirements. The goal of this program is to develop an underground power transmission system suitable for transferring bulk quantities of electricity over distances of 16 to 160 km. Both the capital investment and operating costs must be low enough to make the system attractive to the electric utilities.

The superconducting cable shares the advantages with conventional underground cables of needing only a few feet of right-of-way width rather than the large tracts of increasingly expensive land required for conventional aerial transmission. Recent cost analysis studies[1] show that superconducting cables, although more expensive than aerial transmission, will probably be competitive with other methods of underground transmission at loads greater than 2000 MVA. Initial design studies showed that a flexible, forced-cooled cable offered the best combination of technical and economic features.[2] A helium-cooled cable with an Nb_3Sn superconductor was chosen as the BNL design.

The present goal of the BNL program is the construction of a 100 m outdoor three-phase ac cable rated at 138 kV and 1000 MVA. The refrigerator and the 100 m long dewar are already installed. Terminations and cables are under design, and it is planned to begin installation of the first single-phase cable in 1979. If the results on this model show promise for eventual commercial use, cables of higher voltage and power rating will be developed.

One fundamental phase of this project, the development of the
required insulating materials, is the subject of this paper.

TAPE REQUIREMENTS

Many of the design features of the dielectric are governed by
the necessity to operate the cable at a temperature suitable for
the Nb3Sn superconductor (6 to 8 K). Extruded polymer is not a
viable mode of insulation application since the very large thermal
contraction associated with extruded polyethylene would almost
certainly lead to early mechanical failure of this dielectric.
Instead, the choice was made to lap many layers of plastic tape
and impregnate the butt-gaps with supercritical helium. The major
dielectric, mechanical, and thermal specifications for the dielectric tapes are summarized in Table I.

In order for an experimental cable to be cost-effective, the
materials should be commonly produced and readily available. Also,
fabrication techniques should follow standard practices and

Table I.   Specifications for Dielectric Tapes for
           Use in AC Superconducting Cables.

A. Dielectric (6 to 8 K)

    1. Dielectric constant - 2.5 (maximum)

    2. Dissipation factor - 2×10^{-5} (maximum)

B. Mechanical (293 K)

    1. Yield strength - 10 MPa (minimum)

    2. Tensile strength - 140 MPa (minimum)

    3. Tensile modulus, E_1 - 3.5 to 7.0 GPa

    4. Compressive modulus, E_3 - 10 MPa (maximum)

    5. Friction coefficient, μ_s - 0.250 (maximum)

C. Thermal

    1. Total contraction (293 to 4.2 K) 0.6 to 1.0%

    2. Conductivity (4.2 K) - 5 to 30 mW/m·K

---

technologies wherever possible. The dielectric tapes will be
applied to the cable by high-speed taping machines, designed many
years ago to construct conventional paper-lapped underground cables.
Although precise predetermined winding patterns are used during the
lapping of a kraft-paper cable, even greater care is required during construction of a plastic-lapped superconducting cable. Each
successive layer must be lapped "out-of-phase" with the previous
layer, so that the small butt-gaps between adjacent turns are completely covered by the width of the tape from the very next layer.
Double-thick butt spaces would reduce the inception stress of
partial discharge activity that could ultimately cause tape degradation and dielectric failure. Experience has shown that taping
tensions of 28 to 36 kPa (4 to 5 psi) are required to produce
tight, accurately wound paper cables. The tapes must possess
sufficiently high values of tensile strength and tensile modulus
to ensure that they do not break or deform under these tensions.

Additional requirements on the 293 K tape moduli were revealed
after several experimental test cables were built for BNL.[3] During
cable lapping a component of the taping tension is transformed to
a radial pressure directed radially towards the cable core.[4] These
radial forces present no problem in conventional cables because of
the high compressibility of paper. The more isotropic nature of
polymers transforms taping tensions to extremely high values of
radial pressures in plastic lapped cables. The pressure increases
proportionately to the number of layers of tape applied. During
bending of the completed cable in reeling, the radial pressure can
force tapes to wrinkle rather than slide on one another as they
normally should. Wrinkles will result in a reduced dielectric
strength and can shorten the life of the cable.

The net thermal contraction of the dielectric must be designed
so that it contracts evenly with the conductor during the cooldown
period. Insufficient dielectric contraction would lead to voids
between the inner conductor and the dielectric medium. These voids
would permit harmful partial discharges to occur. Excessive contraction would keep the tapes under tensile and compressive load
while at operating temperature. This stress could either cause
immediate tape fracture, or contribute to accelerated failure owing
to a long-term aging process.

The value of the thermal conductivity is another important
consideration in the design of the dielectric. Too small a value
could cause local heating of certain portions of the dielectric and
produce regions of reduced dielectric strength within the helium
impregnant. An upper limit to the conductivity was also established[5] to prevent excessive thermal coupling between counter-
flowing "go" and "return" helium coolant streams.

In striving for a low-loss cable, we placed a great deal of emphasis on the selection of tapes having very low values of both dielectric constant and dissipation factor. A maximum dielectric constant of 2.5 was chosen both to minimize dielectric losses and to keep the permittivity of the plastic as close as possible to that of the helium impregnant. The loss tangent of 20 x 10^{-6} was set so that the dielectric loss at the likely operating voltage would be no greater than either the conductor loss or the heat leak through the cryogenic envelope. An important design requirement discovered during electrical testing of small cable samples was that the dielectric tapes must be solid rather than of porous construction. Helium-impregnated porous tapes were found to have significantly lower dielectric strengths than solid tapes.[6]

Finally, in order for an underground cable to be cost effective, it should have a life expectancy of 30 to 40 years. Studies of the effects of environmental stress cracking and fatigue failure on the life of dielectric tapes are being conducted.

Initial evaluation of commercially available plastic films disclosed that none simultaneously satisfied all our requirements.[7] Dielectrically acceptable tapes were mechanically weak, and mechanically strong tapes had unacceptable dielectric properties (see Table II). The Teflons, Kaptons, and other exotic tapes had attractive properties but were set aside because of their very high costs. Attempts to reduce the 60 Hz, 4.2 K loss tangents of polysulfone and polycarbonate by altering their chemical construction were unsuccessful. Consequently, the decision was made to modify the dielectric and mechanical characteristics of the less expensive, intrinsically lower loss polyolefins. This development work is described in the following sections.

TAPE DEVELOPMENT

Dielectric Properties

The intrinsic dielectric losses of pure polyethylene and polypropylene are very small at 4.2 K (i.e., $\simeq 5$ x 10^{-6}). The higher values of tan δ measured for commercially produced polyolefins are due to the presence of additives placed in the polymer during the manufacturing process to protect the polymer in its intended air environment. Early work by King and Thomas[8] disclosed that the antioxidant may be one of the major sources of dielectric loss at temperatures of 6 to 8 K. A subsequent study of the effects of antioxidant on tan δ, carried out jointly by Battelle Columbus Laboratories (BCL), the National Bureau of Standards (NBS), and BNL, also showed that the 60 Hz loss tangent of polyethylene, in the region of 4 to 10 K, was strongly dependent upon both type and

Table II. Dielectric and Tensile Properties of Dielectric Tape Candidates.

Polymer type	Dissipation factor* (tan δ x 10)	Dielectric constant*	Yield strength, MPa at 293 K	Tensile modulus, GPa at 293 K	Tensile modulus, GPa at 4.2 K
Polyethylene low density nonoriented (100 μm)	15	2.3	2.1	0.093	5.45
Polypropylene low density nonoriented (125 μm)	9	2.2	6.9	0.248	0.830
Polypropylene biaxially oriented laminated (100 μm)	21	2.3	22.1	1.92	1.44
Polysulfone (125 μm)	60	2.5	39.8	1.90	4.47
Polyimide, Kapton H (100 μm)	90	3.1	44.1	2.84	5.50
Polycarbonate, Makrofol "KG" (60 μm)	55	2.9	49.3	3.48	4.55
Polyester, Mylar (75 μm)	200	2.5	65.3	4.01	4.53

* at 4.2 K and 100 Hz

concentration of antioxidant. One variety of antioxidant, Topanol, in a concentration of 0.1%, was found to result in a loss tangent < 10 x 10^{-6} over the temperature range 4.2 to 10 K.[9]

Mechanical Properties

The most severe problems facing the designer of cryogenic dielectric insulation are those of obtaining satisfactory mechanical properties over a wide temperature range. The dielectric must be able to withstand the variety of forces present during construction, installation, and over the normal life of the cable.

Tensile Measurements. Tape tensile measurements were made at 293, 77, and 4.2 K using an Instron table model testing machine and the associated apparatus shown in Fig. 1. Cryogenic tensile tests were made in the modified 4-liter helium dewar shown in the figure. A pair of grips with cylindrical bearing surfaces were used to hold the specimens. During a cryogenic tensile test, the sample was located inside a 6 cm diameter by 40 cm long compression cylinder, which in turn was submerged in the cryogenic liquid. The compression cylinder was filled with pressurized helium during tests made at 77 K. (Olf[10] has shown that nitrogen gas, at a temperature near its boiling point, can induce crazing and stress cracking in many polymers.) Test specimens were usually 10 cm long by 2 cm wide, and crosshead speeds were (0.05 cm/min). A strip chart record of load versus strain was made of each run. Representative stress-strain curves for both high- and low-density polypropylene samples are illustrated in Fig. 2.

The initial slopes of the curves were used to compute tensile moduli. The maximum stress reached during a run was used to calculate tensile strength, and the strain at fracture was taken as the ultimate elongation. The yield point was defined as the intercept of the stress-strain curve with a straight line parallel to the initial slope and offset 0.2% extension. The tensile data taken at 293, 77, and 4.2 K are summarized in Tables III, IV, and V, respectively.

Examination of Table III shows that the nonoriented polyolefins possess 293 K tensile strengths and tensile moduli well below cable specifications. However, oriented polypropylene and polyethylene tapes have superior tensile properties at this temperature.

Tables IV and V illustrate the pattern of dramatically reduced total elongations and increased embrittlement that accompanies most plastics upon cooling to cryogenic temperatures. These changes are most pronounced with the low-density, low-modulus polyolefins. A high-density, biaxially-oriented laminated polypropylene tape

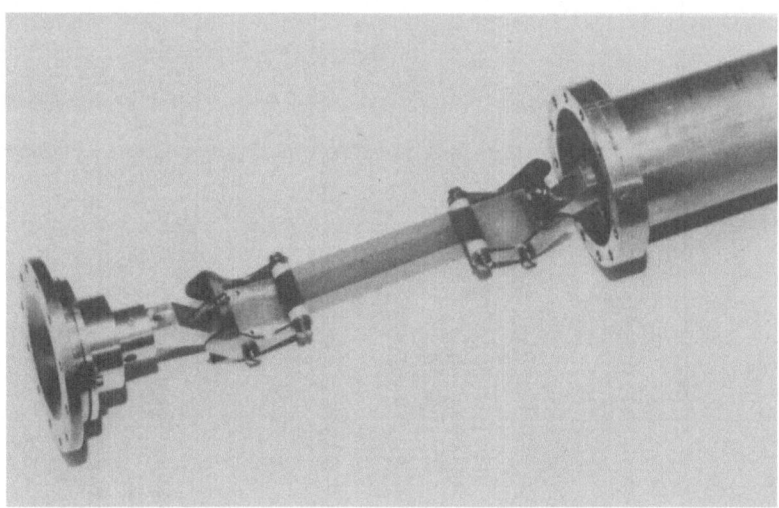

Fig. 1. Apparatus used for tensile measurements.

Table III. Tensile Properties of Dielectric Tape Candidates at 293 K.*

Polymer type	Elastic elongation, %[†]	Total elongation, %	Yield strength,[†] MPa	Tensile strength, MPa	Tensile modulus, GPa
Polyethylene, low density nonoriented (100 μm)	2.48	528	2.1	14	0.093
Polypropylene, low density nonoriented (125 μm)	3.1	1118	6.89	46.0	0.248
Polyethylene, uni-axially oriented, laminated, Valeron (100 μm)	0.64	402	3.94	40.6	0.889
Polypropylene, bi-axially oriented, laminated (100 μm)	1.35	61.8	22.1	169	1.90
Polysulfone (125 μm)	2.25	63.6	39.8	58.8	1.88
Polyimide, Kapton H (100 μm)	1.84	55.2	44.1	182	2.83
Polycarbonate, Makrofol "G" (100 μm)	2.00	72.2	37.1	110	2.13
Polyester, Mylar (75 μm)	2.08	114	65.3	150	4.01

* Measurements made in tape machine direction.   † At 0.2% offset.

Table IV. Tensile Properties of Dielectric Tape Candidates at 77 K.[*]

Polymer type	Elastic elongation, %[†]	Total elongation, %	Yield strength,[†] MPa	Tensile strength, MPa	Tensile modulus GPa
Polyethylene, low density, nonoriented (100 μm)	2.60	3.60	104	104	3.68
Polypropylene, low density nonoriented (125 μm)	2.38	4.55	72.8	74.9	3.40
Polyethylene, uni-axially oriented, Valeron (100 μm)	2.60	10.4	93.1	114	3.88
Polypropylene, bi-axially oriented, laminated (100 μm)	3.28	17.5	183	321	1.52
Polysulfone (125 μm)	2.97	4.40	66.9	85.4	2.38
Polyimide, Kapton H (100 μm)	2.35	12.0	112	260	4.93
Polycarbonate, uni-axially oriented Makrofol "G" (100 μm)	2.85	5.67	93.1	147	3.40
Polyester, Mylar (75 μm)	2.75	5.75	171	276	6.73

[*] Measurements made in machine direction of tape.   [†] At 0.2% offset.

Table V. Tensile Properties of Dielectric Tape Candidates at 4.2 K.*

Polymer type	Elastic elongation, %†	Total elongation, %	Yield strength,† MPa	Tensile strength, MPa	Tensile modulus, GPa
Polyethylene, low density, nonoriented (100 μm)	2.85	2.85	150	150	5.45
Polypropylene, low density, nonoriented (125 μm)	--	1.175	--	38.3	0.830
Polyethylene, uni-axially oriented Valeron (100 μm)	3.1	3.1	125	125	4.23
Polypropylene, bi-axially oriented, laminated (100 μm)	3.26	7.80	207	323	1.44
Polysulfone (125 μm)	2.62	2.98	111	121	4.47
Polyimide, Kapton II, (100 μm)	3.08	5.78	154	256	5.50
Polycarbonate, uni-axially oriented, Makrofol "G" (100 μm)	4.42	7.25	168	248	4.10
Polyester, Mylar (75 μm)	4.52	10.75	190	348	4.53

* Measurements made in machine direction of tape.   † At 0.2% offset.

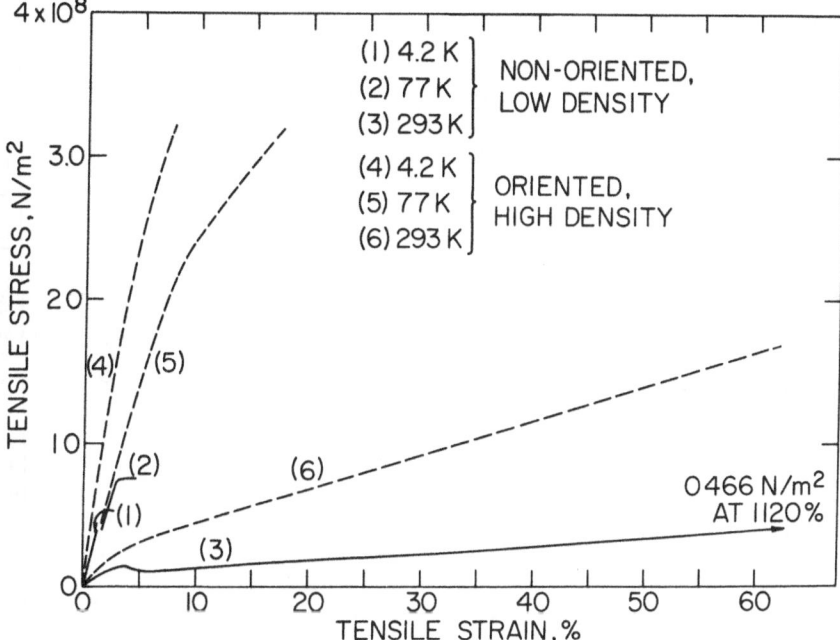

Fig. 2.   Typical stress-strain curves for oriented and
          nonoriented polypropylene films.

under development at BNL has acceptable tensile properties at all
temperatures, while also having dielectric and thermal properties
that meet cable specifications. This film is manufactured in a
thickness of 32 μm and either two or three layers are cemented
together with a 2 μm thick polyurethane binder to produce total
tape thicknesses of either 66 or 100 μm, respectively.

   Very High-Modulus Tapes. The results of tests made with laminated, intermediate-modulus polypropylene show that bending behavior
of cables fabricated with this material should be satisfactory for
cable insulation thicknesses up to at least 1 cm. However, radial
pressure is dependent upon insulation thickness,[4] and higher voltage cables may require higher ratios of tensile modulus, E_1, to
compressive modulus, E_3, to permit reeling without damage to the
insulation. Most standard methods of reducing E_3 would probably
also degrade the dielectric performance of the tape. However, an
increase in the value of E_1 would benefit winding and reeling performance without jeopardizing the electrical properties of the
insulation (see Table VI).

Table VI. Values of Some Factors Affecting Cable Bending Performance.

Polymer type	Tensile modulus, E_1, GPa	Compressive modulus, E_3,* MPa	$n = E_1/E_3$	Coefficient of friction, μ_s†
Polyethylene, uni-axially oriented, laminated Valeron (100 μm)	0.89	17.6	50.4	0.418
Polycarbonate, nonembossed, Makrofol "G" (100 μm)	2.14	18.3	117	0.453
Polypropylene, bi-axially oriented, 2-ply laminate (66 μm)	2.34	16.1	145	0.225
Polypropylene, bi-axially oriented, 3-ply laminate (100 μm)	1.90	7.0	273	0.225
Polycarbonate, embossed, Makrofol "G" (178 μm)	1.71	2.82	607	0.438
Kraft paper, electrical grade (178 μm)	6.40	6.4	1000	0.320

* At 0.14 MPa.

† Cross machine direction to cross machine direction.

Although the tensile moduli of commercial grades of polyethylene are only approximately 1 GPa (1.4 x 10^5 psi), the theoretical modulus of highly crystalline-oriented polyethylene is 0.1 TPa (see Fig. 3). Polymers derive their very high moduli from an orientation because the number of tie molecules connecting crystalline regions is greatly increased during orientation (see Fig. 4). The tie molecules are produced as a result of friction between adjacent blocks of lamellae. This causes chain unfolding at the boundaries of these blocks. In drawn material, the resulting higher modulus is almost directly proportional to the draw ratio (see Fig. 5).

There are two major methods used to produce very highly oriented, high-modulus polymers; cold drawing and hydrostatic extrusion. Although the highest moduli may be reached by way of the cold-drawing technique, hydrostatic extrusion provides better control over the final product dimensions than does drawing.[13]

Working under BNL contract, Battelle Columbus Laboratories (BCL) has begun preliminary work to develop a hydrostatic extrusion process for fabricating ultrahigh modulus polyolefin tapes with a tensile modulus of 7 GPa (1 x 10^6 psi). (Recently, single fibres of highly oriented polyethylene were prepared by Porter[14] that had moduli of 70 GPa (1 x 10^7 psi). Using the die shown in Fig. 6, molten polymer is fed to a rectangular transition zone, which precedes the deformation zone of the die. The polymer is cooled in the transition zone so as to be solid prior to area reduction. The thickness of the transition zone is 1.5 mm and the deformation zone is 125 μm thick. This geometry results in an extrusion ratio or draw ratio of 12:1. The surface of the channel was also Teflon coated to reduce frictional drag. Several tapes 25.4 mm wide, by 125 μm thick, by several meters long have been successfully extruded. Tensile moduli of these tapes were approximately 14 GPa (2 x 10^6 psi). Modifications are planned to improve the die temperature control and to reduce the extrusion ratio to 6:1.

Thermal Properties

Thermal Expansion. Early BNL dielectric strength measurements were made with eight to ten layers of tape helically wound on a 1.27 cm diameter stainless steel mandrel. Most nonoriented polyethylene and polypropylene tape candidates fractured during these tests, which were made at 6 to 8 K. A comparison of the 293 to 4.2 K thermal contractions[15] with the elongation to fracture at 4.2 K revealed the probable reason for this problem (see Table VII). Tapes could not contract while wrapped around the metal mandrel and were forced to stretch beyond their maximum elongations. Tapes possessing elongations considerably larger than required contractions remained intact. To avoid future cryofracture problems,

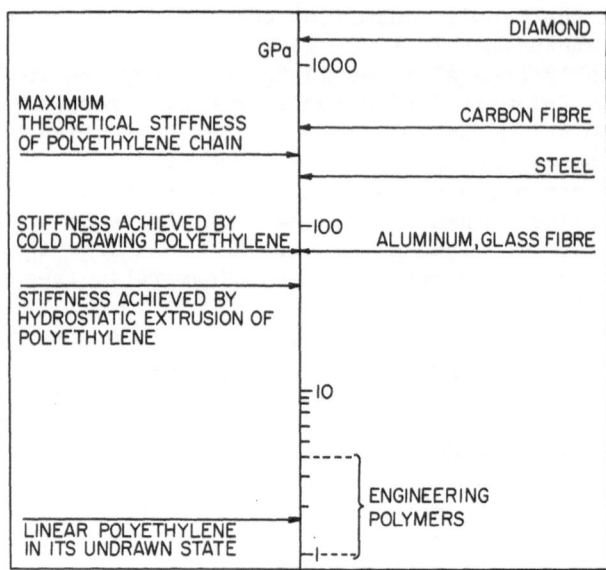

Fig. 3. The comparative stiffness of some typical materials. Data from Ward.[11]

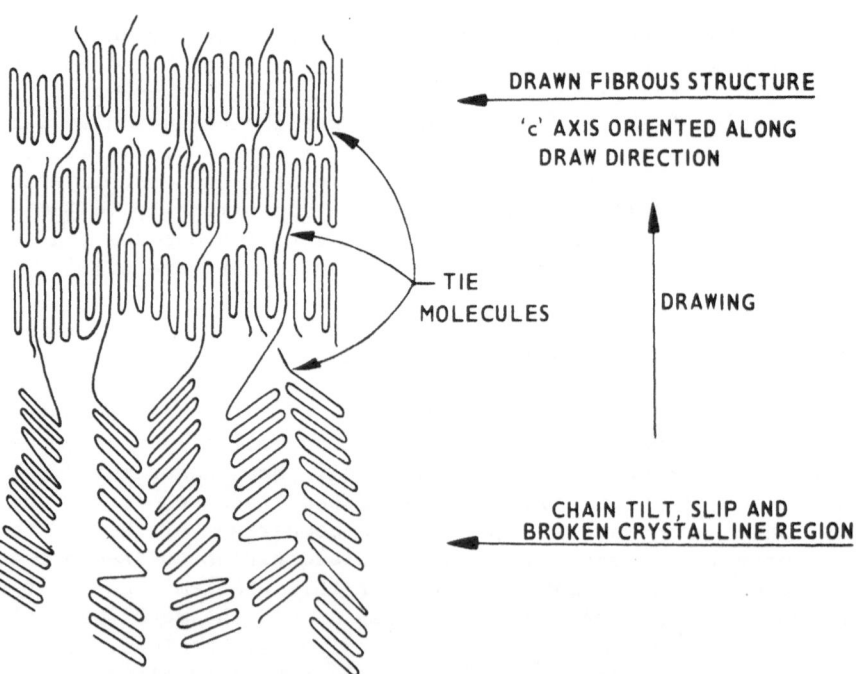

Fig. 4. Structural changes occurring during cold drawing (from Gibbons).[12]

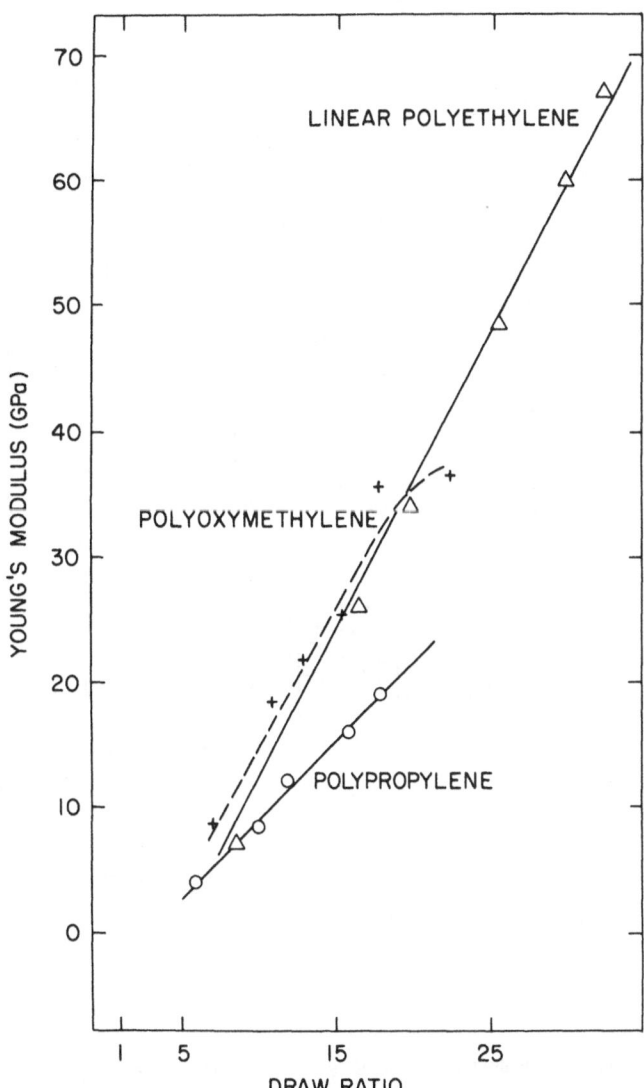

Fig. 5. Young's modulus as a function of draw ratio for ultrahigh modulus polymers. Data from Ward.[11]

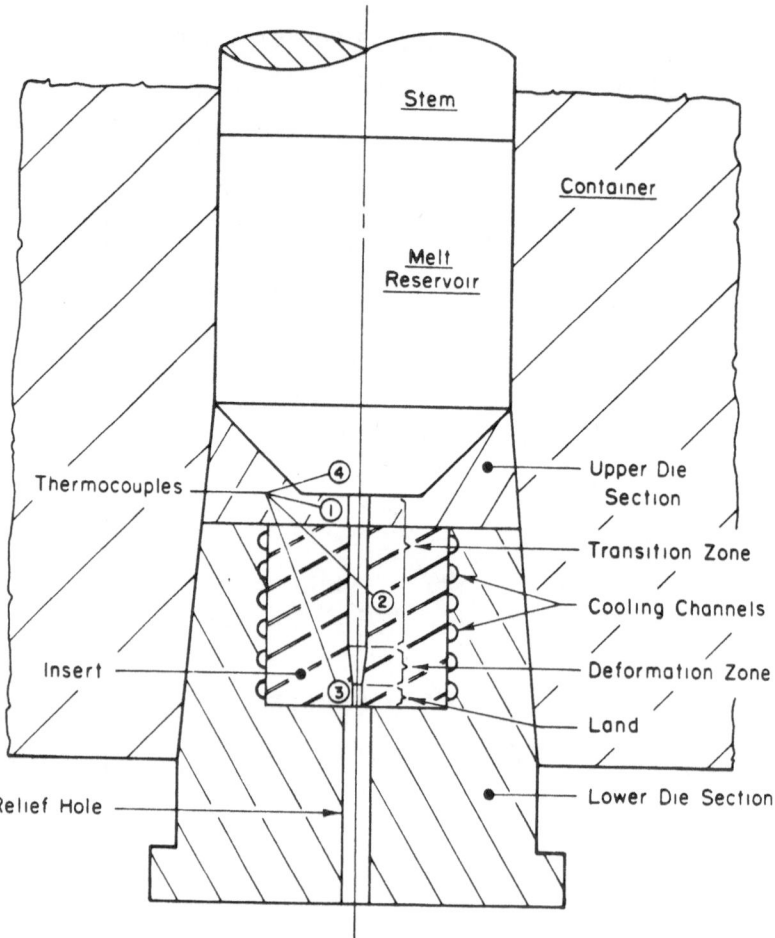

Fig. 6.   Schematic cross section of tape die in
          extrusion container.

Table VII. Thermal Contraction and Tensile Elongation of Dielectric Tapes.

Polymer type	Contraction, 293 K to 4.2 K, %		Elongation to fracture at 4.2 K, %	Elongation/ contraction, %
	Longitudinal	Transverse		
Polyethylene low density, nonoriented	2.74	2.69	2.85	1.02
Polyamide, Nylon-11	1.92	1.85	3.13	1.61
Polyethylene, uni- axially oriented, laminated Valeron	1.70	1.26	3.10	1.80
Polysulfone	1.16	1.07	2.98	2.58
Polypropylene, bi- axially oriented, 2-ply laminate	0.641	--	8.29	12.9
Polycarbonate, Makrofol "KG"	0.474	0.471	10.8	23.0

BNL tape candidates are required to meet an empirical elongation to contraction ratio of $\geq$ 2:1. As a general rule, it was found that the elongation to fracture usually increased as a result of orientation, and the thermal contraction decreased following this treatment. The 293 to 4.2 K contraction of the laminated biaxially oriented tapes was found to be 0.641,[15] and E/C ratios for this material are greater than 12:1. (Details of the BCL thermal expansion technique are described in Appendix I.)

Thermal Conductivity. Conductivity measurements were made of tape candidates to determine whether or not they would satisfy the cable design requirements listed in Table I. The measurements were made for BNL by Jelinek of BCL using the apparatus illustrated in Fig. 7. The method was a modified steady-state conductivity technique where a temperature gradient was established between two copper plates separated by four layers of polymeric film. (Multilayer measurements were always made to approximate the series interfacial resistivity that would be present in lapped cable configurations.) One plate was attached to a controlled heat sink and a measured quantity of heat was added to the other plate by means of an electric heater. With the use of a liquid helium throttling dewar, the ambient temperature could be controlled to within ± 1 K. (Complete details of the conductivity measurement method are included in Appendix II of this paper.)

The results of the conductivity measurements made of several materials in vacuo are shown in Table VIII. All of the polymers showed increasing values of thermal conductivity with temperature over the range of 6 to 300 K. The data in Table IX were taken at 6 K as a function of helium gas pressure at several pressures over the range 0 to 710 kPa (0 to 100 psi). This experiment was designed to more closely approximate the composite heat path present in a superconducting cable impregnated with supercritical helium. Although there is a high dependency of conductivity on pressure between 0 and 178 kPa (25 psi), there is no further dependence of conductivity on pressure above this pressure for the polyethylene and polycarbonate films. The polysulfone conductivity appears to increase again as a function of pressure at 710 kPa. The conductivities of the three materials at 710 kPa meet the specifications of Table I.

CONCLUSIONS

The development of a suitable polymeric tape for use as insulation on ac superconducting cables is a challenging engineering problem for the designer of plastic films. No commercially available tape simultaneously fulfills all of the dielectric, mechanical, and thermal requirements without modification. Porous, paper-like tapes produced dielectrically weak cables.

Table VIII. Thermal Conductivities of Dielectric Tapes.*

Temperature, K	Thermal conductivity, mW/m·K			
	6	20	100	300
POLYMER TYPE				
Polysulfone	2.0	2.5	4.4	11
Polyethylene, uniaxially oriented, laminated Valeron	6.0	7.1	12	81
Polypropylene, biaxially oriented, 2-ply laminate, urethane binder	7.8	11	29	89
Polycarbonate, Makrofol "KG"	9.0	12	15	60
Polypropylene, biaxially oriented, 2-ply laminate, polyethylene binder	9.2	13	43	110

* Measurements made in a vacuum at 0.13 mPa (10^{-6} torr).

Table IX. Thermal Conductivity of Three Polymers as a Function of Helium Gas Pressure at 6 K.*

Polymer type	Pressure, MPa				
	0	178	532	710	
Polysulfone	2.0	8.4	8.4	9.7	
Polyethylene, uni-axially oriented, laminated Valeron	6.0	15	14	16	
Polycarbonate, Makrofol "KG"	9.0	21	25	20	

* Conductivities in mW/m•K.

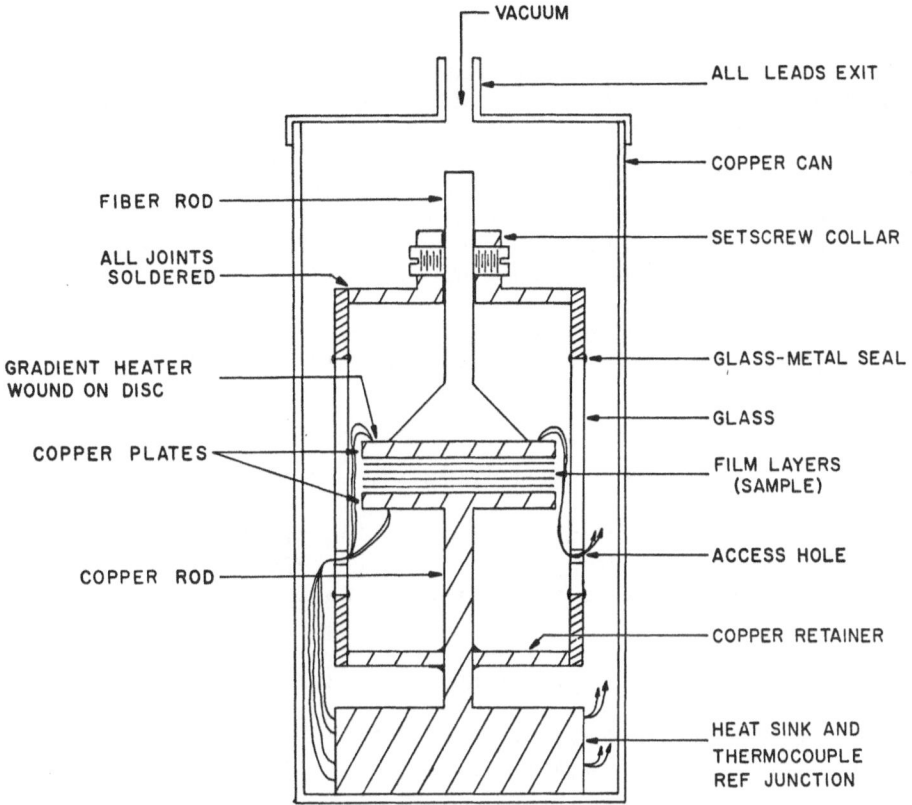

Fig. 7.   Schematic of sample container used for thermal
          conductivity measurements at 480 kPa (70 psi)
          applied pressure on four thicknesses of poly-
          meric film.

Uniaxially oriented polyolefin tapes were found to have tensile properties superior to the nonoriented types, but these
materials often fibrillated during cooldown to operating temperature. Further studies showed that commercially produced, 32 μm
thick, biaxially-oriented polypropylene tapes had acceptable tensile properties at both 4.2 and 293 K. The desired tape thicknesses of 66 μm and 100 μm were obtained by laminating two or three
layers together with a 2.0 μm thick polyurethane adhesive. The loss
tangent and permittivity of this laminate meets design considerations. Biaxial orientation was also found to reduce thermal contraction and increase the 4.2 K elongation of the polypropylene
films. Work has been started to develop a very high-modulus,
single-layer polyethylene tape for use with higher voltage superconducting cables.

ACKNOWLEDGMENTS

The author wishes to acknowledge the technical assistance of
A. L. Minardi and J. Scrofani in obtaining the results reported in
this manuscript. Many other persons have contributed to the design
of the experiments described here, in particular D. H. Gurinsky,
A. J. McNerney, K. F. Minati, and R. Zoller of Brookhaven National
Laboratory, and J. A. M. Gibbons of the Central Electric Research
Laboratory, England. Finally, the author wishes to express his
thanks to E. B. Forsyth for the many suggestions made during
preparation of this paper.

Work was performed under the auspices of the U. S. Department
of Energy.

REFERENCES

1. E.B. Forsyth, "Overview of Electric Transmission (Technology
    and Economics)," paper presented at the 1978 Thirteenth Inter-
    society Energy Conversion Engineering Conference, San Diego,
    California (1978).

2. E.B. Forsyth et al., IEEE Trans., PAS 92, 494 (1973).

3. E.B. Forsyth, A.J. McNerney, A.C. Muller, and S.J. Rigby,
    IEEE Trans., PAS 97, 734 (1978).

4. P.G. Priaroggia, E. Occhini, and N. Palmieri, Fundamentals of
    the Theory of Paper Lapping of a Single Core High Voltage
    Cable, Pirelli, SPA, Milan, Italy (1961).

5. G.H. Morgan and J.E. Jensen, Cryogenics 17(5), 259 (1977).

6. E.B. Forsyth, A.J. McNerney, and A.C. Muller, in: Advances in
    Cryogenic Engineering, Vol. 22, Plenum Press, New York (1977),
    p. 296.

7. A.C. Muller, Revue Generale de l'Electricite 94, 568 (1975).

8. C.N. King and R.A. Thomas, in: 1974 Annual Report of the Con-
    ference on Electrical Insulation and Dielectric Phenomena,
    National Academy of Sciences, Washington, D.C. (1975).

9. F.I. Mopsik, Nonmetallic Materials and Composites at Low
    Temperatures, Plenum Press, New York (1979), p. 85.

10. H.G. Olf and A. Peterlin, J. Polymer Sci. 12, 2209 (1974).

11. I.M. Ward, Phys. Bull., 28(2), 66 (1977).

12. J.A.M. Gibbons, R.M. Scarisbrick, and I.A. Sutherland, Liquid
     Flash Plastic Composites for High Voltage Insulation, Part IV:
     Morphology of Drawn Polythene Films, Internal Report, Labora-
     tory Note No. RD/L/N 31/72, Central Research Laboratories
     (1972), p. 19.

13. T. Williams, J. Mater. Sci. 8, 59 (1973).

14. R.S. Porter and N.J. Capiatti, J. Poly. Sci.-Phys. 13, 1177
     (1975).

15. F. Jelinek and A. Muller, in: Advances in Cryogenic Engineer-
     ing, Vol. 22, Plenum Press, New York (1977), p. 312.

APPENDIX I

Thermal Expansion Measurement Technique

In preparing the tape samples for measurement, the procedure
consisted of cutting enough tape strips (6 mm wide x 38 mm long)
to produce a stack of material 6 mm (¼ in) thick when clamped.
Steel clamping jigs were machined to facilitate this stacking procedure. While clamped, the ends of the stack were trimmed to produce a specimen approximately 32 mm (1¼ in) long, and the longitudinal edge faces of the stack were lightly wiped with Eastman
910 (strain gauge quality) epoxy. The bundles were then unclamped
and end-ground on a metallographic polishing wheel. Thermocouples
were attached to the specimens with thin strips of masking tape
and the beads fastened to the specimen surface with a small drop
of electrical varnish.

Measurement Technique. The thermal expansion measurements
were performed in a fused silica dilatometer with associated LVDT
sensing equipment. The dilatometer has a resolution of 13 nm
(0.5×10^{-6} in). Temperature is controlled at 10° intervals by
utilizing a throttling dewar, which uses the sensible heat resulting from vaporized cryogen and proportional controllers, which
permit temperature control of ±0.1 K in the entire temperature
range.

APPENDIX II

Description of Thermal Conductivity Measurements

In the modified steady-state conductivity method employed in
this experiment, a temperature gradient is established between two
copper plates separated by the polymeric film. One plate is
attached to a controlled temperature heat sink, and a measured
quantity of heat is added to the other plate by means of an electric resistance heater. Conductivity is then calculated using a
form of the Fourier equation:

$$k = \frac{q}{A}\frac{L}{\Delta T} \tag{1}$$

where k is the thermal conductivity, g/A is the heat flow per unit
across section area, L is the film thickness since the conductivity
of copper and the bonding agent are orders of magnitude higher,
and ΔT is the temperature gradient across L measured with Keithley
147 nanovoltmeter.

The gradient heater is a 3-lead unit wound of Evanohm wire,
which has a nearly zero temperature coefficient of resistance. A
constant current source is used to power the gradient heater. The
temperature gradient set up in the specimen after a steady-state
condition has been reached is measured using either gold cobalt
versus "normal" silver differential thermocouples, or miniature
platinum resistance thermometers. The ambient temperature is precisely controlled (± 0.05 K) during a measurement using the output
of a Keithley 150 B null detector, the signal to which arises from
a copper-constantan thermopile mounted on the specimen container,
and a low-temperature modified West controller. The temperature
gradient across the specimen is measured by a Keithley 140 nanovoltmeter.

The measurements are carried out in a liquid helium throttling
dewar. This dewar provides a degree of ambient temperature control
in itself in that by suitably adjusting the throttle value (which
admits helium through a capillary to the dewar sample chamber) and
the vaporization heater voltage (which allows the liquid helium to
vaporize before entering the sample chamber), the cold helium gas
flowing past the specimen fixture (which is highly evacuated) can
be controlled to within ± 1 K. This greatly reduces the burden on
the independent ambient temperature control device used in the
specimen fixture.

For the single film unpressurized experiment, the sample
assembly presented no unusual difficulty. For the pressurized

conditions, a different sample container had to be designed that
could convey the required pressure to the film material and at the
same time prevent excessive heat loss through the mechanical pressurizing structure. Figure 7 illustrates the technique used. The
method was not ideal in that heat losses of 25 to 30% were measured
along the top fiber rod. These losses were taken into consideration in calculating the effective thermal conductivity of the film
using the equation previously discussed. A load of 480 kPa (70 psi)
was applied to the fiber rod using an LVDT load measuring device.
As the load was applied and held, the setscrew collar was tightened.
We estimate that the applied load was retained to within 34 kPA
(5 psi) after cooldown. The glass-metal tube was utilized to
further minimize heat losses during the measurement.

THE MANUFACTURE AND PROPERTIES OF A

GLASS FABRIC/EPOXY COMPOSITE BELLOWS

D. Evans, J. U. D. Langridge, and J. T. Morgan

Rutherford Laboratory
Chilton, Didcot, Oxon, England

INTRODUCTION

This bellows is a small but critical part of a large bubble chamber at present being constructed. Shown in Fig. 1, it forms a gas tight seal between the reciprocating piston and the stationary chamber walls, separating the chamber liquid from the piston backing gas. It is tubular (800 mm diameter), containing only one convolution, and operates at a temperature of 26 K in a 3 tesla magnetic field. For this latter reason, a polymeric material is preferred since this avoids eddy current heating effects.

During normal operation of the bubble chamber, the bellows is required to accommodate the total stroke of the piston (±3.5 mm) and to withstand a pressure swing of ±0.3 MPa (±3 bar) at a frequency of 30 Hz.[1] Detailed consideration of these operating parameters confirms the need for high quality glass-reinforced plastics material and precludes the possiblity of using simple 'wet layup' techniques.

This paper describes the moulding technique, first developed for a 300 mm diameter bellows successfully used in a similar application. It produces a composite of the complex shape required, to close dimensional tolerances and with a glass fabric content in excess of 60% by weight.

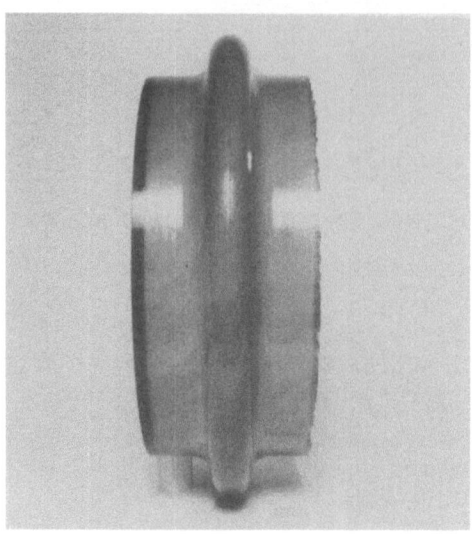

Fig. 1.   Bellows as demoulded with ends untrimmed.

MANUFACTURING TECHNIQUE

General

Apart from shape and size considerations, the prime requirement of the composite is a high and uniform glass content and
this, by definition, excludes resin-rich areas or surfaces.
Matched metal moulding techniques have limitations when used to
process composite materials of this form. In addition, a complicated and costly split mould would be required to remove the male
section from the inside diameter of the bellows after moulding and
curing the epoxy resin. A flexible, self-releasing silicone rubber mould core offers a simple solution.

A general (diagrammatic) cross-sectional view of the mould is
presented in Fig. 2, from which it may be seen that the mould consists of three essential parts, viz., (1) an aluminium alloy split
female mould, (2) an intermediate silicone rubber male core, and
(3) a tapered central aluminium alloy plug. Important features to
note are the spacer inserted between the split halves of the outer
mould, the available space below the central plug, and the channels
moulded into the silicone rubber (top and bottom) to form a reser-
voir for the resin. Prior to use with epoxide resins, the mould
surfaces are treated with a silicone release agent (Tego 290 –
Ambersil Ltd.) and cured for 3 hours at 230°C.

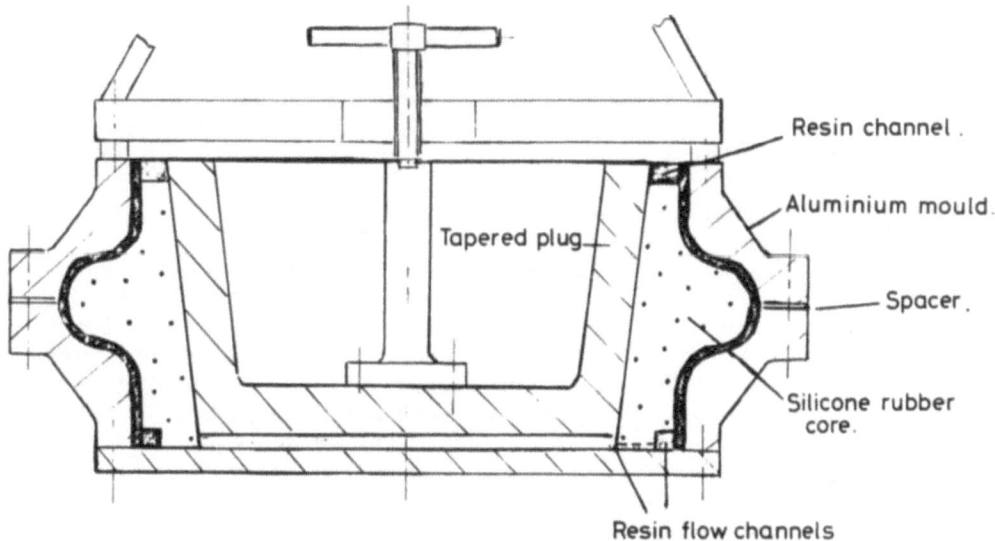

Fig. 2. Cross-sectional view of bellows mould.

Mould Manufacture

The split outer mould and the tapered (5°) central plug are manufactured from aluminium alloy (BS 1470-HS-30) to a conventional design. Provision is made on the mould for a screw jacking system to apply moderate pressure to the plug, used in the later stages of moulding to consolidate the glass fabric/epoxy resin. The split outer mould is provided with an O ring in both top and bottom sections, which, during moulding, bear on a flat spacer ring inserted to compensate for the thickness of the laminate. It is found that a spacer thickness of 5 mm gives sufficient moulding pressure.

Preparation of Rubber Mould

It is important to note that the liquid silicone rubber (Table I) is sensitive to impurities and the cure of the liquid will be inhibited by a range of chemical contaminants.[2] It is therefore necessary to ensure that all moulds and mixing implements are scrupulously clean.

The mould is assembled without a spacer ring separating the mould halves. The resin filler holes are plugged with rubber bungs, trimmed flush on the inside, and sealed over with 'Melinex' adhesive tape. Direct contact with rubber bungs will inhibit cure of 'Sylgard,' as will contact with rubber O rings. With plug extension plate in position, the complete mould is placed in the

Table I. Properties of Sylgard 186[4].

Mixing ratio	10 pbw resin 1 pbw curing agent
Viscosity of base resin	80 Pas
Viscosity of base mixed with 10 phr curing agent added	45 Pas
Useable life of mixed resin	>2 hours
Cure	24 hours at room temperature
+	10 hours at 65°C
Hardness (Shore A Durometer)	32

vacuum tank. The plug is held in a central position using the
crossbeam and screw.

The silicone elastomer is hand mixed initially and then vacuum
degassed. It is then poured into the mould cavity in a manner to
avoid, as far as possible, the inclusion of air.

At the quarter, half, three-quarter, and full stages, the
vacuum tank is closed and sufficient vacuum generated to remove
air bubbles from the casting. Finally, the rubber is cured at
room termperature for 24 hours followed by postcuring for 10 hours
at 65°C. It is necessary to cut or mould channels in the lower
surface of the rubber to allow free access of resin to all areas
of the mould.

In the cured state, Sylgard 186 exhibits high surface friction, which affects the smooth entry of the plug. This is overcome
by lightly dusting the taper bore only with talcum powder, or by
applying a thick coat of liquid resin to the taper plug prior to
insertion in the rubber.

To date, one such rubber former has been used to mould four
bellows, and it is believed that further mouldings are possible.
It must be remembered that although silicone elastomer is self-
releasing from epoxy resins and from clean metal surfaces, it will
bond to silicone release agents. It is therefore essential to
mould the rubber prior to treating any parts of the mould with
release agent.

LAMINATING PROCEDURE

Strips of glass cloth are cut at 45° to the warp and weft
directions from a satin weave fabric having an epoxy-compatible
finish (Marglass cloth 116T with P738 finish). These are wound
around the silicone rubber former, consecutive cloths being
loosely stitched together after removal of the selvedges (Fig. 3),
until 12 complete layers have been applied.

The mould is then assembled around the fabric-covered plug,
care being exercised to prevent trapping of the glass fibres
between mould parts. The spacer is included at this stage
between the upper and lower outer mould parts to allow for the
thickness of the laminate. Rubber O ring seals are contained in
the mould joints, and after assembly these are tested for vacuum
leaks. The tapered plug is then loosely inserted after removal
of the extension plate and the complete mould is placed in a
vacuum chamber.

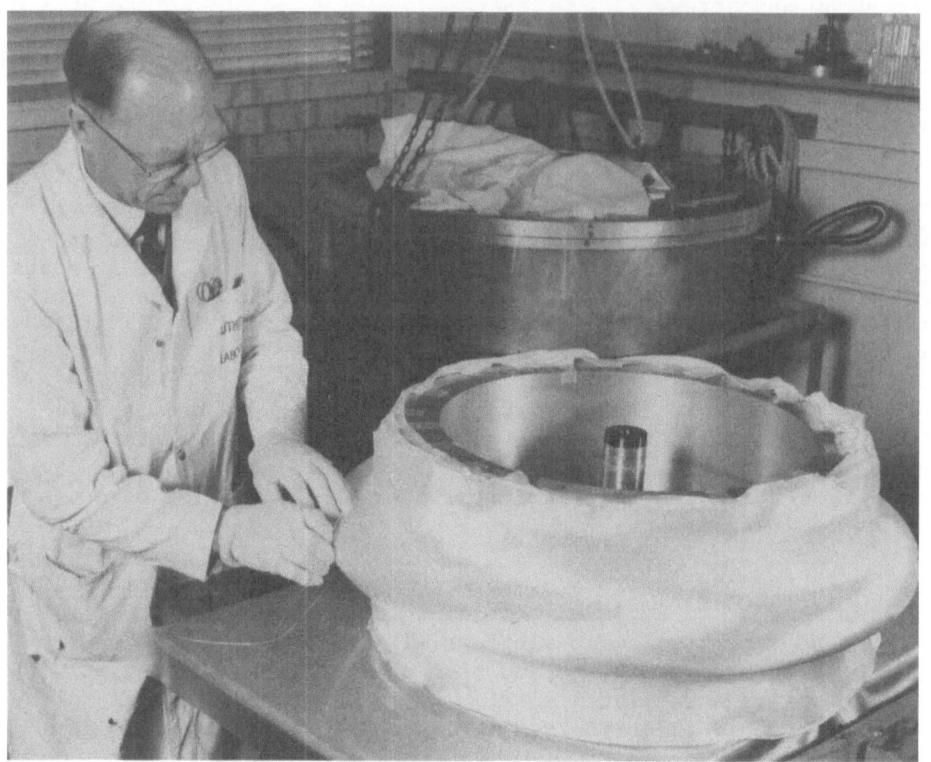

Fig. 3. Dry laying up of glass fibre reinforcement.

IMPREGNATION

The arrangement for the vacuum impregnation is shown diagrammatically in Fig. 4. The vacuum chamber containing the mould
(H, Fig. 4) is heated to a temperature of 30°C and the pressure
reduced to 10 Pa or lower. The mould is maintained under these
conditions for a minimum time of 24 hours prior to impregnation
to establish an even temperature throughout and to completely
degas the glass fabric.

The resin, (Bisphenol A diglycidyl ether with an acid anhydride
curing agent; for formulation see Table II) is metered, hand stirred,
and vacuum degassed at a pressure of 20 Pa prior to transfer into
a separate vacuum chamber (G), which is at a pressure of less than
10 Pa.

Transfer of resin to the mould is initiated by increasing the
pressure in chamber G to 6 kPa. Resin flows into the space below
the plug and, when this is full, appears in the riser pipe, E, which

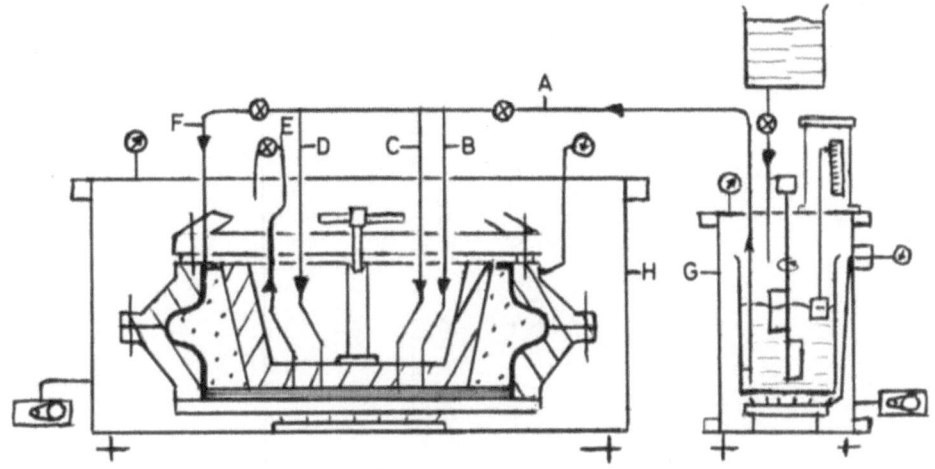

Fig. 4.   Bellows impregnation procedure (diagrammatic).

is then closed.   This pipe E initially serves to evacuate the
space below the plug and the feed pipes.   A reduction in the resin
flow rate indicates that impregnation of the glass fabric is com-
mencing and, at this stage, increasing the pressure in chamber G
to 12.0 kPa maintains the resin flow at 0.2 dm^3/h.   This flow rate
is chosen arbitrarily on the basis that a slow impregnation of the
fabric leads to uniform filling and therefore minimises the risk
of air entering unimpregnated areas when the pressure surrounding
the mould is returned to atmospheric.   The impregnation process
takes several hours and the flow of resin is stopped when the
channel at the top of the mould is completely full.   The penulti-
mate stage of the process is to allow the resin to soak into the
glass for several hours with the mould still under vacuum, and
finally to return the pressure to atmospheric over a period of
about one hour.   The composite is compacted at this stage by the
application of pressure to the tapered plug using the screwed
plunger, excess resin being expelled through the feed pipes in
the bottom of the plug.

The resin is gelled at 60°C, a process which takes approximately 12 hours after the mould has reached this temperature.   The
resin is finally cured for 20 hours at 120°C, the temperature being
raised in six steps over a period of six hours.

A low gelation temperature is used to minimise the thermal
expansion of the rubber part of the mould and so avoid undue
pressure on the composite during the critical gelation period.
Accelerator DY 062 was selected to achieve a long useable life at
30°C, while giving a relatively short gelation period at 60°C.

Table II. Epoxy Resin Formulations.

				Manufactured by	Cure:
For impregnating glass fibre bellows	Resin	MY 740	100 pbw	Ciba-Geigy Ltd.	24 h at 60°C
	Hardener	HY 906	80 pbw		20 h at 120°C
	Accelerator	DY 062	1 pbw		
For bonding bellows to stainless steel	Resin	MY 740	100 pbw	Ciba-Geigy Ltd.	Cure:
	Hardener	EM 308	75 pbw	Thiokol/Chem. Div.	24 h at room
		(for over-wrap EM 308)	50 pbw		temperature

ASSEMBLY AND TESTING

Typical flexural properties of satin-weave glass-fabric-reinforced epoxy resin at various temperatures to 4.2 K are shown in Fig. 5. The results are from test specimens cut from sample laminates prepared from materials proposed for the fabrication of the bellows. To examine the low-temperature flexural fatigue characteristics, the completed bellows is tested in the apparatus shown diagrammatically in Fig. 6. The joints between bellows and both piston and flange are adhesive bonded using the room-temperature-curing epoxy resin system detailed in Table I. Where possible, the joint is assembled dry with a single layer of 0.15 mm thickness glass fabric between the adherends, and the resin is admitted using vacuum impregnation techniques. The joint to the fibreglass piston is overwrapped for added stiffness with 10 mm thickness of glass fabric laminate applied by wet layup techniques, using the same resin system as that used for the bonding. These bonding techniques are similar to those proposed for use in the bubble chamber. The test apparatus closely simulates the conditions present in the bubble chamber so that meaningful fatigue tests may be carried out on both the bellows and the adhesive joints. It also permits tests to be carried out under more severe conditions of piston movement and pressure differential than are encountered under normal working conditions.

The load required to deflect the bellows through the 3.5 mm is calculated to be 20 kN, while a pressure differential of 0.4 MPa (4 bar) across the bellows contributes an additional load of 25 kN on the adhesive. This gives a working shear load on the adhesive joints of 45 kN, equivalent to 21 kN/m of joint length of 0.28 MPa of bonded area.

To date, a prototype bellows has been subjected to more than 7×10^5 cycles of ± 3.5 mm at a temperature of 77 K, whereas the smaller 300 mm diameter version has operated for more than 10^7 cycles of ± 2 mm at 26 K. This gives confidence to the use of fibreglass-reinforced epoxy resins and epoxy-resin adhesives in situations of cyclic stressing at low temperature.

CONCLUSION

The method outlined in this paper is being used to produce high quality glass/epoxy composites of the desired mechanical properties to close dimensional tolerances. Although it is slow and suited only to the production of a small number of components, the technique has proved effective for this application. In view of the small number of bellows to be moulded and the successful test results achieved with the first prototype, little variation

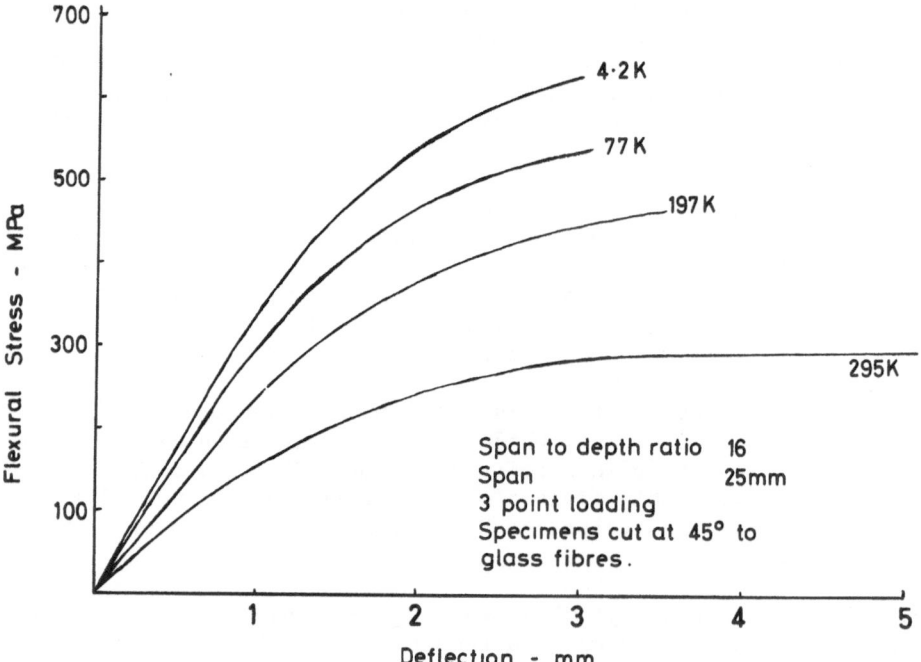

Fig. 5.　Stress strain curves for satin weave glass
fabric laminates at various temperatures.

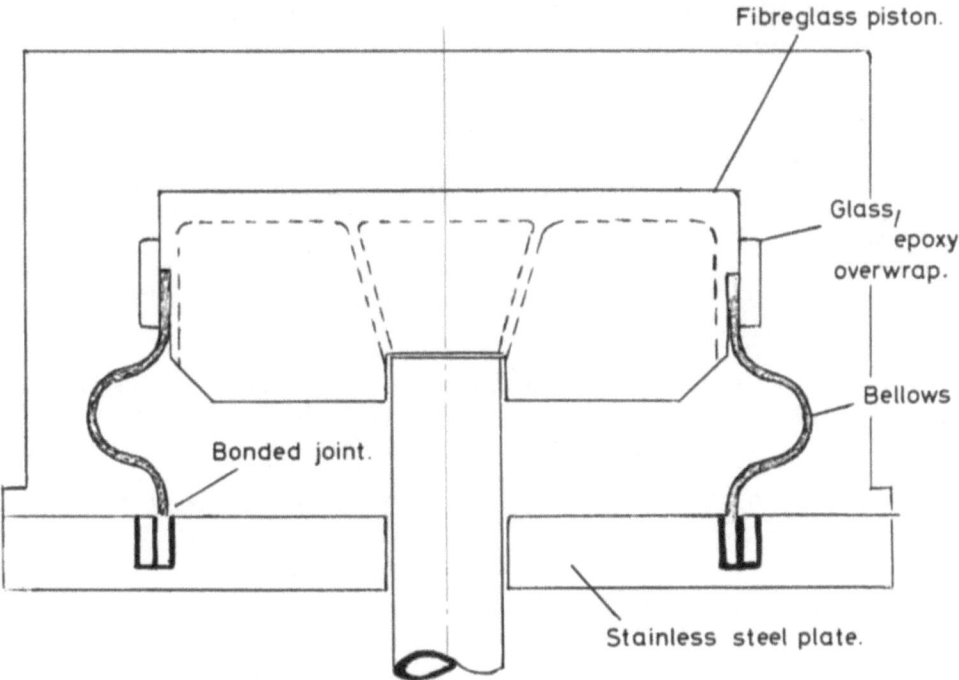

Fibreglass piston.

Glass/
epoxy
overwrap.

Bellows

Bonded joint.

Stainless steel plate.

Fig. 6. Low temperature fatigue testing apparatus (diagrammatic).

in the process has been made and therefore the various moulding parameters are not necessarily the optimum.

ACKNOWLEDGMENTS

Grateful acknowledgment is made to many colleagues at the Rutherford Laboratory intimately involved with aspects of the G.R.P. bellows other than the manufacturing techniques outlined in this paper. We are also indebted to our colleagues in the Chemical Technology Group of the Laboratory for their practical skills and ingenuity, without which the production of these bellows would have not been possible.

REFERENCES

1. B.R. Diplock, Design Note on G.R.P. Bellows, Publ. No. EHS/RCBC/R/T/034/BRD, Rutherford Laboratory, Chilton, Oxon, England.

2. Bulletin 07-312-01, Dow Corning International Ltd., Brussels, Belgium (1969).

3. D. Evans et al., Properties of glass reinforced composites at low temperatures, Report No. RPP/E11, Rutherford Laboratory, Chilton, Oxon, England (1968).

4. Bulletin 07-347-01, Dow Corning International Ltd., Brussels, Belgium (1971).

LAMINATED FIBERGLASS COMPOSITES FOR CRYOGENIC STRUCTURES

IN UNDERGROUND SUPERCONDUCTIVE ENERGY STORAGE MAGNETS

S. G. Ladkany and E. L. Stone

University of Wisconsin
Madison, Wisconsin, U.S.A.

INTRODUCTION

The design and construction feasibility of cryogenic structures (struts) of laminated composites for superconductive energy storage magnets[1] is considered. Energy storage magnets are in the form of large solenoids in deep underground tunnels with stored energies of 1000 MWh to 10 000 MWh. The composite structures support the large solenoid and transmit the magnetic forces to bedrock. Capital cost considerations dictate the choice of low-cost fiberglass composites.

We present the results of several preliminary analytical studies for the supporting strut using glass cloth/epoxy, glass cloth/poly-ester and unidirectional glass/epoxy laminates. The thermal gradient in the strut, the behavior of the rock foundation, the large induced moments, and the high ratio of shearing forces to normal forces produce a very complicated stress pattern in the strut and create major difficulties in its design.

SUPERCONDUCTIVE ENERGY STORAGE MAGNETS

Large superconductive energy storage magnets have radii of 100 to 250 m, heights of 30 to 75 m, and magnetic fields of 2 to 5 T. High currents, 0.1 to 0.3 MA, are carried by round composite aluminum stabilized conductors at 1.8 K.[2] The laminated fiberglass composite struts are designed to support the large solenoid and to transmit the magnetic forces to bedrock. To reduce the circumferential tension, the conductors and the cryogenic dewar system are rippled and discretely supported by the struts at 2 m intervals.

377

The composite struts also act as electric and thermal insulators between the helium dewar and rock.

The large mass of material needed for structural support and the capital investment required for manufacture dictate the selection of commercially available low-cost materials which lend themselves to straightforward manufacturing techniques and ease of installation. Since glass/polyester is one-third to one-half as expensive as glass/epoxy, glass/polyester is an attractive alternative for the strut design.

MATERIAL PROPERTIES

The materials chosen for this preliminary analysis are glass cloth/epoxy, glass cloth/polyester, and unidirectional glass/epoxy. Material properties for these fiberglass composites and others have been surveyed by Kasen and presented in a review article.[3] The University of Wisconsin Energy Storage Group is testing commercially available composites. Room temperature, 77 K, and 4.2 K compressive strength and cyclic fatigue testing is in progress.[4,5] The ultimate strengths of fiberglass composites increase substantially, 40 to 80%, upon cooling to 77 K and below. The compressive and tensile moduli can differ by 20%, but they do not change more than about 10% upon cooling. Shear modulus and strength values vary and depend on the testing method. Average values are used in this preliminary design analysis.

The values of mechanical properties used in this analysis are presented in Table I. The values are obtained from References 3 and 4 and the 3M Company Technical Data Sheet for Scotchply 1002 and are estimates where data is sketchy. The values of thermal properties from Reference 3 are used.

STRUT DESIGN

The loading imposed on the struts varies with axial location within the energy storage magnet. For the preliminary analysis presented in this paper, we consider one support strut configuration in the location that has the maximum shear loading.

The designs presented are for a laminated composite structure fastened to the vertical rock tunnel wall. This structure is 1.5 m wide radially, 3.7 m long axially, and carries magnetically induced compressive radial forces of 17.3 MN and axial shearing forces of -30.5 MN. These forces are applied via the aluminum bulkhead to the free edge of the strut. The bulkhead is a load-carrying member between the strut and the liquid helium dewar. The strut is

Table I. Mechanical Properties

Properties*	Glass cloth/ epoxy		Glass cloth/ polyester		Unidirectional glass/epoxy	
	300 K	77 K	300 K	77 K	300 K	77 K
Strengths, MPa[†]						
σ_1^t	360	660	280	500	1100	1500
σ_2^t	340	610	280	500	20	29
σ_1^c	380	710	180	350	620	900
σ_2^c	340	610	180	350	140	300
S	41	76	31	38	30	41
Moduli, GPa[†]						
E_1	25.5		22.0		41.5	
E_2	23.0		22.0		12.0	
G	4.45		3.75		11.0	
Poisson's Ratio						
ν_{12}	0.25		0.25		0.25	

* 1,2 = material directions; t = tension; c = compression
[†] Pa = Pascal = N/m^2

cantilevered to the rock tunnel walls and is fastened by rock bolts as shown in Fig. 1.

The strut is covered by layers of superinsulation to minimize the radiation heat leak and is cryogenically shielded by two discrete shields along its length. For practical purposes, the strut can be considered adiabatic between the heat shields and between the heat shields and the ends. We chose two equally spaced intermediate cooling stations at 11 K and 70 K based on refrigeration power optimization studies carried out for similar composite materials.[6]

The first two preliminary designs utilize slabs of glass cloth/ epoxy and glass cloth/polyester. For the third design, we chose a

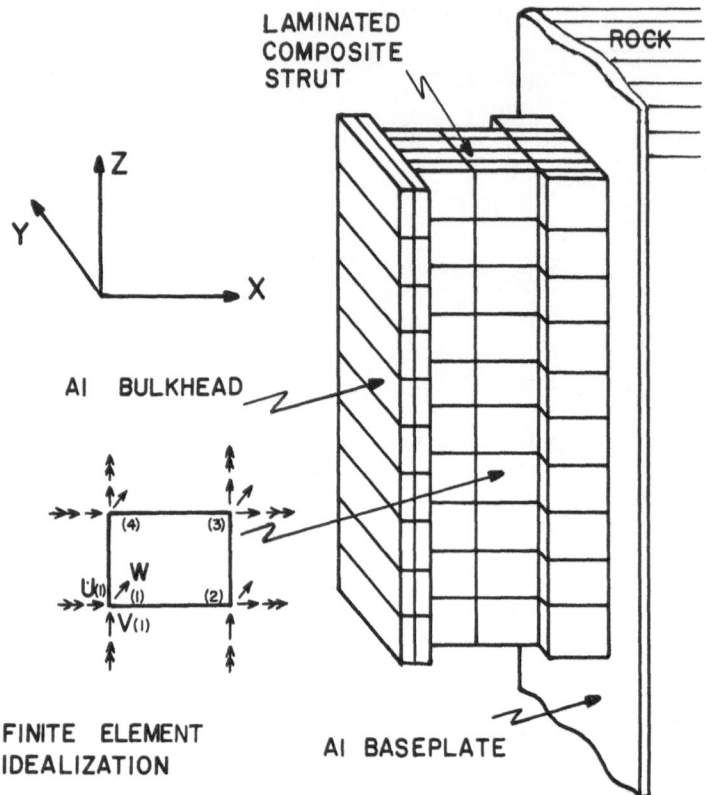

Fig. 1.   Schematic of a composite supporting structure
           showing the finite element mesh layout of the
           strut and a representative plate element
           having four nodes.

symmetric 20 ply unidirectional glass/epoxy laminate having angu-
lar arrangements with the radial direction of [0°, 10°, 18°, 52°,
-57°, 83°, -72°, 80°, 45°, 90°]$_s$. The effective thermal conductivities for this variable angle composite strut and the heat
fluxes and temperature profiles for all three designs are presented elsewhere.[7]

STRESS ANALYSIS

To analyze the complicated stress field, which is created in
the strut by the magnetic and thermal loadings and by the boundary
conditions, finite element analysis is used. Highly specialized

elements are needed to account for the orthotropic behavior of the individual plies and the anisotropic behavior of the laminate or strut. The elements need to account for the sizable shear deformation in the strut, especially since at the bulkhead the axial shearing forces, F_z, are 1.76 times larger in magnitude than the radial compressive forces, F_r.

The strut design uses new, hybrid finite elements[8] for the plane stress and plate bending analysis of layered structures. The rectangular elements, which have four corner nodes and five degrees of freedom at each node, assume both a stress field inside the elements and linear displacement functions around their boundaries. The element stiffness matrices are derived from a complementary energy functional. Figure 1 shows a schematic representation of the finite element mesh used in the analysis of the strut. The magnetic loading is distributed along the outer nodes of the aluminum bulkhead.

Using finite element analysis on a representative 0.1 m wide strut, radial and axial displacements of the bulkhead and the strut-rock interface are calculated and shown in Fig. 2. The effect of the large induced moment can be seen in the radial displacements, U_r. For all three materials, the displacements at the rock-strut interface are the same. The radial stiffnesses are nearly the same; however, the axial stiffness, or resistance to shear deformation, of the cloth/epoxy or cloth/polyester is considerably less than that of the variable angle composite.

FAILURE ANALYSIS

The three designs are analyzed for failure using the Tsai-Hill failure criterion:

$$\sigma_1{}^2/X^2 - \sigma_1\sigma_2/X^2 + \sigma_2{}^2/Y^2 + \tau_{12}{}^2/S^2 = c \leq 1 \tag{1}$$

where σ_1, σ_2, and τ_{12} are the parallel, normal, and shear stresses for an individual ply or glass cloth slab, and X, Y, and S are the strengths. In our analysis, the strengths used are one-half of the ultimate strengths given in Table I. Room temperature and cryogenic tests conducted at Wisconsin[4,5] show that the combination of fatigue and sample variability dictate the lowering of the design strengths. Stress outputs from the finite element analysis are transformed from the global x-z coordinates of the strut to the 1-2 coordinates of the ply. Failure of the ply occurs if $c \geq 1$. In this preliminary analysis, the thickness of the ply or slab is increased until failure no longer occurs.

Using this analytical procedure, it is found that a glass cloth/epoxy strut must have a minimum thickness of 0.29 m for the

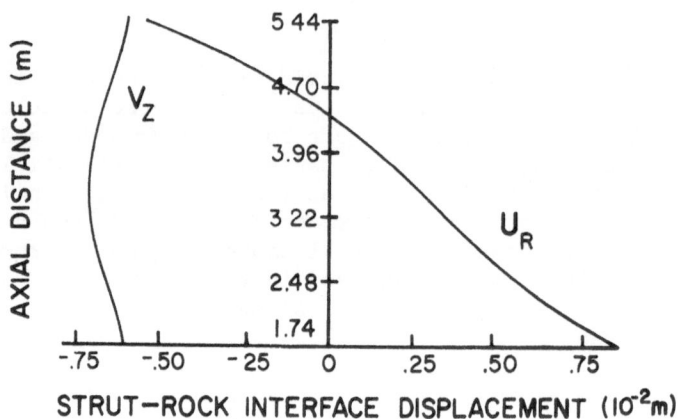

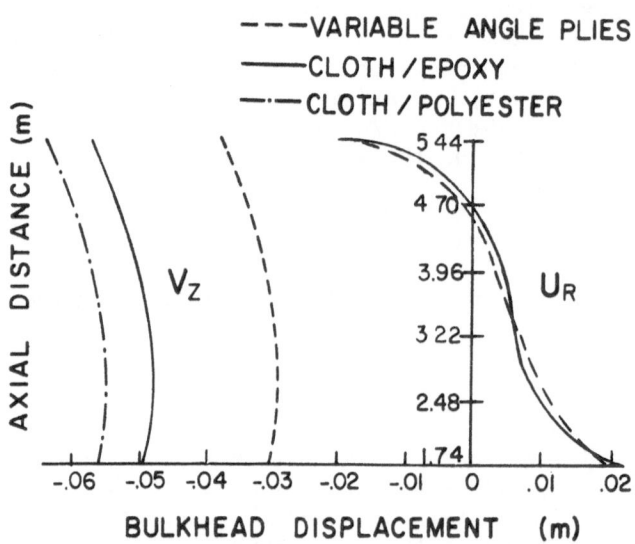

Fig. 2.   Radial displacement, U_r, and axial displacement,
V_z, at the strut-rock interface and the bulkhead for 0.1 m wide composite struts.

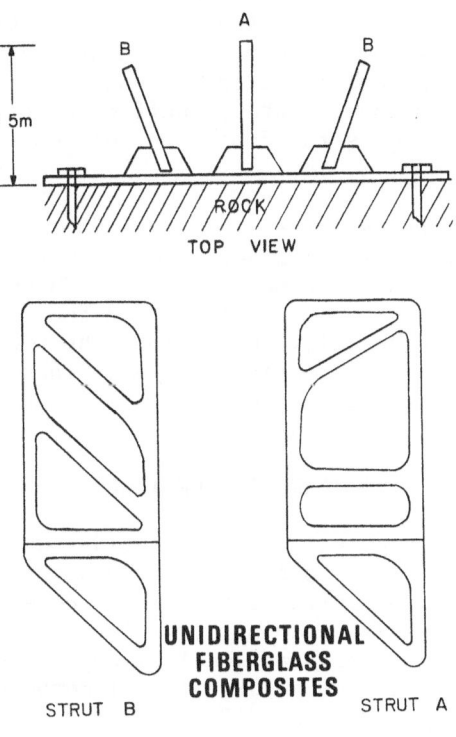

Fig. 3. Design of a modular strut having two
continuously wound frames of unidirectional fiberglass composite.

length between the 1.8 K helium dewar and the 77 K shield and a
thickness of 0.47 m between the 77 K shield and the rock. A glass
cloth/polyester strut would require thicknesses of 0.58 m and
0.61 m, respectively. The 20-ply laminate requires thicknesses
of 0.58 m and 1.0 m. In this analysis, some plies are allowed to
fail in tension normal to the fiber direction in cases where the
other plies could carry the additional stress.

An examination of the relative stresses in the struts and
the material properties shows that shear strength is the governing factor in determining the minimum thicknesses required. The
proportionately high axial loading on the strut results in high
shear stresses and, consequently, inefficient usage of the composite materials. The strut design is not complementary to the
loading mode.

Analytical results for the stresses in the composite strut
that are due to the thermal gradient are, on the whole, an order
of magnitude lower than the magnetically induced stresses. However, at the 77 K cryogenic shield and at the rock base, the
thermal stresses, especially the maximum shearing stresses, could
reach one-third of the magnetic stresses. Thus, thermal stresses
cannot be neglected.

ALTERNATIVE DESIGNS

Alternative conceptual designs consider different strut configurations and loading conditions. Some strut designs have braces
below the strut to more efficiently carry the axial loading. One
design to be examined, a modular, jointless strut of continuously
wound glass fibers or unidirectional plies, is shown in Fig. 3.
The frame-truss structures allow uniaxial members to carry stresses
in their preferred directions. Studies of reinforced plastic
trussed-web girders[9] showed that joints can reduce the strength of
a composite member by over 50%. Additional areas of investigation
are the interfaces between the bulkhead and strut and between the
strut and bedrock.

ACKNOWLEDGMENTS

The authors are grateful to M. A. Hilal and W. C. Young for
their many helpful discussions and to V. A. Koch and K. A. Sakr
for their effective help in the preparation of this manuscript.
This work is supported by the U. S. Department of Energy.

REFERENCES

1. R.W. Boom et al., Wisconsin Superconductive Energy Storage
    Project, Vol. I (1974), Vol. II (1976), Annual Report (1977),
    University of Wisconsin, Madison, Wisconsin.

2. S.G. Ladkany, in: Advances in Cryogenic Engineering, Vol. 24,
    Plenum Press, New York (1978), p. 375.

3. M.B. Kasen, Cryogenics 15(6), 327 (1975).

4. E.L. Stone and W.C. Young, Proc. 7th Symposium on Engineering
    Problems of Fusion Research II, 77CH1267-4-NPS (1977), avail-
    able from IEEE Service Center, Piscataway, New Jersey.

5. E.L. Stone, L.O. El-Marazki, and W.C. Young, in: Nonmetallic
    Materials and Composites at Low Temperatures, Plenum Press,
    New York (1979), p. 283.

6. M.A. Hilal and R.W. Boom, Advances in Cryogenic Engineering,
    Vol. 22, Plenum Press, New York (1977), p. 224.

7. S.G. Ladkany, Proc. ICEC 7, IPC Science and Technology Press,
    Guildford, England, to be published.

8. S.G. Ladkany, Five Hybrid Elements for the Analysis of Thick,
    Thin or Symmetric Layered Plates and Shells, Ph.D. Disserta-
    tion AWB.L1572.5303, University of Wisconsin, Madison, Wiscon-
    sin (1975).

9. F.C. McCormick, Further Studies of a Trussed-Web Girder Com-
    posed of Reinforced Plastics, Report No. VHTRC 76-R16, Virginia
    Highway and Transportation Research Council, Charlottesville,
    Virginia (1975).

SELECTION OF MATERIALS AND MANUFACTURE OF A

SUPERCONDUCTING 0.60 MJ PULSED ENERGY STORAGE COIL

Z. N. Sanjana and M. A. Janocko

Westinghouse R & D Center
Pittsburgh, Pennsylvania, U.S.A.

INTRODUCTION

This report presents information on the selection of nonmetallic materials that were used in the construction of a superconducting solenoidal, pulsed energy storage coil of nominally 300 KJ stored energy capacity. The coil was built by Westinghouse Electric Corporation for the Magnetic Energy Transfer and Storage (METS) program of the Los Alamos Scientific Laboratory.[1] It consisted of four concentric layers of a liquid helium cooled, unpotted, non-compacted, cabled superconductor wound on glass-reinforced epoxy coil formers. Apart from the superconducting cable and its steel mandrel, which are discussed more completely in Reference 2, the entire structure of the coil was made from nonmetallic materials. These comprised the four coil formers, fiberglass packing and glass-epoxy banding tapes, Kapton[(R)] electrical interlayer insulation, and G-10 laminate endplates and terminal supports.

NONMETALLIC COMPONENTS

Coil Formers

Geometry. Dimensional constraints, coupled with the use of a flat-wound, porous conductor of somewhat low current density,[3] necessitated a four-layer coil design to obtain the desired coil inductance. Each layer of the conductor is supported by a separate coil former shell, though the conductor itself is continuous from one coil terminal to the other. Together these formers provide the main

structural body of the coil. The innermost coil former is shown
in Fig. 1. The other three formers were similar in appearance.

The coil formers had average diameters of 0.46, 0.49, 0.52,
and 0.55 m. As shown in Fig. 2, the formers had a circumferential,
helical, machined groove of 1.9 cm width to accept the superconductor and axial slots of approximately 0.6 cm width for liquid helium
flow. After machining, the second, third, and fourth formers were
axially slit so that they could be slipped onto the next smaller
former after completion of its superconductor layer. Transition
slots were provided, as seen in Fig. 2, for the continuous winding
of the conductor from one layer to the next.

Material Selection. The coil formers were made of a filament-
wound glass-epoxy material. The selection of the resin system was
based on a limited amount of cryogenic testing and on the ability
of the resin system to form composites of the mass required by the
size of the cylinders. The latter constraint is an important practical consideration because the heat of exotherm and the curing
shrinkage can result in large internal stresses and even flaws in
composites of large mass. The Micarta Division of Westinghouse
Electric Corporation recommended their Grade HY-1457 filament-wound
cylinder as being the best for manufacturing in the sizes involved.

The resin system of this Grade HY-1457 was evaluated along
with a resin system, which appeared to be the best filament-winding
resin for cryogenic applications.[4,5] This is the Aerojet 2 resin
system developed by Soffer and Molho.[4] Modifications of the two
resins were also evaluated.

The various formulations examined are presented in Table I
along with room temperature tensile strength and tensile strength
at 77 K after 5 cycles from room temperature into a dewar containing liquid nitrogen.

The specimens used in these tests were standard ASTM "dogbone"
specimens cast to shape (ASTM D638-72, Type I Specimen). Apart
from the specimens that were tensile tested, evaluations were also
made of thermal shock behavior.

It was found that the Aerojet 2 system and the modified Aerojet 2 were both marginally satisfactory. If they were inserted
rapidly into liquid nitrogen, they tended to crack at the first
immersion. This tendency to crack on rapid immersion into liquid
nitrogen was considerably reduced if the edges of the "dogbone"
specimens were sanded smooth. If the specimens were inserted slowly
($\sim$3 cm/s) into liquid nitrogen, the tendency to crack was much less
and, if the specimen edges were sanded smooth, cracking was totally
eliminated during slow immersion.

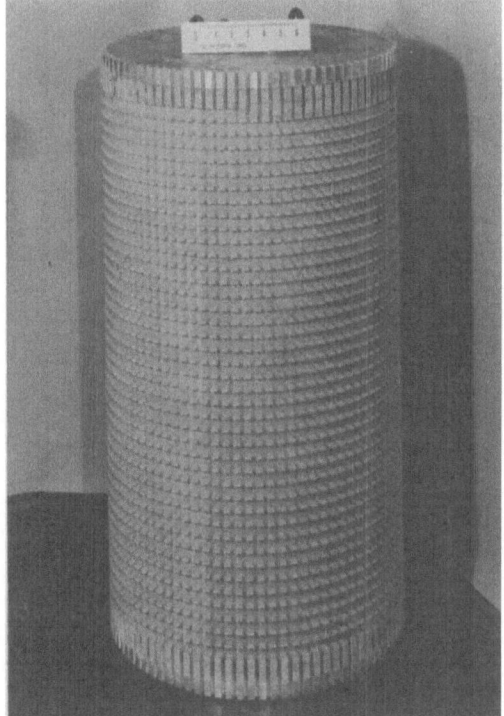

Fig. 1.   The innermost coil former after machining.

Fig. 2.   End portion of the outermost coil former, showing toothed
machined surface and conductor transition slot.

Table I

Type resin system	Components	Formulation (parts by wt.)	Tensile strength at room temperature, MPa		Tensile strength at 77 K after 5 cycles between 77 K and 300 K, MPa	
			Average*	Standard deviation	Average*	Standard deviation
Soffer, Aerojet 2	Epon 828 DDSA Empol 1040 BDMA	100 115 20 1	45.0	--	--	--
Modified, Aerojet 2 (lower viscosity for easier filament winding)	DER 332 DDSA Empol 1040 BDMA	100 126 20 1	48.7	0.2	52.1	11.1
IMD G-10 filament winding system	Grade HY-1457	--	58.9	7.5	71.2	11.7
IMD G-10 filament winding system with rubber additive	Grade HY-1457 + Hycar CTBN (5%)	--	57.9	15.7	92.8	8.8

*Average of 5 samples for all tests.

The Micarta resin system was found to be much less prone to cracking during rapid immersion, and, with edges sanded smooth, there was no cracking even with rapid immersion (for 15 cycles, 5 specimens). This superior thermal shock behavior may be caused by the higher tensile strength of this resin (see Table I). The Micarta filament winding system with a toughening rubber additive, Hycar CTBN, was found to have good shock behavior - all 10 of the samples were tested without sanding the edges and only one cracked during 15 cycles. The presence of the rubber additive also resulted in a 30% improvement in tensile strength at 77 K. The addition of Hycar CTBN to the standard Micarta filament winding system could be advantageous for other structural components fabricated from this type of material. For the present coil structure, this modification was not attempted.

Additional thermal shock tests were performed on ring samples cut from a filament-wound tube of Micarta Grade HY-1457. After 10 cycles of rapid immersion in liquid nitrogen, the rings were examined for damage, such as micro- and macro-cracking in the matrix and fiber-to-resin separation. Samples from cycled and control rings were cut and the cross sections examined by a scanning electron microscope at magnifications of 1 000X and 2 000X for damage or flaws created by the cycling. Past experience with several resin systems has indicated that microscopic examination for signs of damage, such as micro-cracking in the resin phase and fiber-resin separation, permits early detection of property degradation. No difference was found between the thermally stressed and the unstressed ring samples.

Based on these evaluations, the selection of Grade HY-1457 filament-wound cylinders supplied by the Westinghouse Micarta Division was made. These cylinders incorporated E-glass reinforcement filaments wound at a 55° angle to the axis. The design properties applying to these cylinders, which were used in the coil design, are given in Table II. Details of mechanical design and stress calculations are given in References 2 and 6.

Machining. The machining of a complicated tooth structure (as shown in Fig. 1) into a material of this type is not straightforward. Using a mockup, a satisfactory machining plan was tested. To avoid excessive heat generation and thermal degradation of the matrix, moderate speeds were used throughout, and a cooling mist of distilled water was used. After turning the cylinders to the required inside and outside diameters, the helical conductor slot was machined. Multiple passes were made on a numerically controlled milling machine using special carbide end mills. The axial cooling slots were then cut. Some tooth breakage was experienced on the mockup until proper feed rates were ascertained. Only about 10 teeth of a total of about 14 000 were broken on the four formers.

Table II. Properties of Coil-Former Cylinders.

	300 K	4 K
Axial compressive strength	172 MPa	241 MPa
Axial tensile strength	86 MPa	96 MPa
Circumferential tensile strength	379 MPa	552 MPa
Interlaminar shear strength	34 MPa	48 MPa
Circumferential modulus of elasticity	20.6 GPa	33.1 GPa
Specific gravity	1.99	
Circumferential thermal expansion coefficient (average over the range 300 to 77 K.)	6×10^{-6}/K	

ENDPLATES AND TERMINAL SUPPORTS

It was decided to febricate these parts from epoxy-glass laminates, as their properties are more than adequate for the intended
application.[7] In order to select between NEMA grade G-10 and G-11
types of epoxy-glass laminates, thermal shock tests were performed.
Samples of the two types of laminates 2.5 cm thick were obtained
from the Micarta Division of Westinghouse Electric Corporation.
To simulate effects of machining, 0.3 cm wide grooves were cut into
the sample blocks, which were then cycled by liquid nitrogen immersion. The G-10 type of laminate was found to be much more resistant
to crack formation and extension than the G-11 type, particularly
when the groove was cut in the plane of the laminate. The G-11
type samples produced cracks emanating from the machined groove
after the first thermal cycle, and these cracks extended rapidly
upon subsequent cycling. The G-10 type showed no cracking after
10 cycles, and on this basis G-10 was selected for the endplates,
terminal supports, and other miscellaneous structural parts.

Thermal expansion coefficients were measured in all three
directions and were found to be as follows: for the range 300 to
77 K, the overall coefficient in the warp direction was $9.75 \times
10^{-6}$/K; in the fill direction, 13.4×10^{-6}/K; and in the thickness
direction, 41×10^{-6}/K.

Conductor Restraint Components

Design Philosophy. The conductor in a high field superconducting coil must be restrained from motion caused by high Lorentz
forces, lest the heating produced by such motion lead to a local
normalization of the wire, propagation of the normal zone, and a
complete coil quench. The design of the conductor restraint and

cooling systems of the Westinghouse coil, explained in greater detail in Reference 3, is based on the total elimination of adhesives or potting compounds in direct contact with the conductor and on the maintenance of helium coolant access to every part of the conductor. The resulting method of conductor support is shown schematically in Fig. 3. The conductor wound onto the former is shown in Fig. 4.

Fiberglass Tape Packing and Glass Roving Filler. To fill the space between the conductor and the top of spacer teeth, layers of fiberglass tape were used. They are Carolina Narrow Fabric Company Type 15106-2, 1.9 cm wide and 0.25 mm thick, and Type 4536-1, 1.9 cm wide and 0.076 mm thick.

To fill in the fillets between the top corners of the conductor and the teeth, glass fiber roving was used, as supplied by Owens-Corning Fiberglass, designated Type 30 Roving No. 431X7225.

Circumferential Banding Tape. Each layer was banded with a unidirectional glass reinforced epoxy tape before application of the Kapton$^{(R)}$ interlayer insulation. The purpose of the tape was to restrain radial conductor motion, and secondarily, to bridge between teeth tops. The tape used was 3M Company 4 cm wide, Scotchply$^{(R)}$ Type 1003. The first application of the tape was half-lapped and the next layer was butted, thus providing 3 layers of tape or approximately 0.76 mm of tape buildup. To cure the banding tape, the whole coil was placed in a forced air oven and heated to 423 K for 16 h. Four thermocouples were inserted in the various layers to different depths to make sure that the coil interior reached the required temperature.

Interlayer Insulation

A maximum interlayer voltage of one-half the terminal-to-terminal voltage can be imposed on a four-layer coil. The design terminal voltage was 40 kV, and thus interlayer insulation capable of withstanding 20 kV was required. Disregarding the additional insulation provided by the helium coolant, wire insulation, and coil formers, a sufficient barrier of continuous electrical insulation was provided by 0.6 to 1.0 mm of Kapton$^{(R)}$. Seven or eight layers (depending on permissible coil diameter build) of 0.127 mm thick Kapton$^{(R)}$ sheet were wound, jelly-roll fashion, around each coil layer after the Scotchply$^{(R)}$ banding tape had been applied. Because of "walking" of the material off of one end of each former, it was necessary to apply the insulation in two pieces, with seams atop one another. Thus, a minimum of five Kapton$^{(R)}$ layers resulted. Kapton$^{(R)}$ is a polyimide material having excellent dielectric strength from 4 to 573 K and is supplied by the Film Department of

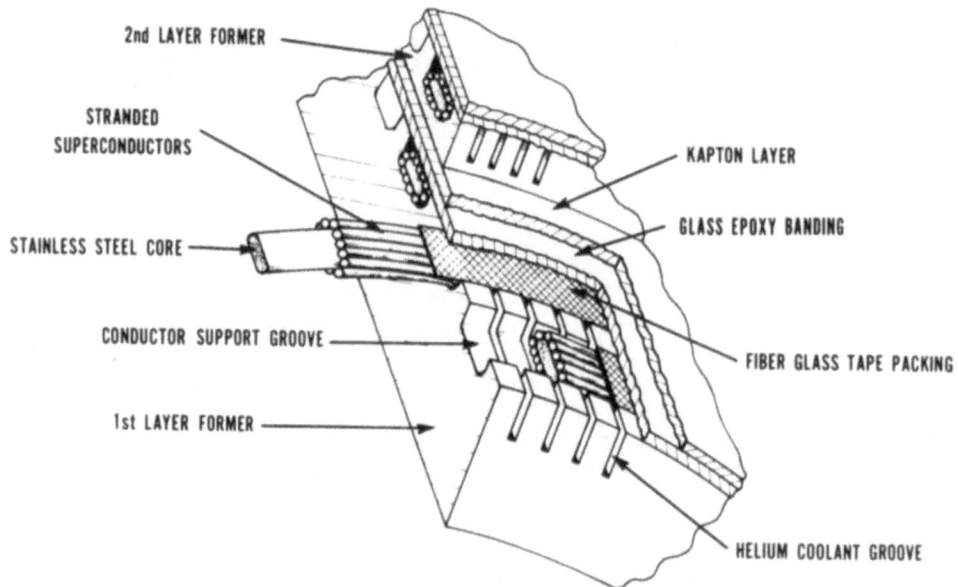

Fig. 3.   Conductor cross section, showing conductor
          support and insulation components.

Fig. 4. View of conductor wound onto the former.

E.I. duPont de Nemours & Company. Dielectric strength of the
0.125 mm thick Type H film used is 120 kV/mm. Fig. 5 shows the
insulation and the transition from one former to the next.

Outer Wrap. To prevent the Scotchply$^{(R)}$ banding tape on the
outermost cylinder from becoming loose during the curing operation,
the outermost layer was banded with a heat shrinkable polyester
[Mylar$^{(R)}$] tape.

Adhesives. During the machining of the first, innermost coil
former, about 10 teeth broke off due to a flaw (a dry spot delami-
nation) that was present in one area of the cylinder. These teeth
were joined to the former using a polyurethane adhesive. Polyurethane adhesives are, in general, the best adhesives for use at
cryogenic temperatures. The adhesive used was a 3M Company product, 3549 B/A. The lap shear strength of G-10 to G-10 joints
using this adhesive was tested and found to be 17.5 MPa at 77 K.

Other convenience adhesives used during assembly were Hysol
Epoxy Patch Kit No. 608, a clear quick-setting adhesive taking
about 4 min to set up, and No. 309, a filled paste taking about
9 min to set up. These were used to join the end of one roll of
fiberglass tape to the start of the next roll and to hold down
the ends of these tapes.

Fig. 5. The third former being slid onto the completed second
         layer.

COIL PERFORMANCE

The initial cooldown of the coil to 70 K was performed by
Westinghouse in the very rapid time of 4 h. No untoward events
occurred. The coil was further tested at Los Alamos Scientific
Laboratory and far exceeded all of the design specifications.
The specifications called for a stored energy of 300 kJ at 10 kA
current, a 2.3 T maximum field, and 40 kV transfer voltage. The
actual test levels reached by the coil to date are 0.60 MJ at
14.1 kA and a 3.2 T peak field.[8] Maximum transfer voltage of
58 kV was achieved with a fast transfer of energy in a 1 ms period.

The successful tests of the coil indicate that the materials
performed their intended structural and electrical insulation
functions more than satisfactorily. Westinghouse is at present
building a 0.4 MJ 25 kA, lower loss, 3-layer energy storage coil
of similar design for LASL.

ACKNOWLEDGMENTS

The design, construction, and testing of this coil involved the efforts of teams at LASL, headed by J. D. Rogers, and at Westinghouse, headed by C. J. Mole, whose individual members contributed in so many overlapping ways as to preclude individual acknowledgment. The authors' main contributions were in materials selection and coil assembly, respectively, and we gratefully acknowledge the contributions of the other members of the teams in these two areas. E. Mullan was the project manager.

This work was sponsored by the University of California and Los Alamos Scientific Laboratory, Subcontract No. XN 4-32767-3.

REFERENCES

1. J.D. Rogers et al., Advances in Cryogenic Engineering, Vol. 23, Plenum Press, New York (1978), p. 48.

2. C.J. Mole et al., Advances in Cryogrenic Engineering, Vol. 23, Plenum Press, New York (1978), p. 57.

3. E. Mullan, M.A. Janocko, and D.C. Litz, Proceedings of the Seventh Symposium on Engineering Problems of Fusion Research, IEEE Publ. No. 77CH1267-4NPS, IEEE, New York (1978), p. 327.

4. L.M. Soffer and L. Molho, Cryogenic Resins for Glass-Filament-Wound Composites, NASA CR-72114, available from NTIS, Springfield, Virginia (1967).

5. J.V. Larsen and R.A. Simon, Carbon Fiber Composites for Cryogenic Filament-Wound Vessels, AD 743470, available from NTIS, Springfield, Virginia (1972).

6. E. Mullan, Final Report to Los Alamos Scientific Laboratory, Subcontract #XN4-32767-3, Westinghouse Electric Corp., Pittsburgh, Pennsylvania (1977).

7. M.B. Kasen, Cryogenics 15, 327 (1975).

8. J.D.G. Lindsay, P. Thullen, and D.M. Weldon, Los Alamos Scientific Laboratory, private communication.

APPLICATION OF GRAPHITE-EPOXY TO CRYOGENIC TELESCOPES

L. D. Michelove

Itek Corporation
Lexington, Massachusetts, U.S.A.

INTRODUCTION

The basic driving forces for materials selection for aerospace
optical systems are weight, stiffness, and stability. These systems, whether cameras, telescopes, spectrometers, or heliographs,
have used and helped to drive the state-of-the-art of beryllium,
beryllium-aluminum, magnesium, magnesium-lithium, other experimental alloys, and also materials in the composites field. The weight
and stiffness criteria are self-explanatory, but the stability needs
require some further explanation before we examine the use of composites in these systems.

OPTICAL CONSIDERATIONS

Figure 1 shows in schematic detail an idealized reflective
telescope for aerospace applications. The focal plane is a cryogenically cooled series of photo-optical sensors operating below
100 K. The walls and baffles of the system also must operate in
that temperature region to prevent their becoming extraneous inputs
to the sensor system. The restraints this puts on the materials of
construction and how these problems are met are the subject of this
section.

The structure is used to provide the light-tight enclosure
needed for optics, to protect the glass and sensors from damage,
and to precisely locate the various components with relation to
each other. A typical system may be 0.5 meters in diameter and 1.5
meters long. With multiple mirrors, the total light path through
the system may be as long as 7.5 meters.

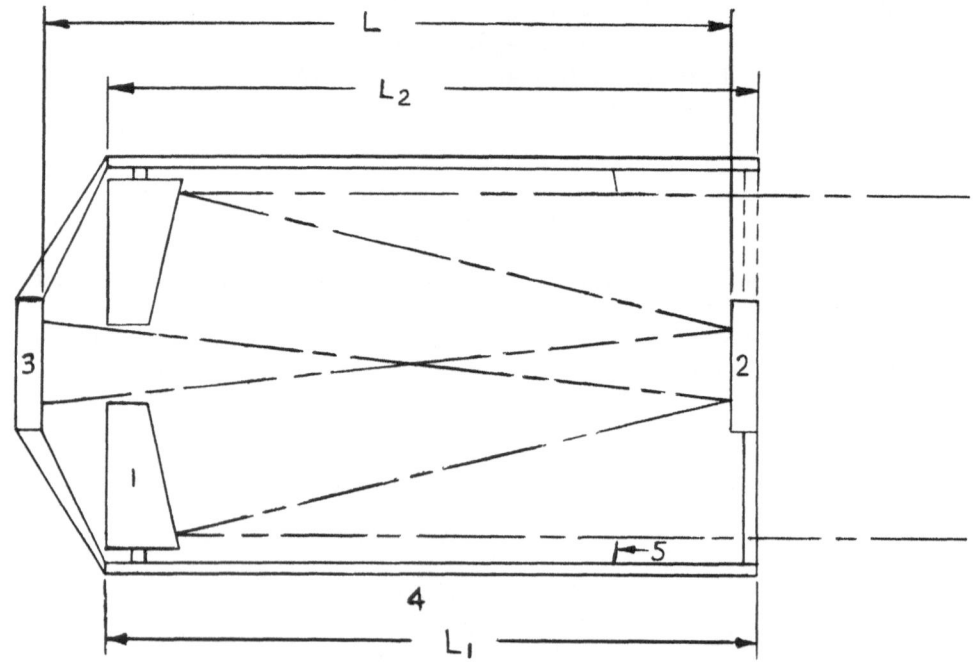

Fig. 1.   Idealized space telescope.

1 - Primary mirror; 2 - Secondary mirror and centering spider;
3 - Focal plane; 4 - Composite metering structure; 5 - Baffles.

The first and most obvious parameter of interest is the length
change between room temperature (fabrication, testing, and optical
alignment) and the operational environment of cryogenic temperatures and space vacuum. Of primary concern is the spacing between
the mirrors (L) and the change with temperature (ΔL). Additionally,
of great significance is the ΔL_1 and ΔL_2, or the tipping and tilting
of each optical element, including the focal plane, with respect to
the optical axis. The basic length change can be controlled by
fabricating the structure from low thermal expansion material.
Table I illustrates some of the traditional materials as well as
composites that have been applied to these systems. The changes in
ΔL_1 or ΔL_2 are less easily come by.

Additional considerations are the mechanical aspects derived
from testing and fabricating at 1 g on the earth and operating in
free flight on orbit. Understanding of these aspects comes from an
analysis of the structure based on a knowledge of tensile strength,
microyield strength, launch loads, and the relaxation of these and
other second-order inputs via multiple load paths upon gravity release. This understanding requires that the design engineer, the
materials engineer, and fabrications people communicate closely on
all details of design and construction of individual parts as well
as the system.

We talk of despace, (ΔL) decenter, tip, and tile, (ΔL_1, ΔL_2,
$\Delta L_1/\Delta L_2$) over several hundreds of degrees Kelvin in terms of a few
parts per million total. In standard systems, length changes of
millimeters are tolerable; in optical systems, total length changes
from all inputs of 50 parts per million over 2 meters may make the
difference between advancing man's understanding of the universe
and another expensive failure.

MATERIALS PROPERTIES

As can be seen from the data in Table I and Table II, graphite-
epoxy composites have the properties necessary to meet the demanding
needs of aerospace optical systems. Several companies have developed,
either on contract to NASA or other agencies or on their own, computer programs and analytical techniques that allow them to design
multi-directional layups of unidirectional graphite epoxy tapes and
fabrics that produce near-zero thermal expansion. These materials
have the strength and stiffness required and, with the proper fabrication care and control, the uniformity essential to optical needs.
These composites are lightweight and many have resin systems that
meet the outgassing and vacuum stability requirements of space systems as well.

Table I. Materials Properties.

Material	Density, kg/m²	Young's modulus, E, GN/m²	UTS*, MN/m²	MYS†, MN/m²	CTE‡ α, $10^{-6}\cdot K^{-1}$	$\Delta\alpha$, 10^{-8}	Thermal conductivity, KW/m·K
Fused silica	2.2	73	0.5	1.8	.56	5	1.37
ULE	2.2	68	0.5	1.8	.03	15	1.31
Beryllium (1–70)	1.8	304	2.4	5.5	11.2	40	220
Aluminum (6061–T6)	2.7	69	3.1	12	23	< 6.0	171
Invar	8.1	141	4.49	25	1.3	5.0	145
Titanium	4.5	110	11.04	82.5	10.4	40	120
GY70/X 30 Uni	1.8	293	5.5	13	1.0	‑‑	35 (inplane)
GY70/X 30 Quasi ISO	1.8	103	1.93	~10	0.1	10	20 (inplane)
GY70/X 30 4040 Satin Quasi ISO	1.8	97	2.0	13.8	0.2	2	12–20

* UTS = ultimate tensile strength.
† MYS = microyield stress.
‡ CTE = coefficient of thermal expansion.

Table II. Property Changes with Temperature.

Material	Ultimate tensile strength Θ, GN/m²			Microyield, MN/m²			Coefficient of thermal expansion α, $10^{-6} \cdot K^{-1}$		
	300 K	150 K	77 K	300 K	150 K	77 K	300 K	150 K	77 K
Fused silica	50			0.18			0.56	-0.1	-0.1
Beryllium	240			0.55			11.2	4.0	0.5
Invar	449	562	730	2.5	2.8	3.2	1.3	2.7	3.1
GY70/X 30, unidirectional	550	550	500	1.3	1.0	0.3-0.5	1.0	1.2	1.5
GY70/X 30, quasi-isotropic	193	190	155	1.0	0.8	0.2-0.3	0.1	1.4	4.2
GY70/X 30 4040 Satin, quasi-isotropic	200	189	89	1.38	1.37	0.1-0.8	0.2	0.22	0.53

Several questions need further probing, however, before whole-sale application of composites to these particular systems' needs can be approved:

1. Long-term stability of mechanical properties.
2. Long-term vacuum stability, including moisture effects.
3. Property changes with large ΔT.
4. Cryogenic survival.

The question of cryogenic survival is the most germane, since the others need not ever be considered if the material cannot with-stand exposure to temperatures below 150 K, the arbitrary upper limit for this area of materials technology. As can be seen in Table III, many composites of graphite-epoxy show a distinct break in the uniformity of their properties as they cross into the cryo-genic region. Not all properties change drastically, but the ones of most interest to optical instrument manufacturers are severely affected. Coefficient of thermal expansion (CTE) and microyield strength are severely altered. For the GY70/X 30 system of inter-est to our programs, the microyield drops from over 20 000 psi to below 8 000 psi after one cycle below 150 K and continues to lower during several additional cycles. The CTE decreases and then stabilizes at the new value after about 10 cycles. The phenomenon that causes this is called microcracking, or translaminar stress relief. Microscopic cross sectioning of samples clearly displays a matrix of fine cracks perpendicular to the fiber direction and apparently not interconnected.

Table III. Microyield Properties of Graphite-Epoxy Laminates.

System	Microyield, MN/m^2	
	Room temperature	After 77 K exposure
HMS/934	27.6	3.1
HMS/3501	3.6	3.6
HMS/759	0.7	1.4
HMS/339	16.1	0.7
GY70/X 30-Uni	$\sim$ 10	$\sim$ 8
GY70/X 30-4040	13.8	13.8

This microcracking has been associated with the stress developed by the differential shrinkage during cure between the resin and the high modulus fiber. The release of these stresses at low temperature is accomplished with little or no damage to the fibers and seems not to affect the fiber-to-resin bond. Since the cracks propagate perpendicularly to the plane of the laminate, they can induce cracks in coatings or surface layers and may cause damage to seal coatings and foils. The major effect appears to be on the microyield properties, as noted in Table III, but there is little change in other mechanical properties, as shown in Table II. The exact nature of the phenomena is not yet well understood, but, as will be shown, the problem can be avoided.

Two approaches to this problem, both reasonable from a user's point of view, are possible. One is to design the systems to accept the properties after cryogenic exposure, build the components, cycle them to a stable point, and accept the results. Because the mechanical loads are generally low in optical systems, this is not out of the realm of possibility. The same approach is often taken in the use of metals in similar situations, where multiple cycles to low temperatures are used to stabilize fabricated components of beryllium, aluminum, fiberglass (training), etc. The risk here is that gravity unloading or further cycling will produce unknown effects. The risk appears small, but larger than most others taken in this field.

The second approach is to develop a technique or techniques that will allow the designer to use the full potential of the material and avoid the microcracking phenomenon. This is the approach taken at Itek through its vendors, and the test data presented in Tables II and III indicate that it has been successful. GY70/X 30 material, woven into a 2-dimensional fabric, a 4040, 8-harness satin weave, when laminated as a balanced composite (0, 30, 60, 90, 120, 150)s produces a product with the properties required. A complex structure similar to that shown in Fig. 1 is in production, and the test data for samples from this are included in the statistical base of the tables and figures.

Property changes with the large ΔT from the 500 K fabrication temperature to the 100 K operational temperature appear to be advantageous, based on present test programs. Mechanical properties improve: tensile strength and shear strength both show a 20 to 30% rise at 100 K; microyield properties show little change, and the thermal expansion stays between 0 and 0.5 ppm/K over the entire range.

MOISTURE EFFECTS

The effect of moisture absorbed into the composite is potentially as important to the overall system as the cryogenic

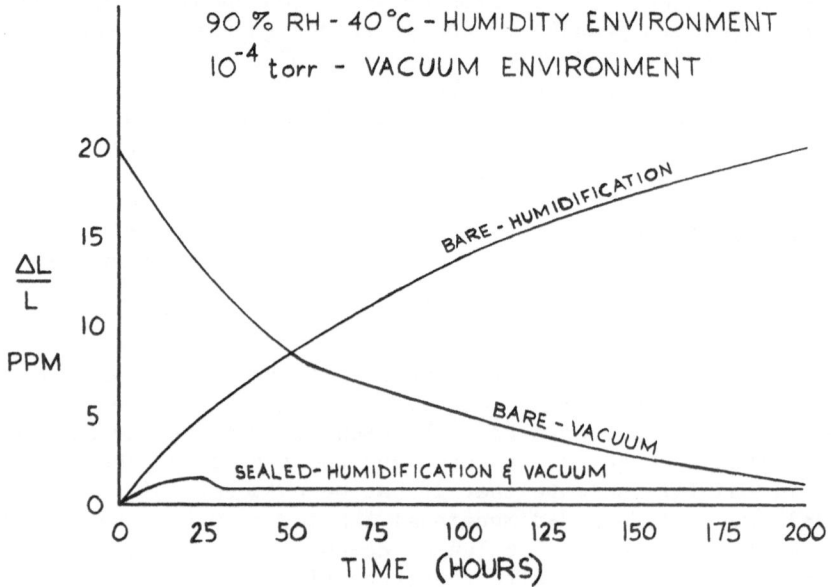

Fig. 2

microcracking. Changes, downward, of mechanical properties of 10 to 40% have been noted. CTE changes of 20% or more have been noted. In addition, subsequent exposure to vacuum and low temperatures show time-dependent changes in mechanical and physical properties and a long lifetime degassing of the absorbed water, complicating the specific environment of the optics. It is, therefore, most desirable to eliminate these variable phenomena. Two approaches have been taken. One is to maintain a dry environment, below 15% relative humidity, from some point in fabrication and assembly until the system is in orbital operation. This can be difficult in practice and painful for the test and assembly personnel involved. The second technique is to dry and then seal-coat the composite structure to maintain a constant internal moisture content. Many processes have been applied, with varying success. Cryogenic exposure of most coatings where the substrate composite is subject to microcracking has proved impractical, as the cracks appear to propagate through the coating. However, a multilayer, well-adhered metallic coat, of electroless nickel and indium-based solder applied to the woven composite has proven effective. As measured by the sensitive CTE changes, Fig. 2 shows the reduction in moisture pickup and release attainable with this seal. Long-term tests are in progress to establish the total effectivity of the process.

A third, presently experimental, process is being evaluated. This involves the addition of moisture barrier compounds to the basic resin system of the composite, eliminating the need for coatings of any sort.

CONCLUSION

As can be seen from the data presented, the application of composites to cryogenic space optics is advantageous and practical. Care must be taken to accommodate in the systems design and the materials fabrication all of the potential hazards. This has been accomplished at Itek and elsewhere and opens the potential for larger systems and even greater usage of composites in difficult environments.

ACKNOWLEDGMENTS

Most of the actual fabrication and test work presented was accomplished by Mr. J. Stumm and Dr. G. Pynchon of Composite Optics of San Diego, California. Analytical techniques and mechanical design procedures were developed by Mr. J. Pepi at Itek, OSD. Historical data was compiled by W. Barnes, Jr.

Work was performed under a series of internally sponsored programs, F30602-76 and FO4701-77-C-0107.

BIOGRAPHICAL DATA

L. D. Michelove is Supervisor of Materials and Processes at Itek, Optical Systems Division in Lexington, Massachusetts. He has been responsible for this activity through the Apollo Camera Programs and the Viking Lander Camera System.

VACUUM IMPREGNATION WITH EPOXY OF

LARGE SUPERCONDUCTING MAGNET STRUCTURES

M. A. Green, D. E. Coyle, P. B. Miller, and W. F. Wenzel

Lawrence Berkeley Laboratory
Berkeley, California, U.S.A.

INTRODUCTION

The Lawrence Berkeley Laboratory (LBL) has been building a series of thin solenoid magnets for use in high energy physics experiments.[1,2,3] The LBL thin-magnet concept integrates a super-conducting coil, a shorted secondary and a tubular cooling system into a single-coil package. The magnet coil is vacuum impregnated with a low viscosity epoxy which, after casting and curing, forms an integrated crack-resistant package.

This paper talks about the engineering realities of potting large superconducting magnet structures. The epoxy LBL has chosen was selected for its vacuum impregnation properties rather than its thermal contraction properties. The LBL experience suggests that the fabrication technique could be more important to the success of large coils impregnated with epoxy than the formulation of the epoxy itself. The apparent success we achieved in building solenoids up to two meters in diameter (no training up to the coil critical current) shows that this view may have merit. The LBL epoxy formulation, the technique for using the epoxy, and the LBL epoxy casting and curing method in large coils is described.

THE LBL EPOXY FORMULATION

The epoxy formulation used by LBL was developed about 10 years ago to vacuum impregnate large room temperature magnet coils. The successful vacuum impregnation of large coils not only requires the proper epoxy, but it also requires proper casting and curing technique. The general properties of the epoxy used by LBL are:

1. Low viscosity (~500 centipoise at 25°C).

2. Long pot life (~4 hours at 50°C).

3. Good wetting power.

4. A reasonable cure schedule and temperature (24 hours with a maximum temperature at 80°C).

5. Good crack resistance. (This applies at room temperature.)

6. Good vacuum properties. [It has been checked down to 0.7 mPa (5 x 10^{-6}torr).]

There are a number of formulations that yield the properties listed above. The formulation used for the LBL test coil epoxy is:

1. 50 parts by weight of unmodified low-viscosity resin (EPON 826),

2. 50 parts by weight of polyglycol diepoxide resin (Dow Chemical Company DER-736),

3. Silicone antifoam fluid - about 1 drop per kilogram of mixture (Dow Chemical Company DC-200), and

4. 28 parts by weight aromatic - amine hardener (TONOX, Huagatuc Chemical Division of Uniroyal).

Other formulations are suggested by Turner. The formulation given here is the only one actually used in our test coils. The formulation is important, but mixing technique is just as important. The procedure used by LBL is as follows:

1. Mix the two resins together (the EPON 826 and DER-736) and heat to about 45°C. Deaerate prior to Step 3.

2. Melt the TONOX slowly at 80°C. When the TONOX is melted, strain it through a cheese cloth strainer and cool down to about 45°C.

3. Pour the TONOX mixture into the resin mixture slowly while stirring with a mechanical stirrer that stirs in as little air as possible. Stir at a temperature of 40 to 45°C. Continue stirring until the mixture is completely mixed.

Once the epoxy has been mixed, it is deaerated at a temperature of 40 to 45°C. Deaeration of the mixture is accomplished by pumping on the mixture until large bubbles cease to form. Deaerate down to a pressure of 30 Pa (200 μm) or lower. In order to have the maximum pot life while pouring, much of the

deaeration should occur before mixing the resin and the TONOX.
The pot life of the resulting mixture will be about 4 hours (at
50°C). At 45°C, the pot life can be extended to about 6 hours.
Note that the pot life varies according to ambient temperature
and the size of the batch. For more information see Reference 4.

SMALL SAMPLE TESTS

LBL has not, in general, gone through the extensive cryogenic
epoxy studies other laboratores have.[5,6,7] However, measurements
of cracking resistance and total thermal contraction coefficients
have been made at LBL on a number of samples of filled and unfilled epoxy as well as other plastics and composite materials.
Two groups of measurements were made during the last three years,
and they are in agreement with measurements made at other laboratories. The LBL measurements include small sections of superconducting coil. The LBL cracking and thermal contraction
measurements were all done at liquid nitrogen temperature. (These
yield enough information to make reasonable engineering decisions.)

Table I shows the results of various measurements made of
thermal contraction and cracking resistance at LBL. The materials
used in the thin superconducting solenoid magnets are included in
the table. The crack resistance of unfilled epoxy sections is
often rather poor. Experience indicates that when these epoxy
sections are filled, the crack resistance improves greatly.

VACUUM IMPREGNATION OF LARGE SUPERCONDUCTING COILS

The three high-current-density LBL test superconducting
solenoids were built using the vacuum impregnated epoxy described
previously. These magnets use a copper-based niobium-titanium
superconductor, which operates at current densities as high as
1.24 GA/m^2. These coils are cryogenically stable. Therefore,
relatively little energy deposited locally is required to drive
them normal (about 10 to 40 kJ/m^3, depending on current density
and magnet induction).[9] Since epoxy cracking can release energy
locally, it is important that a good epoxy technique be used in
these coils.

The epoxy chosen for impregnation of the large superconducting test coils was chosen for its impregnation properties.
Impregnation of conventional coils has been done in the LBL shops
for over a decade.

The LBL epoxy formulation appears to crack at cryogenic
temperatures if large epoxy volumes are unfilled. Care has been

Table I. Total Thermal Contraction from 295 K to 77 K and the
          Crack Resistance for Various Materials Which Are Used
          in Thin Superconducting Magnets.

Material	Total thermal contraction coefficient 295 K to 77 K	Crack resistance
Copper *	2.9×10^{-3}	Good
Aluminum *	3.8×10^{-3}	Good
Sn 62, A 2, Pb 36 Solder	5.8×10^{-3}	Good
Epoxy EPON-820, 50 parts Versamid-140, 50 parts unfilled	12.0×10^{-3}	Poor
Epoxy EPON-826, 50 parts[†] DER-736, 50 parts TONOX, 28 parts unfilled	8.8×10^{-3}	Moderate
NEMA-G10 rod along the length	$2.3\text{--}2.7 \times 10^{-3}$	Good
NEMA-G10 Sheet along the grain ⊥ to the grain	$1.7\text{--}2.2 \times 10^{-3}$ $7.0\text{--}7.8 \times 10^{-3}$	Good Good
ESCAR superconducting coil section along the conductor ⊥ to the conductor	2.5×10^{-3} 2.7×10^{-3}	Good Good
TPC superconducting coil section[†] along the conductor ⊥ to the conductor	2.4×10^{-3} 3.6×10^{-3}	Good Good

* Standard values.

† 1978 measurements; all other values were measured in 1975
   by D. Hunt.

taken to avoid cracking within the epoxy. (In theory, the coil
assembly is designed for rapid cool down. One should be able to
cool the coil in liquid nitrogen and have the coil assembly survive.) Cracking of the epoxy within the coil can be controlled
by the following procedure:

1. The epoxy should be kept in compression. The coil
    is pretensioned to minimize the amount of tensile
    stress the epoxy sees. (This technique is analo-
    gous to the technique used to make prestressed
    concrete.)

2. The amount of epoxy within the structure should be
    minimized. This reduces the time needed to intro-
    duce the epoxy and the time under vacuum, and it
    minimizes the potential for cracking.

3. The unfilled spaces within the epoxy should be
    minimized; this reduces the size of potential cracks.
    (Thus, the energy released due to cracking is mini-
    mized.)

4. Voids and bubbles should be eliminated as much as
    possible.

5. Sharp corners within the mold and on surfaces
    exposed to the resin should be eliminated.

The basic structure of a magnet coil is shown in Fig. 1.
This figure shows a cross section of the one-meter diameter
coils. The two-meter diameter coil has a slightly different but
similar structure. The spaces in the coil were either filled
with glass cloth or with dacron cord. The superconducting wire
was wound on the coil form under considerable prestress (about
0.2 GN/m^2). This prestress guarantees that the relative thermal
contraction between the coil and aluminum bore tube is matched.
In addition, the coil has enough prestress so that a minimum
amount of tensile stress is put into the epoxy.[10]

The cooling tube, which is wound outside of the magnet coil,
will carry two-phase helium to cool the coil. During potting
it is used as a means for heating the coil while the epoxy is
poured and cured. (This second use for the cooling tube is
extremely important.) The thick, nearly pure aluminum bore tube
also distributes the heat to the epoxy during pouring and curing.
As a result, the epoxy temperature within the coil can be controlled to about $1°C$.

The aluminum bore tube is made vacuum tight so that it can
be used as part of the vacuum impregnation mold. (A vacuum tight
mold is essential.) The outer shell of the vacuum mold consists
of sheet steel, which is wrapped and clamped around the coil. The

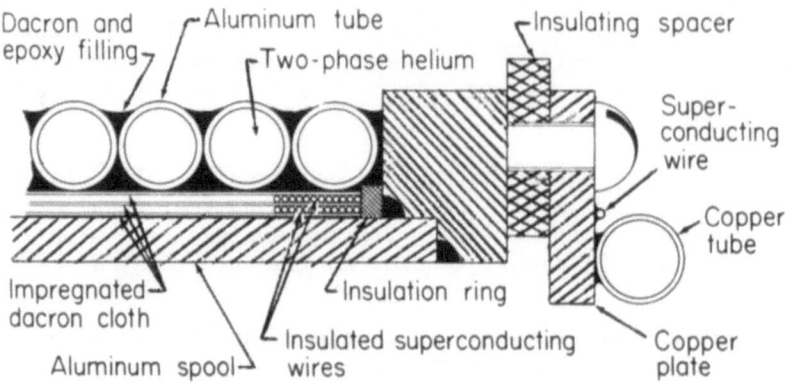

Fig. 1. A cross section of the one-meter diameter test solenoid
magnet vacuum impregnated coil package.

vacuum tightness of the vessel is ensured through the use of a
generous amount of dux-seal, RTV (silicone rubber), and good
clamping of the steel onto the coil cooling tube. The vacuum
space within the coil was pumped to about 1.3 mPa (10^{-5} torr) while
the cooling coil was heated by warm water at a temperature of
about 80°C. Once the mold assembly had been leak checked with a
mass spectrometer, it remained on the pump for 24 hours at 80°C
to outgas water and other contaminants. Then the temperature was
dropped to around 45°C. The finished mold assembly for the two-
meter diameter test coil is shown in Fig. 2.

Once the coil mold was made vacuum tight, pumped down, and
outgassed, it was backfilled with dry nitrogen to a pressure of
about 270 Pa (2 torr). The temperature of the epoxy, which had
been mixed and deaerated, was carefully controlled over a temperature range of 45 to 50°C. (The epoxy pot life at 45°C is about
6 hours; at 50°C the pot life goes down to 4 hours.) The coil
mold was temperature controlled by flowing hot water through the
cooling tube. The water used to maintain the epoxy pot and mold
temperature was heated by a cam controlled temperature controller,
which regulates the water temperature during deaerating, potting,
and curing.

The epoxy flow within the coil is extremely important. The
LBL epoxy will flow through channels as small as 25 μm, but the
flow rate is slow. Care must be taken to ensure good epoxy flow
without large sections of unfilled epoxy. In the LBL solenoid,
the bond between the coil and the aluminum bore tube is very

Fig. 2.   The two-meter diameter test solenoid vacuum impregnation
          shroud showing the multiple pouring posts and vacuum
          pumping port.

important. The bore tube was sand blasted and treated to ensure
a good bond. The mold on the outside was treated with mold release
so that separation could be achieved. In general, it has been
found that multiple entry ports for the epoxy results in a much
faster pour. The A coil (a one-meter diameter test coil) took 7.5
hours to fill through one port. The B coil (also a one-meter
diameter solenoid) took about 1.5 hours to fill because 4 ports
were used. The C coil (a two-meter diameter solenoid) took about
7 hours to fill through multiple ports. The C coil has longer
distances for the epoxy to flow; it has almost three times the
epoxy mass; and it had more than one batch poured into the mold.
Therefore, the fill time was longer.

The cam controlled temperature controller sets the cure cycle for the epoxy. Once the epoxy is introduced, the temperature is held for about 12 hours; then the temperature is slowly raised to 60°C (this takes about 3 hours). The cure at 60°C takes 8 hours. Then the temperature is raised to 80°C. (This takes about 2 hours.) The magnet and the mold are maintained at 80°C for at least 8 hours. The outer molding shell can be removed after the magnet temperature has dropped to room temperature. The pour and cure temperature schedule for the A magnet (one of the one-meter diameter solenoids) is shown in Fig. 3. (Note the epoxy was not held at 50°C for 12 hours before starting the cure.)

After potting and curing, the steel potting shell was broken away. Rough edges on the coil surface were buffed in preparation for a surface glaze. The surface glaze consists of 60 parts by weight of low viscosity resin (EPON 826) and 40 parts by weight of polyamide resin (Versamid 140). The surface glaze was applied

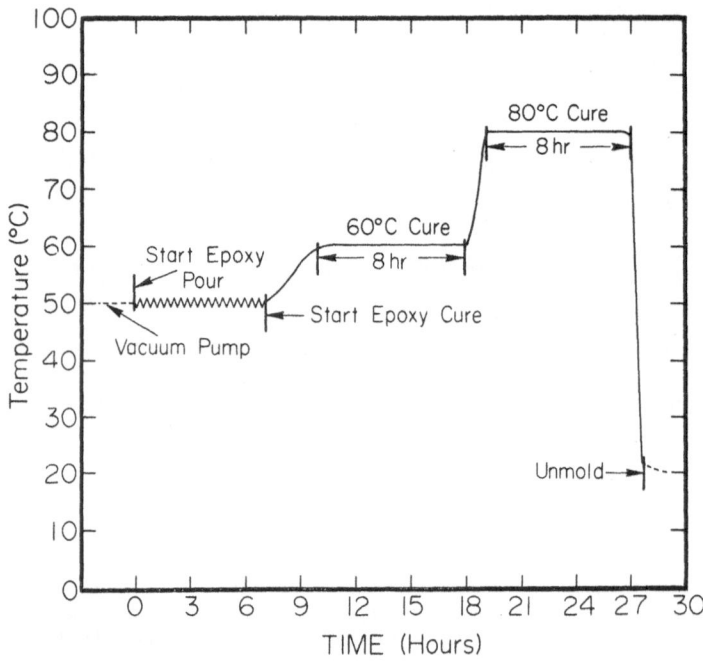

Fig. 3.   The cure cycle (temperature vs. time) for the A and B one-meter diameter test solenoids.

with a brush and almost wiped dry to form a layer as thin as possible. (This epoxy is prone to cracking, but it is well away from
the superconductor.) The glaze epoxy was cured at room temperature
or under heat lamps until it was hard and dry to the touch. Fig. 4
shows the finished C coil solenoid magnet.

TESTS OF THE SUPERCONDUCTING COILS

The tests of the three superconducting solenoid magnets show
that the vacuum impregnated epoxy technique is viable in large
solenoid magnets. The three coils do not train like many coils,
which have been built elsewhere. Table II compares the three
large LBL solenoids.

Fig. 4. The finished two-meter diameter test solenoid (coil C).

Table II. The Characteristics of the Three LBL Large Vacuum
           Impregnated Superconducting Solenoid Magnets.

	Coil A	Coil B	Coil C
Magnet diameter, m	1.035	1.035	2.004
Magnet length, m	0.464	0.461	0.697
Conductor prestress, GN/m^2	0.16	0.21	0.23
Epoxy mass, kg	9.2	8.5	25
Epoxy potting temperature, C	50	50	47
Epoxy pour time, h	7.5	1.5	7.0
Magnet critical current, A	804	1036	1440
Maximum current achieved, A	804	972*	1440
Maximum superconductor matrix current density achieved, GA/m^2	1.044	1.237	0.814
Maximum magnet stored energy, MJ	0.255	0.374	1.921
Number for training quenches	6	0	0
Current for first training	597	—[†]	—[†]

* This magnet did not train. 972 A was the limit of the
   power supply.

† There was no training observed.

     The A coil trained because of sudden coil motion, which was
caused by the breaking of the epoxy bond between the coil and the
bore tube.[11]   There was no evidence of training owing to energy
release caused by epoxy cracking or fractures within the niobium
titanium. The three magnets operated at high current densities
and at high stored energies. It is felt that the bad epoxy bond

in coil A was due to epoxy that had begun to set before it was
poured. (The part of the coil that was poured last was the part
that failed.)

The A magnet was cooled down 6 times; the B magnet was cooled
down 5 times; and the C magnet was cooled down 4 times. All three
magnets show evidence of cracking on the surface. The cracks go
into the layer of glass just below the surface. There was no
degradation of performance with subsequent cooldowns. Once the
magnet achieved its critical current, there was little subsequent
training. Tests on coil B and coil C showed no degradation of
the bond between the coil and the bore tube.

CONCLUDING REMARKS

The epoxy impregnation technique used by LBL appears to work
despite rather poor cryogenic properties of the basic epoxy formulation. It appears that technique is more important than epoxy
formulation in determining the performance of large potted superconducting solenoid coils. It should be noted that dipole or
quadrupole magnets may not perform as well as these solenoids
because the force distribution within the coil is different. The
key to using epoxy in large high-current-density solenoids is
minimizing the epoxy mass, filling the epoxy, and prestressing
the coil so that the epoxy is kept in compression.

As a result of the LBL test coil experiments, the laboratory
has started construction of a 2.16 m superconducting thin solenoid,
which is 3.3 m long. The current density in the superconducting
matrix will be about 0.7 GA/m^2; the magnet stored energy will be
10.9 MJ. The design current is set at 70% of the superconductor
critical current along the load line. The magnet is expected to
contain over 70 kg of epoxy resin.

ACKNOWLEDGMENTS

The authors wish to thank D. Hunt for his measured data on
70 samples of various composite materials and also for
his advice on performing the measurements. They thank A. P.
Barone, of the LBL Assembly Shop, for preparing the samples for
thermal contraction coefficient measurements.

This work was performed under the auspices of the United
States Department of Energy.

REFERENCES

1. M.A. Green, P.H. Eberhard, and J.D. Taylor, "Large High Current
    Density Superconducting Solenoids for Use in High Energy Physics
    Experiments," Proceedings of ICEC-6, IPC Science and Technology
    Press, Guildford, England (1977).

2. M.A. Green, Cryogenics, 17(1), 17 (1977).

3. M.A. Green, "Large Superconducting Detector Magnets with Ultra
    Thin Coils for Use in High Energy Accelerators and Storage
    Rings," paper presented at the 6th International Conference on
    Magnet Technology, Bratislava, Czechoslovakia, August 1977,
    Publ. No. LBL-6717, Lawrence Berkeley Laboratory, Berkeley,
    California (1977).

4. J.O. Turner, "Flexibilized Epoxy Formulation, Unfilled, and
    Its Use in Vacuum Impregnation of Magnet Coils," LBL Specifi-
    cation M20C, Lawrence Berkeley Laboratory, Berkeley, California
    (1970).

5. Materials for Use in Superconducting Magnet Construction, a
    report of the Group European Superconducting Synchrotron
    Studies (GESSS) Collaboration, GESSS-3 (April 1974).

6. D. Evans et al., "Epoxy Resins for Superconducting Magnet
    Encapsulation," Report No. RHEL/R251, Rutherford Laboratory,
    Chilton, Didcot, England (1972).

7. G. Hartwig, IEEE Trans. Magn. MAG 11(2), 536 (1975).

8. D. Hunt, private communication concerning his measurement of
    total thermal contraction from 300 K to 77 K. D. Hunt made
    measurements on 70 different samples of various filled and
    unfilled plastic formulations. D. Hunt's work is unpublished.

9. C. Schmidt, "The Introduction of Normal Zone (Quench) in a
    Superconductor by Local Energy Release" to be published in
    Cryogenics, Report SUPRA/78-26 EG, CEN-Saclay, Gif-sur-Yvette,
    France (1978).

10. M.A. Green and J. D. Taylor, "Construction of the A Coil,"
    Internal Report UCID-3835, Lawrence Berkeley Laboratory,
    Berkeley, California (1976).

11. M.A. Green, "The Development of Large High Current Density
    Superconducting Solenoid Magnets for Use in High Energy
    Physics Experiments," doctoral thesis, Publ. No. LBL-5350,
    Lawrence Berkeley Laboratory, Berkeley, California (1977).

AN OUTLINE DESIGN FOR A CRYOGENIC INTERNALLY INSULATED

LIQUEFIED NATURAL GAS PIPELINE FOR ARCTIC GAS RECOVERY

E. W. Johnson and G. Walker

The University of Calgary
Calgary, Alberta, Canada

INTRODUCTION

A technological assessment of long distance transport of natural gas in liquefied form was carried out for the Government of Canada, Department of Energy, Mines and Resources.[1] This was done to determine its viability as an alternative to vapour phase pipelines, with special application for use in the Arctic.

A cryogenic pipeline has two major components: a pressure pipe to contain the contents and insulation for thermal isolation of the low temperature liquid. The insulation may be inside or outside the pipe; external application predominates in installations now in operation.

Externally Insulated Pipelines

External insulation permits the use of pressure pipe of a minimum diameter and thus of a minimum wall thickness but requires that the pipe be constructed of cryogenic-rated material, such as stainless steel, 9% nickel steel, or 5083-H113 aluminium. These materials are considerably more expensive than ordinary carbon-steel pipe, ranging from ten times for stainless steel to three times for nickel steel or aluminium. The external insulation, usually urethane foam, is secured to the pipe with a metal or plastic vapour barrier and a protective jacket. Such insulation is relatively weak mechanically and could not impose any significant axial restraint on the pipe. Since such a pipe will be constructed at ambient temperatures, but operate at about 194 K, there will be a considerable range of expansion (0.15% for steel and 0.3% for aluminium), which will have to

421

be accommodated by expansion loops or bellows. Where the pipe is
buried, these will have to be constructed in suitable chambers.

External insulation also requires that the entire mass of the
metal pipe must be cooled to working temperature on each occasion
that the pipe is loaded. This represents a significant cooldown
thermal load, although it would not affect continuous operation.

Internally Insulated Pipelines

Internally insulated pipelines are necessarily of larger external diameter and wall thickness, but may be constructed of
conventional pipeline carbon steel, which is considerably cheaper
than cryogenic materials. The range of operating temperatures will
be less than for the externally insulated pipe so that large savings may be made by eliminating expansion loops or bellows. Since
the steel pipe is on the outside, ground or pipe anchors and bend
supports may be readily attached. Conventional soil movement restraint practices for buried pipelines may be used without the need
to protect the fragile insulation.

A final advantage lies in the fact that the pipe does not need
to be cooled when bringing the line into operation. The refrigeration load to cool the insulation is roughly the same for both
methods of insulation, so that there is a significant saving in
refrigeration load during startup. This results, in turn, in savings in the production and storage of the inert cooldown and purging
fluid, liquid nitrogen.

INSULATION MATERIALS

Conventional insulation materials are of the closed-cell foam
type and are unlikely to be used in this application, because the
pressure will crush the cells, increasing the thermal conductivity.
Yates[2,3] investigated open-cell foams for use with liquid hydrogen
and reported best results with polyphenylene oxide (PPO). This is
an anisotropic open-cell material with elongated cells in the depth
direction, as developed by General Electric. From Yates' data on
the thermal conductivity of this material and the known properties
of methane, Walker, Stuchly, and Read[4] have estimated that the effective thermal conductivity for PPO foam filled with methane at 7.0
MPa ($\approx$ 1000 psi) would be of the order of 70 mW/m·K. It is, however, very important to note this value is speculative and experimental verification is urgently required.

COMPARATIVE MATERIAL COSTS

An estimate of the relative amounts of materials used in the two types of construction may be obtained using the classical equation for the stress in a thin-walled cylinder:

$$\sigma = pd/2t \qquad\qquad (1)$$

where σ is the hoop stress, p, the internal pressure, d, the pipe diameter, and t, the wall thickness.

For a given system with fixed operating pressure and design stress, it can be seen that t and d are in direct ratio. If suffixes E and I refer to external and internal insulation, respectively, it may be shown that the ratio of the masses is given by

$$M_I/M_E = (d_I/d_E)^2 \qquad\qquad (2)$$

The diameters for the two cases are connected by the relation

$$d_I = d_E + 2K \qquad\qquad (3)$$

if it is reasonably assumed that the same thickness of insulation, K, is used in each case.

Figure 1 is derived from the above equations and indicates the range of parameters for which it is advantageous to use internal insulation. For the proposed pipeline, for which the fluid flow diameter is 914 mm (36 in) and the insulation is 102 mm (4 in), $d/K = 9$, and the internally insulated pipe weighs 1.44 times that for the external insulation. At the other extreme, if $d/K = 1$, then the weight ratio is 9:1, and it would be uneconomic for lines of any length. Point z indicates the break-even point, if it is assumed that cryogenic materials are three times as expensive as standard pipeline material.

Figure 2 indicates the gross pumping and cooling load as a function of fluid flow diameter for two insulation thicknesses. It will be seen that both requirements are reduced by increases in pipe diameter.

Figure 3 shows the effect of insulation thickness on the cooling power requirements: pumping energy is virtually independent of the insulation thickness. Whereas cooling power requirements are rapidly reduced by increases in insulation thickness, the mass of the steel required will increase linearly.

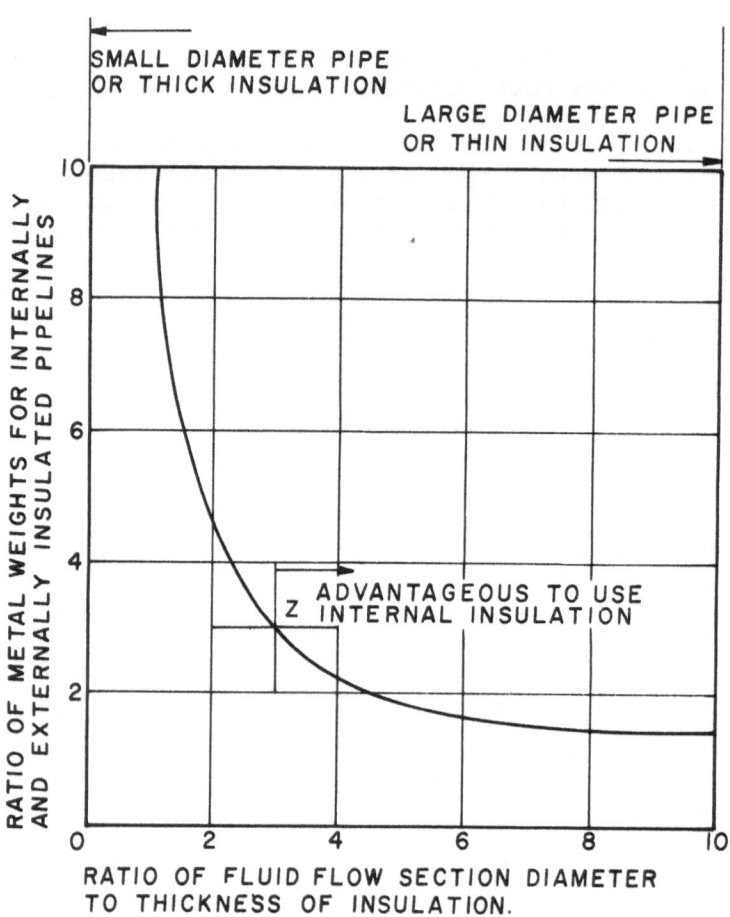

Fig. 1.   Ratio of metal weight for internal and external insula-
          tion as a function of insulation thickness.

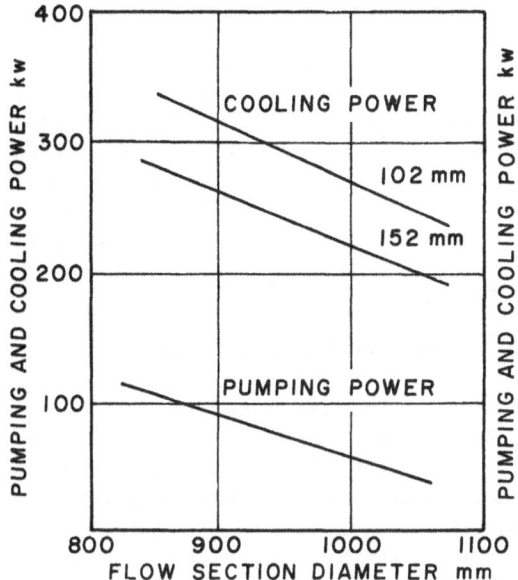

Fig. 2.   Gross power requirements as a
          function of flow section diameter
          (from Ref. 4).

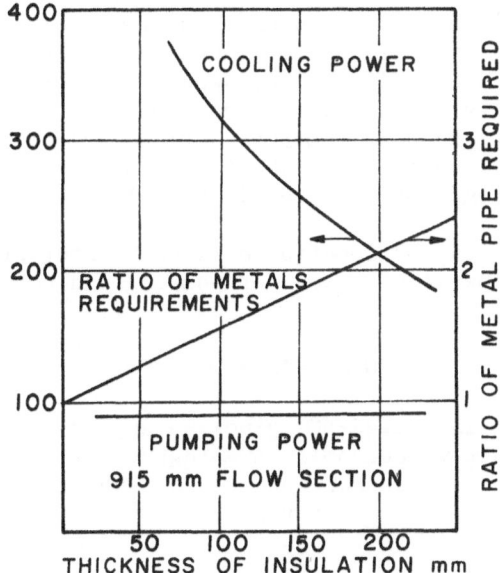

Fig. 3.   Gross power requirements as a function
          of insulation thickness (from Ref. 4).

FIELD CONSTRUCTION

The following procedure would probably be used for field construction. The pipe would be delivered to the site with a sleeve
insert of insulation somewhat shorter than the pipe length. After
the pipe is welded in the normal way and inspected, the insulation
would be pushed axially along the pipe until it connects with the
previously installed section. At intervals, an extra insulated
sleeve would have to be inserted to compensate for the joint allow-
ances.

COMPUTER MODEL OF PIPELINE

The theoretical design was based on the proposed 2300 km
Mackenzie Valley pipeline running south from the shores of the
Arctic Ocean to a maximum elevation of 1200 m and terminating in
central Alberta at a junction with existing systems. A computer
programme developed by Canuck Engineering of Calgary was used to
optimise the design. This programme solved the two sets of equations governing the pressure drop owing to friction and differences
in altitude and the temperature rise owing to friction and conduction through the insulation.

A pump and cooling station was assumed to be constructed
wherever the pressure was calculated to drop to a specified minimum (3.72 MPa). The horsepower was calculated to raise the pressure back to the design maximum (5.45 MPa) and to cool the fluid
to the minimum specified (185 K) in preparation for entry to the
next segment. Fuel gas required for pumping and cooling the fluid
was assumed to be withdrawn from the suction side of the station
and amounted to just under 7% of the flow. A complete description
of this computer programme is given in Ref. 1.

Table I indicates the location of each booster station, its
elevation and the installed power for pumping and cooling.

CONCLUSIONS

The internally-insulated LNG pipeline appears to be a viable
solution with attractive economic advantages. It is particularly
attractive in areas such as Northern Canada, where low ambient
temperatures prevail for much of the year.

ACKNOWLEDGMENT

The work described in this paper was carried out at the request
of the Government of Canada, Department of Supply and Services, DSS
file No. 17SQ 2344D-6-9014.

Table I

Station number	Distance from Arctic, km	Elevation, m	Flow hm^3/day	Pumping power, MW	Cooling power, MW
0	0	15	70.79	0	0
1	93.6	148	70.57	4.21	19.35
2	208.2	181	70.32	4.23	22.79
3	346.7	95	70.03	4.26	27.13
4	464.7	110	69.77	4.21	23.69
5	587.8	109	69.50	4.20	24.70
6	707.7	127	69.24	4.18	24.33
7	817.3	201	68.99	4.15	22.55
8	942.5	202	68.72	4.16	25.22
9	1050.2	291	68.47	4.12	22.10
10	1169.1	339	68.21	4.12	23.66
11	1238.1	616	68.04	4.02	14.73
12	1379.8	559	67.76	4.12	26.20
13	1529.3	477	67.46	4.12	27.68
14	1596.4	770	67.29	3.97	14.20
15	1735.3	741	67.01	4.08	26.61
16	1858.6	800	66.74	4.05	24.70
17	1985.2	841	66.47	4.04	25.45
18	2033.8	1222	66.33	3.89	10.85
19	2211.9	1048	65.98	4.09	33.56
END	2301.4	1181	65.98	--	--
			Totals	78.22	439.50
			Total Power	517.72	

Maximum operating pressure: 5.45 MPa
Minimum operating pressure: 3.72 MPa
Inlet or suction temperature: 185 K
Input flow: 105 000 m^3/day (1.215 m^3/s)
Pipe external diameter: 1118 mm (44 in)
Insulation thickness: 102 mm (4 in)
Flow-section diameter: 914 mm (36 in)
Roughness: 0.0003
Pipe wall thickness: 7.9 mm
Pump efficiency: 80%

REFERENCES

1. J.M. Stuchly and G. Walker, Technology Assessment of Long Dis-
    tance LNG Gas Pipelines, Government of Canada, Department of
    Supply and Services DSS file No. 17SQ 2344D-6-9014 (1977).

2. G.B. Yates, in: Advances in Cryogenic Engineering, Vol. 16,
    Plenum Press, New York (1971), p. 128.

3. G.B. Yages, in: Advances in Cryogenic Engineering, Vol. 20,
    Plenum Press, New York (1975), p. 327.

4. G. Walker, J. Stuchly, and M.J. Read, The Internally Insulated
    Liquid Natural Gas Pipeline for Arctic Gas Recovery, ASME
    77-WA/HT-17, American Society of Mechanical Engineers, New
    York (1977).

CONTRIBUTOR INDEX

MATERIAL INDEX

431